Focus on Man

Focus on Man

SYLVIA CHAPLIN

ARNOLD-WHEATON
A Division of Pergamon Press

Arnold–Wheaton
A Division of Pergamon Press
Butterley Street, Leeds LS10 1AX
and
Hennock Road, Exeter EX2 8RP

Pergamon Press Ltd
Headington Hill Hall, Oxford OX3 0BW

Pergamon Press Inc.
Maxwell House, Fairview Park, Elmsford, New York 10523

Pergamon Press Canada Ltd
Suite 104, 150 Consumers Road, Willowdale, Ontario M2J 1P9

Pergamon Press (Australia) Pty Ltd
P.O. Box 544, Potts Point, N.S.W. 2011

Pergamon Press GmbH
Hammerweg 6, D-6242 Kronberg, Federal Republic of Germany

First published 1982

Typeset by Oxprint Ltd, Oxford
Printed in Great Britain by A. Wheaton & Co. Ltd, Exeter

ISBN 0 08-024981-7

Contents

ACKNOWLEDGEMENTS

The author and publisher wish to thank the following individuals and organisations who kindly provided photographs and data for use in this book:

AGA Infra-red Systems Colour Plate 8, Fig. 15.3; Aerofilms Ltd Fig. 4.3; Alfred Hospital, Melbourne, Australia Colour Plate 11; B.B.C. Figs. 17.9, 17.11; Beecham Proprietories Fig. 7.7; BICC Fig. 18.9; Boulton Hawker Films Fig. 1.12; British American Optical Company (AO Safety Products) Fig. 17.3; British Arkady Co. Ltd Colour Plate 3; British Diabetic Association Fig. 14.4; British Epilepsy Association Fig. 18.25; British Telecom Fig. 3.10; Camera Press Ltd Fig. 7.5; Cancer Information Association Fig. 4.19; Central Office of Information Figs. 6.2, 6.3, 6.4, 6.8, 9.1, 9.6; Cheshire County Fire Brigade Fig. 4.2; Clement Clarke International Ltd Fig. 4.17; Copydex Adhesive Fig. 2.10; DEB Chemical Proprietories Ltd Figs. 7.6, 15.11; Department of Health and Social Security Fig. 9.6; Droylsden Training Centre 18.29; E.M.I. Medical Fig. 13.4; Family Planning Association Fig. 20.12; Dr E.M. Forster Table 9.1; Dr Theodore Friedman (*Scientific American*) Fig. 1.18; Geigy Pharmaceuticals and Dr P.M. Jeavans Fig. 18.24; Gene Cox Figs. 1.2, 1.3, 1.4, 1.5, 1.6, 3.3, 3.6, 3.17, 6.1, 7.1, 15.12, 16.3, 18.14; General Dental Council Colour Plates 4, 5, 6; Gibbs Oral Hygiene Service Fig. 11.15; Gillette Industries Fig. 19.7; Granada TV (*Facts For Life* – Schools TV programme) Figs. 2.4, 20.6, 20.9; Greater Manchester Police Table 3.1; R. Harris Esq. Fig. 5.7; Health Education Council Figs. 2.5; 6.12; 7.10; 9.9, 10.20; Heather Angel Fig. 4.10; Help the Aged 20.14; HMSO Fig. 7.17; ICI Figs. 8.7, 8.13, 15.14; Japan Information Centre Fig. 4.4; John Topham Figs. 10.3, 10.5; Keeler Optical 17.21; London News and Feature Service Fig. 16.15; Manchester Evening News Figs. 6.4, 6.5, 6.9, 7.8, 11.8, 12.20, 13.2, 16.7, 16.17, 18.22; Manchester Royal Infirmary Figs. 1.9, 1.10, 1.11, 2.7, 3.5, 5.15, 12.3, 12.5; Medical Research Council (*The Composition of Foods* – McCance and Widdowson, HMSO) Table 9.1; Messenger Newspapers Ltd Fig. 3.8; Morningside Training Centre Fig. 18.29; Mothercare Ltd Fig. 2.9; Mouth and Foot Painting Artists, E. Twisington-Higgins Fig. 18.16; National Childbirth Trust Fig. 20.4, 20.10; National Portrait Gallery Fig 20.1; National Society for Mentally Handicapped Children Fig. 18.28; North West Water Authority Figs. 6.10, 10.8–10.12, 12.12, 12.18, 12.19; Office of Population Censuses and Surveys Table 7.9; Olympus Optical Company Colour Plate 2; Optical Information Council Fig. 17.1, 17.8, 17.9; Oxfam 9.3; Pharmaceutical Society of Great Britain Fig. 7.9; Philips Figs. 2.6, 3.4; Procter and Gamble Ltd Colour Plate 7; Rentokil Ltd Figs. 10.1, 10.16–18, 12.8, Tables 12.3, 8.1; ROSPA Fig. 3.7; Royal College of Surgeons (Edinburgh) Figs. 3.12, 7.4; Royal College of Surgeons (London) Colour Plate 1; Royal Marsden Hospital, London Fig. 14.3; Royal National Institute for the Deaf Fig. 16.9; St John Ambulance Brigade Figs. 3.11, 4.21; Shandwick PR Company Ltd Fig. 15.17; Sherleys Fig. 8.11; Silvaflame Co. Ltd Fig. 3.9; A.W. Swallow Esq. Figs. 3.26–8; Professor Sydney Selwyn, Westminster Medical School Colour Plate 9; Tesco Fig. 9.2; Thames Water Authority Figs. 12.11, 12.14, 12.15; Vision International Fig. 1.12, 10.13, 10.14, 10.15; Weight Watchers Fig. 9.14; Wellcome Trustees Fig. 7.2, 7.14; The White Helmets, R.C.S. Motor Cycle Display Team Fig. 4.5; Women's Royal Voluntary Service Fig. 9.13; World Health Organization Figs. 7.13, 7.16, 8.1, 8.2, 8.3, 8.5, 8.6. The author provided Figs. 2.8, 5.17, 16.2.

The author would like to thank Cliff Gould, Anthony Redford and Alan Taylor who read the manuscript at various stages and who made many valuable contributions to the text. Thanks also to Ann Rowland, Science Department, Knutsford County High School, Cheshire, who gave assistance on the classroom practicability of the material contained in draft sections of the book. The author is most grateful to Jill Burton, George W. Rogers and Richard Neave for particular photographic help and advice, and to Barbara Roddy who typed the manuscript with speed and accuracy.

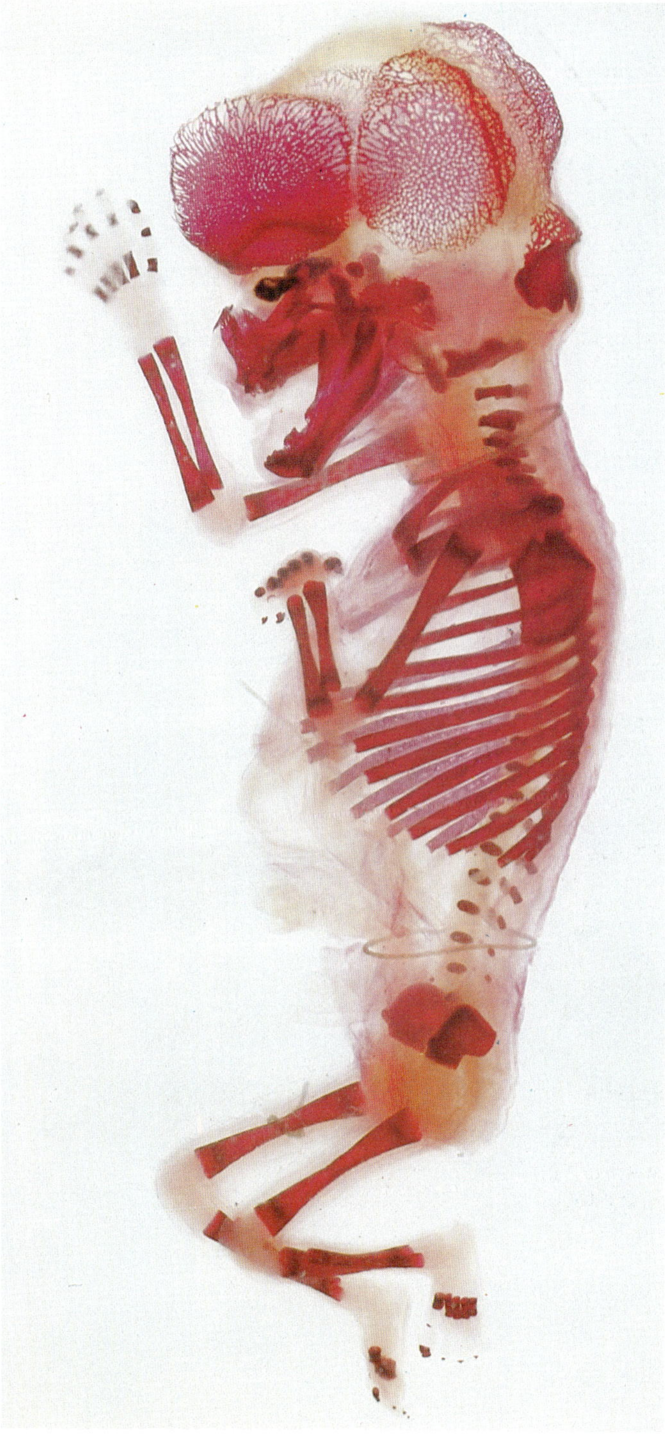

1 *The red areas show where calcium has been laid down in the bones of this 16-week-old foetus*

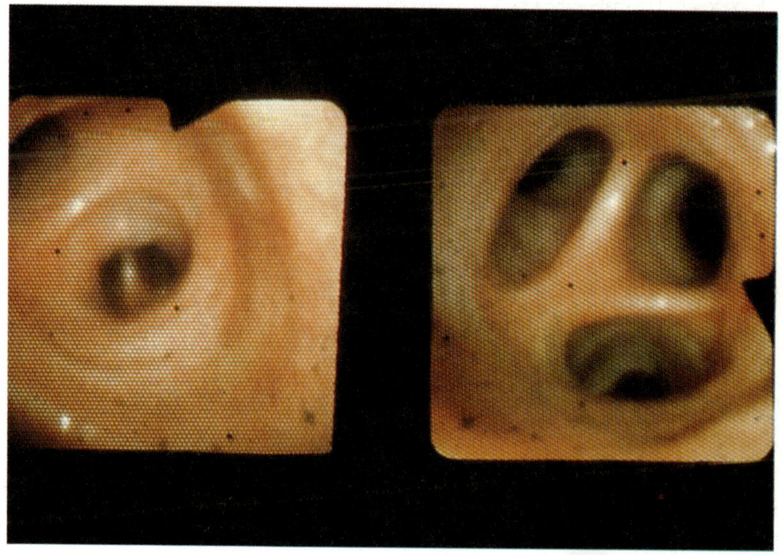

2 *A bronchoscope allows the doctor to explore the respiratory passageways*

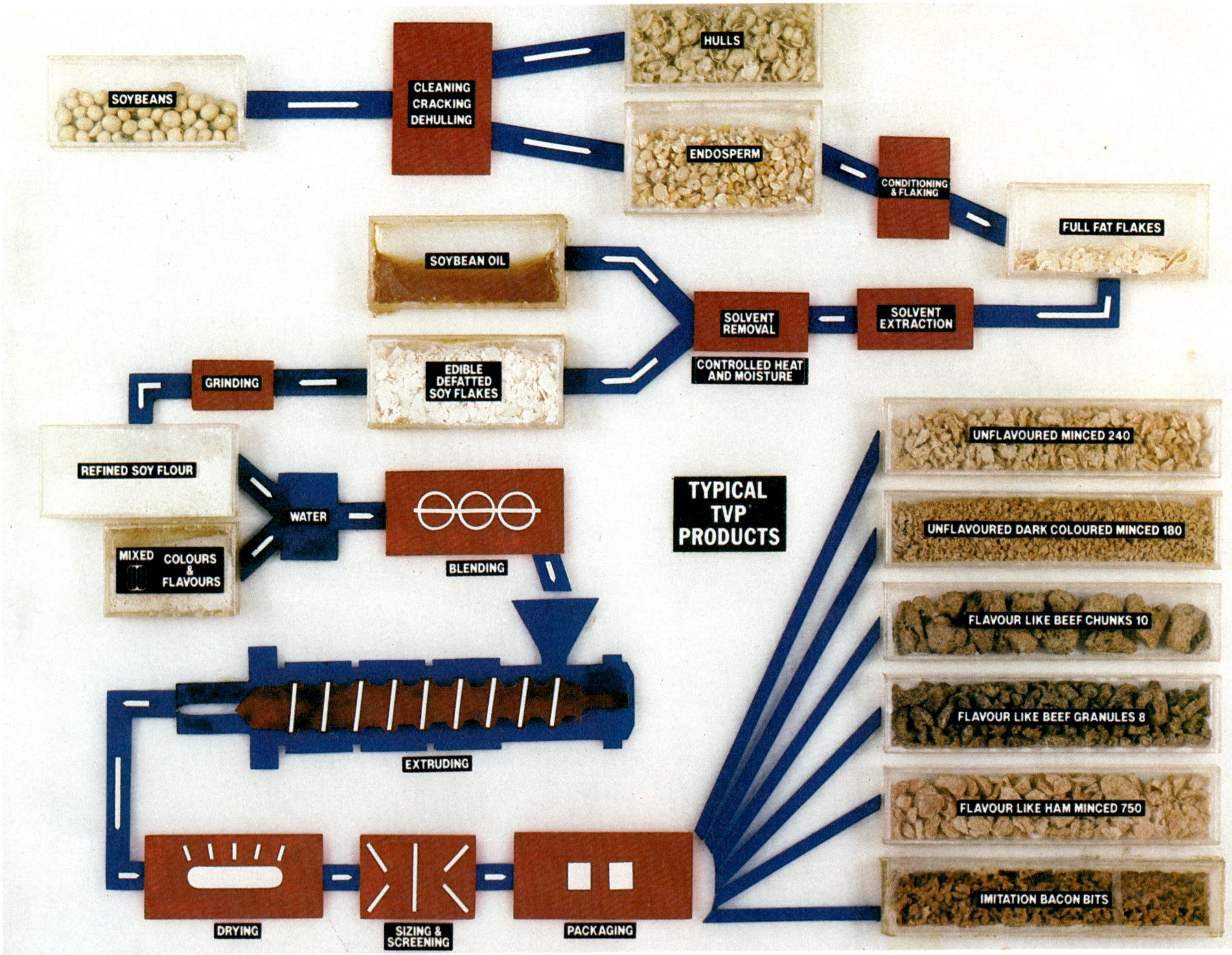

3 *The processing of soya beans to yield textured vegetable protein (TVP)*

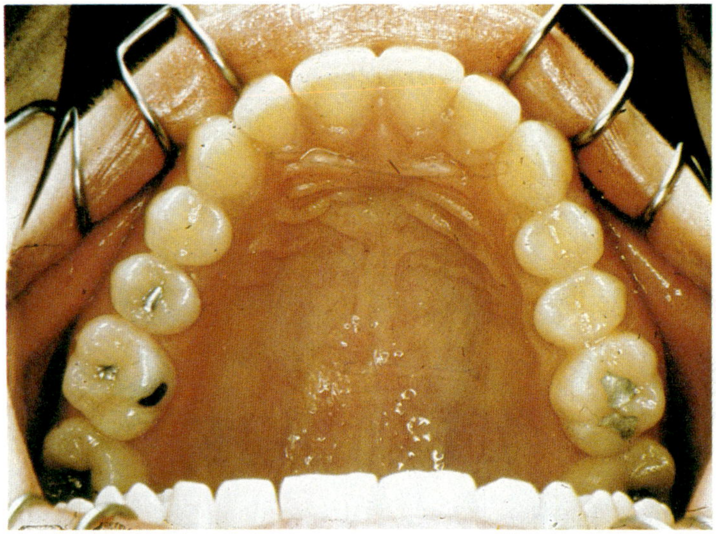

4 *The lower jaw with back teeth that have had their decay treated. The decayed region is drilled out and the cavity is then filled with an amalgam*

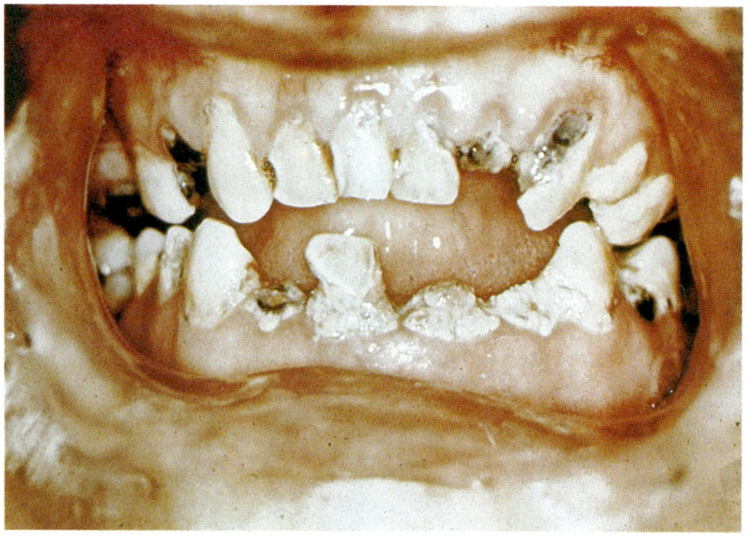

5 *Rotten and decaying teeth*

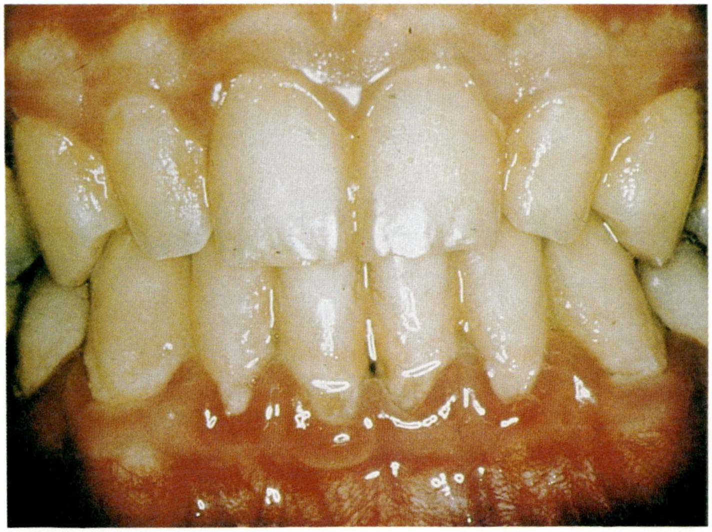

6 *Gum disease*

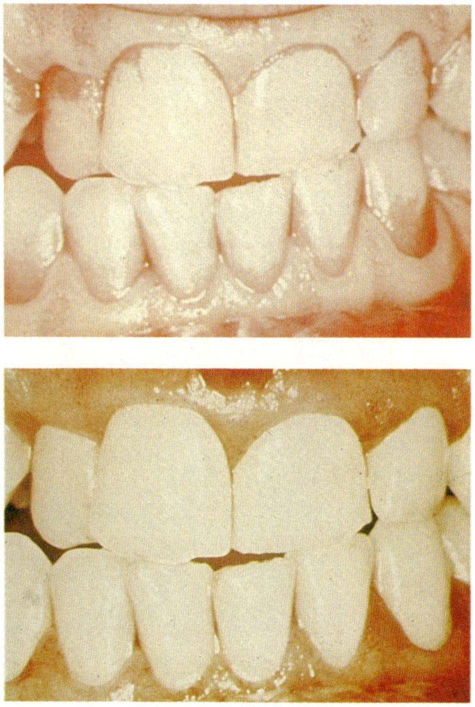

7 *(a and b) After the teeth have been brushed properly no traces of plaque can be seen*

8 *(right) A colour thermogram illustrates a special use for the infra-red camera. The hand on the right has been severely burnt. Dead flesh does not give off heat, so any area that is black is flesh burnt beyond recovery. From this information the doctor knows quickly which areas are most likely to need a skin graft. Dark blue and dark green indicate cool areas. Magenta, red, orange and yellow indicate progressively warmer regions. The hotter the burnt area, the greater is the chance of recovery*

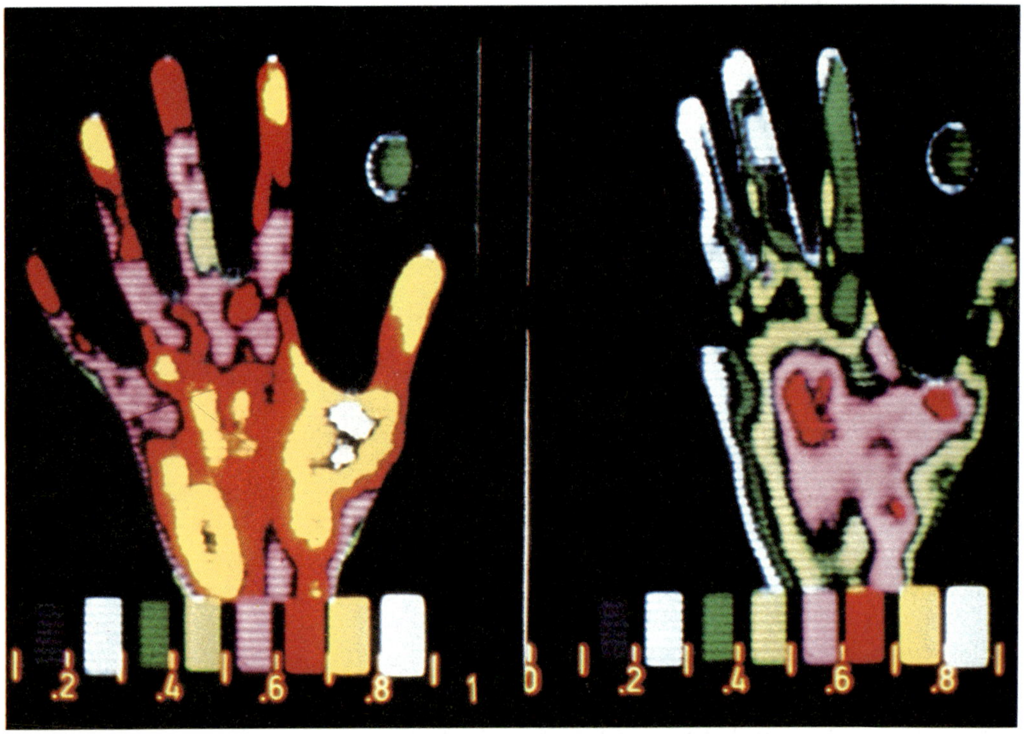

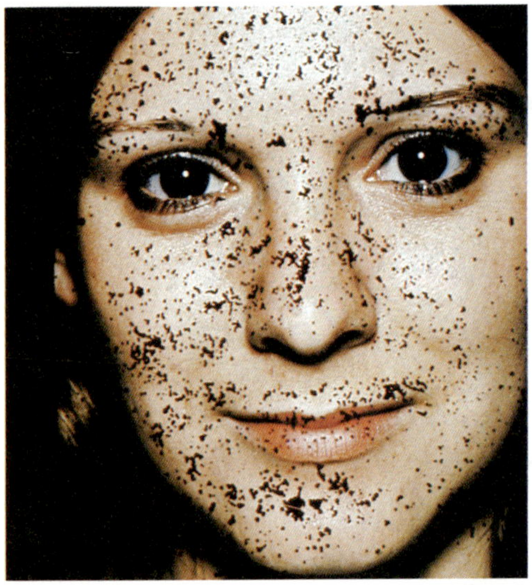

9 *Evidence of bacteria on the face*

10 *A woman about to fit her contact lens*

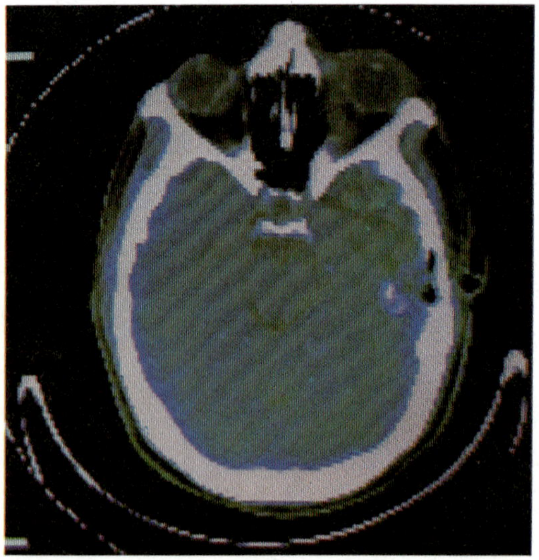

11 *The result of a brain scan. The brain tissue, orbits and optic nerves can be seen clearly. With a normal X-ray all this detail would be hidden by the skull*

1 Alive!

1.1 **Building units**

The man building the house in Fig. 1.1 is using small building units called bricks. What about the man himself? Is he made up of similar building units? We cannot see them with the naked eye, but if we use a microscope they become very clear (Fig. 1.2).

The building units of the human body are called *cells*, and because we consist of many cells man is called a *multicellular* life form.

The cells shown in Fig. 1.2 have been gently scraped from the inner surface of a person's mouth. Looking at these cells might make us think that all body cells are flat. However cells, like bricks, are three-dimensional; that is, they have depth as well as length and breadth. Whether we see them in three dimensions or as flattened objects depends upon how we look at them. The surface cells inside the mouth are very thin. They are easily removed by gentle pressure as single cells. Similar cells flake off from our body surface all the time. They even rub off as we wash and dry our skin (page 192).

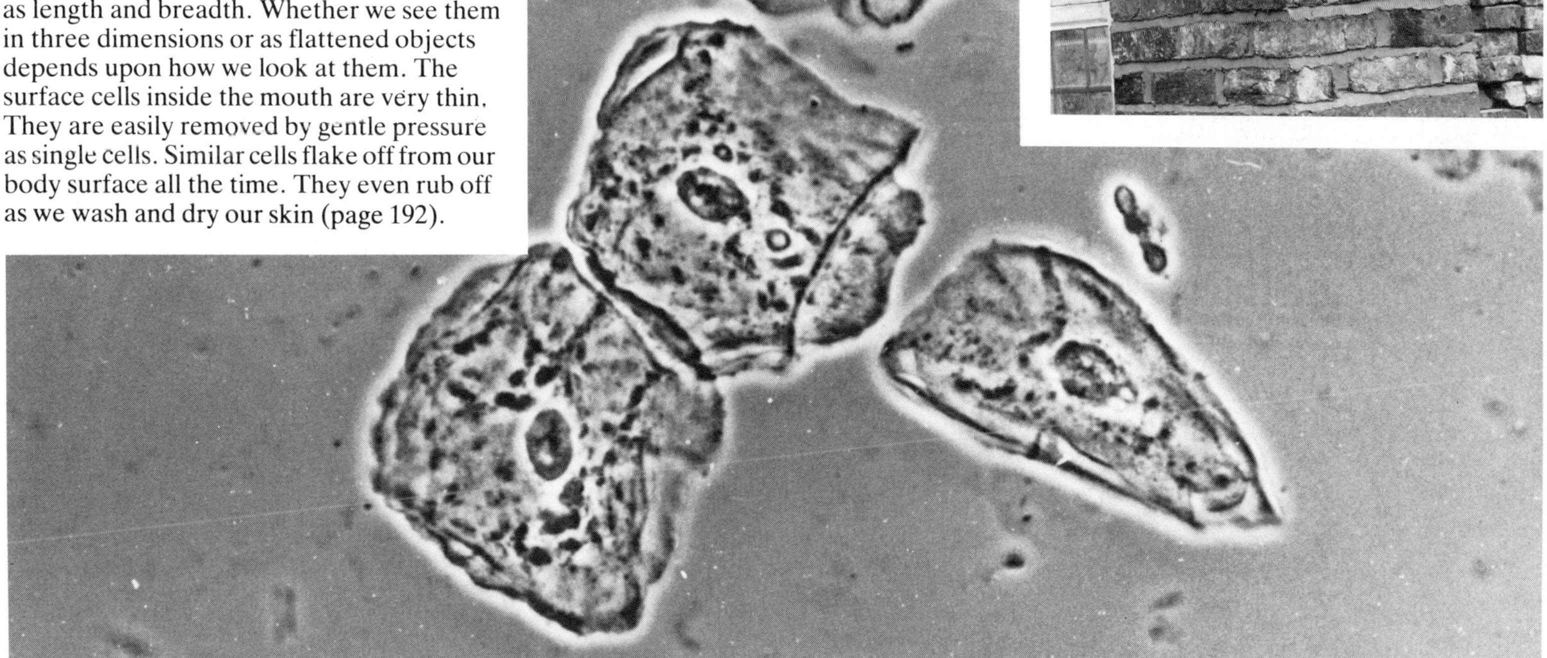

Fig. 1.1 *The building units of a house are bricks and* **Fig. 1.2** *those of the human body are cells*

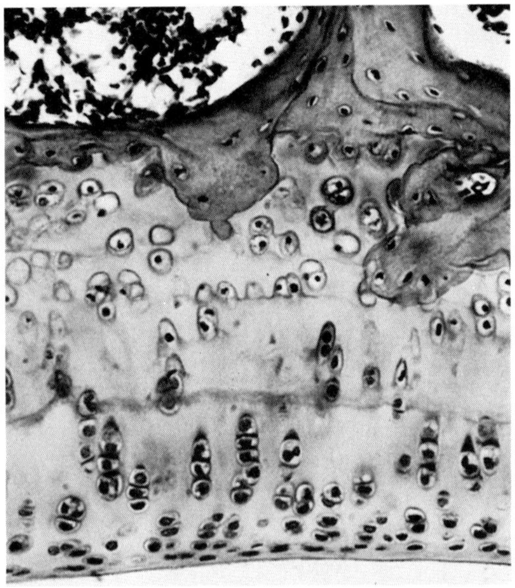

Fig. 1.3 *A very thin slice (section) of cartilage*

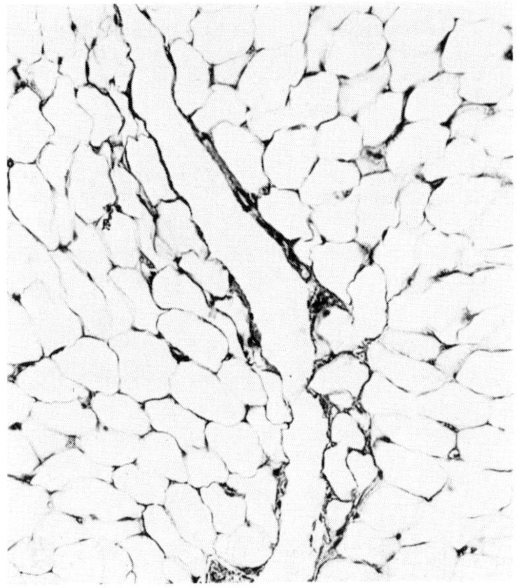

Fig. 1.4 *Body cells that store fat (adipose tissue)*

Figs 1.3 and 1.4 are photographs, taken through a microscope, of cells that are found in other regions of the human body. These cells are not shed from the surface of the body, so it is more difficult to obtain them as single cells. In these circumstances we have to start with a fragment of living material which probably contains hundreds of cells. This makes it hard to see clearly a single cell. It is like trying to see a single colour slide when it is bundled together with a number of others. When the bundle is held up to the light, no light passes through and it is impossible to see the detail on any single slide. As the slides are removed one by one, the light can pass through the thin film and we can see each of them quite clearly. To see cells more clearly a small fragment of living material is taken and sliced up like a loaf of bread. The slices or *sections* are so thin (thousandths of a millimetre) that the details of individual cells can then be seen. Figs 1.3/4 show such sections. To reveal the three-dimensional structure of the cells, the slices can be taken through the material from top to bottom (a longitudinal section) or from side to side (a transverse section).

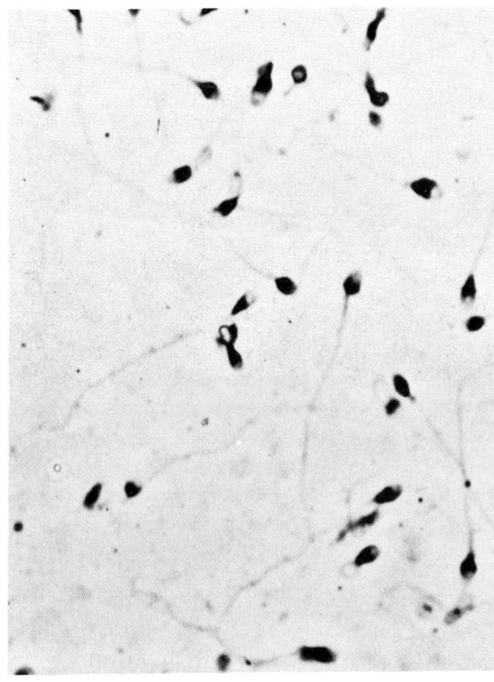

Fig. 1.5 *A man's reproductive cells – sperm*

1.2 Tissues, organs and organisms

By looking at Figs 1.2, 1.3 and 1.4 you may have noticed already that the cells in our bodies are not all the same. When building a house a builder uses special bricks to do different jobs. Foundation bricks are used for strength, air-bricks to let in air. Similarly, in the human body there are different cells which specialize in carrying out different tasks or *functions*.

Although our bodies contain these different types of cell, like all multicellular life forms we begin life as a single cell which is smaller than a full stop on this page. This single cell is formed when a specialized reproductive cell from a man joins up with a reproductive cell from a woman inside the woman's body.

The man's reproductive cell is called a sperm (Fig. 1.5). The woman's reproductive cell is called an ovum (Fig. 1.6). The fusing together of a sperm and an ovum is called *fertilization* (page 251). The single fertilized

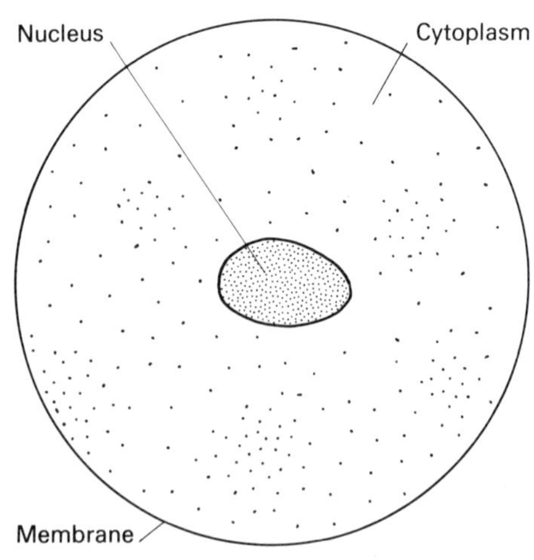

Fig. 1.6 *A woman's reproductive cell – an ovum*

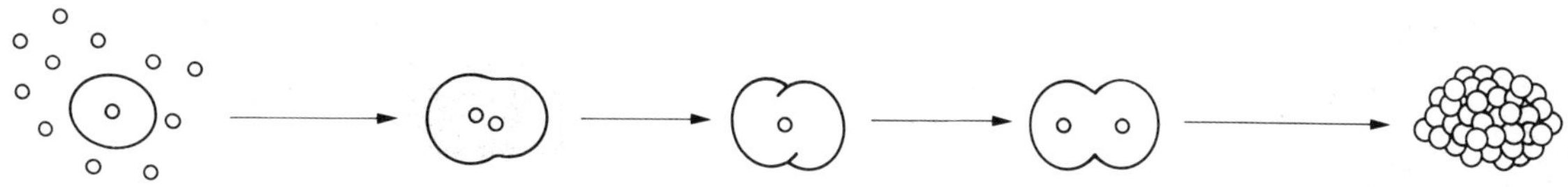

Fig. 1.7 *Fertilization and cell division*

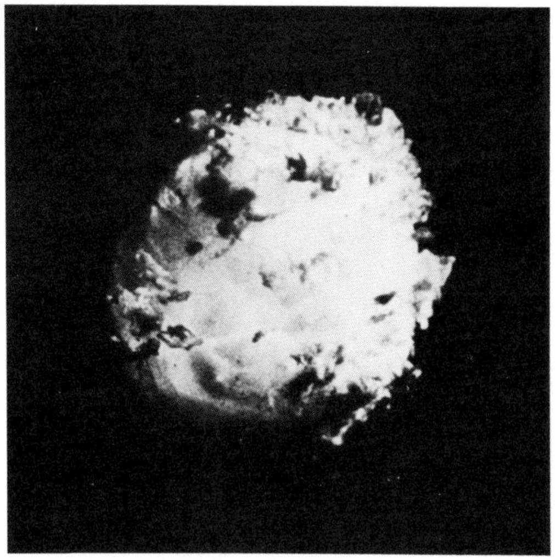

Fig. 1.8 *The embryo at three weeks*

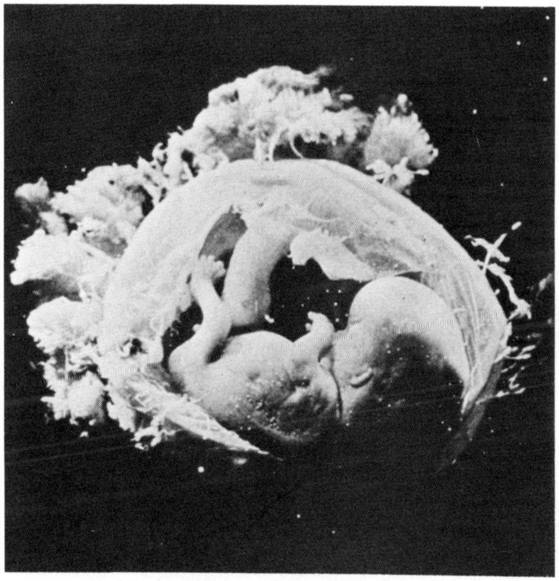

Fig. 1.9 *The foetus at nine weeks*

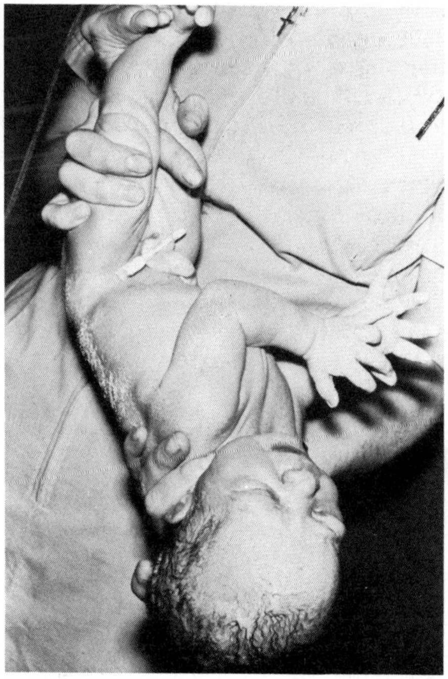

Fig. 1.10 *A newborn baby*

cell that is formed by this union is called a zygote. Within approximately thirty hours this zygote divides into two. After about three days it has become a cluster of cells (Figure 1.7). After 28 days the total mass of cells has become about 3000 million times heavier than the original mass of sperm plus ovum (Figs 1.8–1.9). If you think that this change in size is amazing, look again at Figs 1.8–1.10 and try to see what else has happened. The cells do not remain alike, nor do they form a bigger and bigger ball of cells. Instead, different types of cell are produced and a particular shape is formed.

The development of different types of cell happens as the cells grow and divide. This development is called *differentiation*. The other process that takes place is one of organization. Cells which are alike in structure group together in various regions and work together on a particular activity.

TABLE 1.1 THE MAIN TISSUES OF THE HUMAN BODY

Type of tissue	Purpose	Structure	Examples	Where they may be found – some examples
Epithelial	To line and cover external and internal surfaces	Cells are closely packed. Can consist of a layer one cell thick, or several layers	Pavement (squamous) epithelium 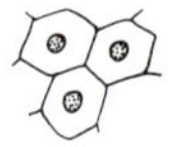 Flat cells (see Fig. 1.2) Ciliated epithelium 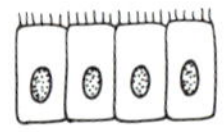 Cells like columns Cubical epithelium 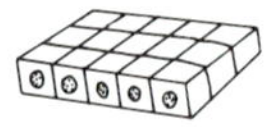 Cells like cubes	Lining of the mouth Passageways of the respiratory system (page 47) In glands
Connective	To bind other tissues together. To fill spaces	Cells are separated from one another like currants in a cake mixture. The mixture between the cells is called the *matrix*. The matrix can contain two types of fibre: white fibres and yellow elastic fibres	Areolar tissue 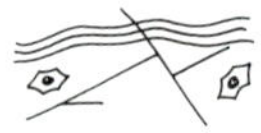 Adipose tissue (fatty tissue)  Fat stored in the cell (see Fig. 1.4) Blood tissue  Blood cells are in a liquid matrix (see page 67)	All over the body as packing material Lower layers of the skin In the circulatory system
Skeletal	To support	Very similar to connective tissue, but the matrix is much harder	Cartilage tissue (see Fig. 1.3) 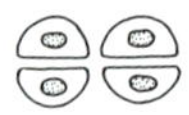 Bone tissue (page 25) 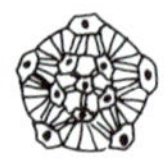	At the ends of bones In the framework of the body – the skeleton
Muscular	To create movement	Made up of a number of fibres which can contract, i.e. become shorter	Striped muscular tissue (see page 32)  Unstriped muscular tissue (see page 35) 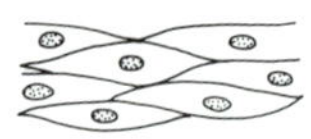 Heart muscle (see page 57) 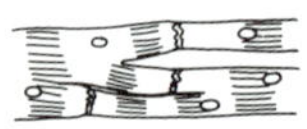	Leg and arm muscles. Walls of the digestive system. We cannot control this muscle Only in the heart
Nervous	To form a communication network within the body	Each cell has many branches leaving its surface. One branch may be much longer than the others	Brain tissue Spinal tissue (see page 228) 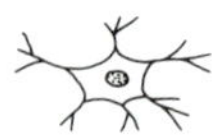 Nerves (see page 223) 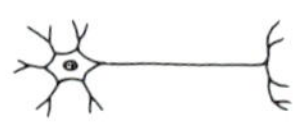	Central nervous system Nerve network all over the body

Cells forming a group in this way are called a *tissue*. Table 1.1 describes the main types of tissue that are found in our bodies.

The growth of tissue leads to the formation of *organs*. Table 1.2 traces the development of some of our organs during the first nine months of life. Each organ is constructed from several tissues. For example, a muscle consists of muscular tissue, blood tissue, nervous tissue and connective tissue. Each tissue supports the activity of the others so that the organ as a whole is able to carry out a certain task. It is like a machine that can carry out a task only because of the special design and action of all of its parts.

TABLE 1.2 **DEVELOPMENT OF ORGANS**

	Time from fertilization	**Development**
FOETUS	30 hours	Fertilized egg is dividing into two cells
	3 days	Now a cluster of cells. In the next few days it will settle down on the wall of the womb (uterus)
	2 weeks	Embryo can be seen as a small dot. Specialization of cells has already started
	4–6 weeks	'Head' and 'tail' regions can be seen. The embryo is developing rapidly. Its heart, in a simple state, starts pumping. The spine and the placenta are both developing
	7–8 weeks	About 1.25 cm long. Head much larger than the rest of the body. Face starting to take shape. Eyes visible as black dots. Limb buds of arms and legs can be detected. The placenta is growing and the umbilical cord is about 2.5 cm long
EMBRYO	10–11 weeks	The embryo is now called a foetus. About 6.5 cm long and 20 g in weight. All body systems are present. Time is now needed for growth and development. Limbs grow longer. Eyelids close over eyes.
	14 weeks	About 12.5 cm long and 100 g in weight. The placenta is fully formed and the foetus fills the uterus, which begins to stretch. First hair appears. Later a softer hair called *lanugo* grows all over the body. Also an oily substance called *sebum* protects the skin
	20–24 weeks	Over 30 cm long. The mother feels movements as the foetus exercises its limbs. Can be seen sucking its thumb. Nails on hands have grown
	28 weeks	About 35 cm long and 1 kg in weight. If it were born now it might survive, but would need special nursing because some systems would not be able to cope, for instance it could not regulate its body temperature, nor suck sufficiently to feed. More time needed
	40 weeks (full-term)	About 50 cm long and 3 kg in weight. Fatty tissue has been laid down under the skin. Lungs are ready for breathing, and the digestive system ready to cope with food. Lanugo is shed. Finger-nails and toe-nails are long

TABLE 1.3 THE NINE SYSTEMS OF THE BODY

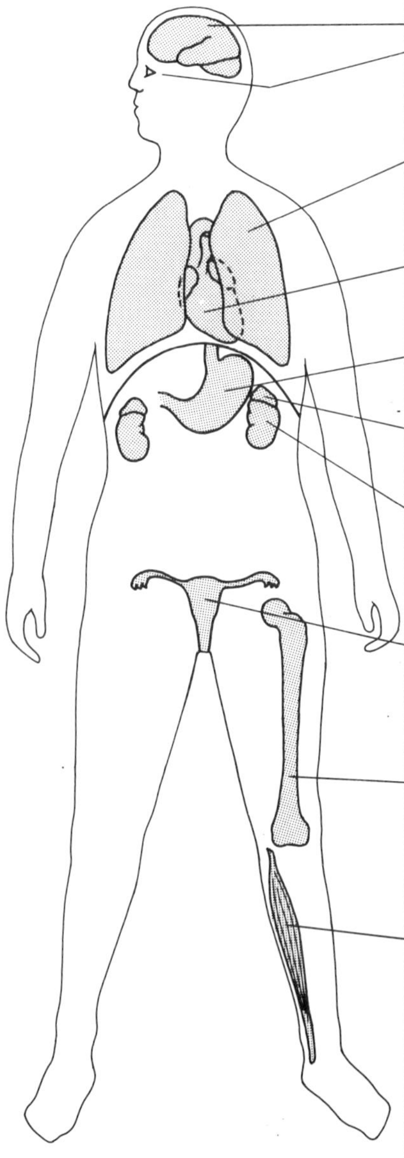

SYSTEM	Some organs working within the system	How the system serves the body
Nervous	Brain Eye	Communication and co-ordination
Respiratory	Left lung	Takes oxygen into the body
Circulatory	Heart	Transports materials
Digestive	Stomach	Processes our food
Endocrine	Adrenal gland	Co-ordination
Excretory	Left kidney	Removal of waste materials
Reproductive	Womb (uterus)	Survival of the species
Skeletal	Thigh-bone (femur)	Supports and protects other systems
Muscular	Calf muscle	Allows movement of the body

A number of machines make up the assembly line of a car factory. They each have their own activity, but they are put to work within a system to produce a particular result – a car. In our bodies organs co-ordinate their activities within nine basic systems (Table 1.3). Each system of organs serves the body in one particular way. For example, the circulatory system (composed of the heart, blood-vessels and blood) deals with the transport of materials around the body.

So in the first nine months of life we develop from a single cell into a complex *organism*. This organism, the body, can exist only when all of the systems, organs, tissues and cells work together (co-ordinate) and support each other's activities. The activities of a cell arise because it consists of *protoplasm*. A cell's activities are the activities of its protoplasm.

1.3 **Cells are 'bags of life'**

Bricks are made from clay, cells are formed from *protoplasm*. This is a complex substance chemically. Its composition varies greatly both within each cell and from one cell to another. Protoplasm contains a large amount of water (about 75 per cent) and a variety of other chemical substances, particularly *proteins* (see page 114). Protoplasm can exist in a semi-solid state, like a jelly, or in a more liquid form. It can even change its state from one form to the other. Whichever state it is in, it forms a 'skin' on its surface called the *cell membrane*. This cell membrane is only a few molecules thick, yet it forms a barrier which can control the movement of materials into or out of the cell. (This action of the cell membrane is discussed in greater detail in Chapter 5.)

The most important fact that separates protoplasm from the clay of the brick and other materials is that it is *alive*. A cell and its protoplasm are rather like a factory. Within both of them a large number of chemical processes are continuously taking place.

These activities occur in anything which is alive, from *unicellular* to *multicellular* life forms. The amoeba shown in Fig. 1.11 is a typical unicellular life form. It is a single cell which is able to exist on its own. We can observe its activities of life quite easily. The activities of life that we see in ourselves are the total of the activities of all the cells in our bodies. (No single cell in our bodies can perform all the activities of a unicellular creature like an amoeba.)

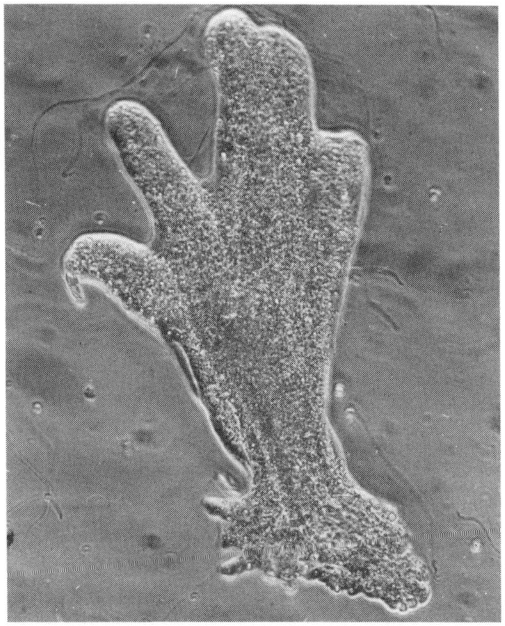

Fig. 1.11 *Amoeba – a unicellular (single-celled) life form*

TABLE 1.4 **THE ACTIVITIES OF LIFE**

ACTIVITY	Description	
Feeding and nutrition	The food which we eat is processed and then sent to our body cells. The cells take in molecules of food to fuel their activities	
Respiration	We breathe in air to supply oxygen to the cells (external respiration). Our cells use the oxygen to release energy from food by *tissue respiration*	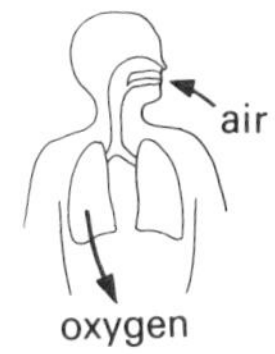
Excretion	The activities of our cells produce waste substances. Our body gets rid of these waste materials by excretion	
Sensitivity	We react to changes that take place in and around our bodies	
Movement	We can move parts of our bodies or our bodies as a whole	
Growth	Our body grows from birth to adulthood. Also, our cells constantly grow and divide to replace worn-out cells	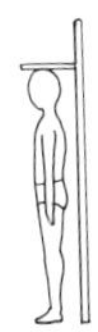
Reproduction	The production of one or more individuals similar to the parents	

1.4 **The activities of life** (Table 1.4)

Feeding

A factory needs to be supplied with energy if it is to keep its processes going. Similarly, the protoplasm of a human body cell needs energy for all its activities. It obtains this energy from the food that passes into the cell through the cell membrane. This taking in of food is one of the activities of living cells. A multicellular form of animal life has to *feed* to supply food to all of the cells in its body.

Respiration
The energy that human body cells need is 'locked' in the chemical bonds that hold together the large molecules of food. To release this energy the protoplasm of the cells acts upon the food, using chemical reactions to break the chemical bonds. This takes place in several stages. As more chemical bonds are broken, energy is released and smaller molecules are formed.

The term *respiration* is used to describe the activity of releasing energy from materials inside the (cellular) protoplasm. You have probably studied burning in other science lessons and you will have discovered that oxygen must be present for burning to take place. Similarly, oxygen is vital to certain stages or types of respiration known as *aerobic respiration*. For example, sugar is oxidized during aerobic respiration, releasing energy, water and carbon dioxide. If we use the 'chemist's shorthand' we can represent this reaction as follows:

$C_6H_{12}O_6$ +	$6O_2$	= $6CO_2$	+ $6H_2O$ +	energy
sugar (glucose)	oxygen	carbon dioxide	water	(2.8 MJ)

The left-hand side of the equation shows the materials that are present at the beginning of the reaction. The right-hand side gives what is produced at the end of a whole series of reactions. Also, in a cell the energy is not released in a flash of flame or puff of smoke! Some of the energy is given off as heat (page 186) but the rest is passed on to other molecules within the protoplasm. These other molecules may use the energy in further reactions or it may remain stored within the molecules for future use. ATP (adenosine triphosphate) acts as a type of energy store, as we shall see later (pages 41 and 186). It is able to make energy available very rapidly.

Excretion
One of the problems of any factory process is the disposal of waste. Waste products are produced by the activities of the protoplasm. If the waste products remained in the cell they would interfere with its activities or even harm its structure. So, like a factory, a cell must rid itself of its waste. The discharge of waste from living material is called *excretion*.

Sensitivity (Irritability)
The survival of a single-celled creature like an amoeba is threatened when there is a change in its surroundings, that is, in its *external environment*. For example, there may be a large change in temperature or the water in the pond may dry up. If the amoeba is to survive it must be able to monitor these changes and take action in response to the situation. Our bodies, like other multicellular life forms, have two environments. One is outside the body, the other inside the body. The development of this *internal environment* means that our body cells are kept in conditions which assist their survival. They are like plants in a greenhouse where the temperature is constant and all the food and water needed is supplied. As we shall see in Chapters 14 and 18, our bodies monitor and respond to changes in both their internal and their external environment. This enables man to survive in a wide variety of conditions.

Movement
A response or a search for food can involve movement. Movement of part or the whole of the body is another sign of life.

Growth
Not all of the materials taken into the protoplasm are used to provide energy. Some are used to form the building materials of the cell itself. This building process is called *assimilation*. Adding to the cell's structure in this way can lead to an increase in size, i.e. *growth*. The growth of a cell leads to another development. For example, think of a canteen in a school or factory. The canteen staff's activities are geared to the number of people they have to cater for every day. If they were suddenly asked to cater for another hundred people they might very well find it impossible to meet the extra demand. To cope with this increase in numbers another unit might have to be built on to the canteen. For similar reasons there is a limit to the growth of a cell. When this limit is reached the cell will divide. Cell growth and division take place all the time in our bodies, so that cells which are worn away or destroyed by injury can be replaced. Growth may mean not only an increase in size, but also a *change* in form which can be quite complex.

Reproduction
The reproduction of one or more individuals similar to the parent is a positive sign of life.

1.5 The structure of cells

The variety of substances that are present in protoplasm gives rise to the many different structures and chemical properties that are seen in cells. Yet when we look at our body cells very closely we see that they do share certain features. A typical cell may be drawn (Table 1.5) to show those parts which are found in most of our cells. The cell's protoplasm can be divided into two main regions: the *cytoplasm* and the *nucleus*. Within the cytoplasm other structures are found. These are responsible for certain activities within the cell (see third column in Table 1.5). However it is the nucleus that appears to rule over the cell's activities, structure and formation. We say that the nucleus is in charge of the cell's *metabolism*.

To see some of the cell's structures more clearly a section is usually passed through dyes or stains. Structures such as the nucleus absorb more stain than other areas of the cell. All of the cells you have looked at so far have been treated this way.

TABLE 1.5 STRUCTURE OF A TYPICAL ANIMAL CELL

	Name	Form	Function
The size of most cells lies between 10 and 100 μm (1 μm = $^1/_{1000}$ mm)	**Cell membrane**	Very thin outer layer of cytoplasm	Controls movement of substances in and out of cell. Appears to 'stick' cells together
	Cytoplasm	Transparent jelly-like substance	Most of the cell's chemical reactions take place inside the cytoplasm
	Mitochrondia	Thin rod-like structures in the cytoplasm	Play an important part in the release of energy within the cell
	Nuclear membrane	Thin layer. Separates nucleus from cytoplasm	Without a nucleus a cell cannot divide or grow. It appears to control the cell's *metabolism*
	Nucleus	Spherical or oval body in the cell. Stains more darkly than the cytoplasm	
	Chromosomes	Solid bodies in the nucleus	Genetic material (see Section 1.6)

1.6 The cell nucleus and genetic material

The builder in Fig 1.1. has a plan which tells him how he is to build the house. The 'plan' that a cell works to is contained within its nucleus and is called *genetic* material. This genetic material can be seen through a microscope in a cell that is about to divide. It shows up very clearly as a collection of thin ribbons if the cell is stained with a special dye (see Fig. 1.12). These ribbons are called *chromosomes* and there are forty-six of them in the nuclei of normal human body cells. (The cells of other animals and plants contain a different number of chromosomes and this gives them a different plan to work to.)

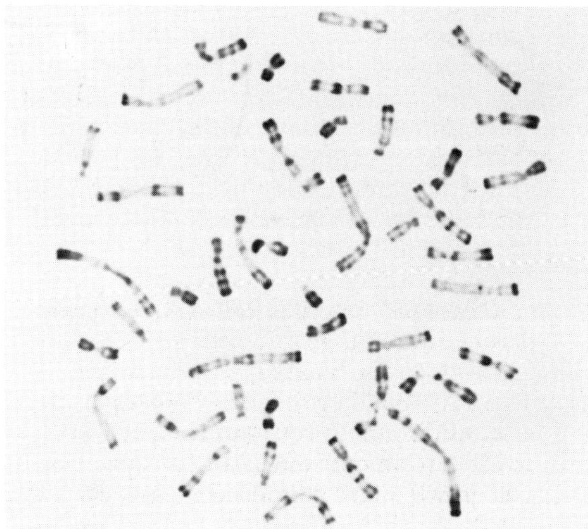

Fig. 1.12 *Chromosomes*

Genetic material is formed from protein and a substance known as *DNA* (deoxyribonucleic acid). Before the structure of chromosomes and DNA was investigated, scientists who studied characteristics within a family could only guess at the existence of something in the cell that 'gave the instructions' for a certain characteristic, such as hair colour or eye colour. They used the word *genes* to describe these unknown 'instructors' and imagined them to be lined up on the chromosomes like beads on a necklace.

When the structure of DNA was investigated it was discovered that there was an orderly arrangement of certain chemicals along the DNA molecule. Most of us are familiar with Morse code, in which series of dots and dashes are used to represent letters of the alphabet (Fig. 1.13a). In a chromosome it is the arrangement of the chemicals in the molecule which forms a code that the cell can 'read' (Fig. 1.13(b)). A gene is now regarded as a chemical group in the chromosomes that gives a 'word' of instruction. The genes instruct the cell on the manufacture of its protoplasm and its activities. That is how our cellular structure and our body structure are determined by the code in the cells' genetic material.

(a)
If we know Morse code we can translate a message like this, and act on its instructions

(b)
A cell is able to translate the coded message of its chromosomes and act on these instructions

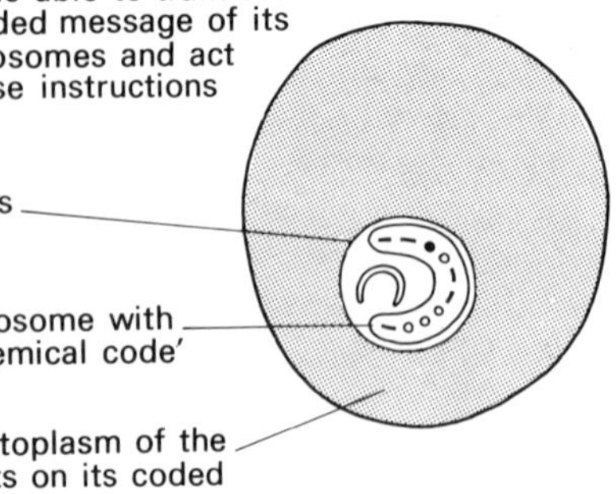

A group such as o•— could form one gene

Fig. 1.13

1.7 Cell division – mitosis

When our body cells are damaged or wear out they must be replaced by new cells. This takes place automatically as other cells divide. Cell division also provides the body with the ability to grow. However, if the new cells are to have the correct structure the body's genetic code must be handed on when the cells divide. To make sure that the new cell contains the same forty-six chromosomes a process of division called *mitosis* takes place.

Mitosis When a cell is about to divide, each chromosome appears to become shorter. At the same time each chromosome makes a copy of itself (its double). It appears to have two halves which remain joined at one point (Fig. 1.14(a)). As division continues, the chromosomes move into the centre of the cell (Fig. 1.14(b)). Then the doubles of each chromosome are drawn apart and end up at opposite ends of the cell (Fig. 1.14(c)). A nucleus membrane forms around each group of chromosomes. The cell becomes waisted and soon splits into two (Figs 1.14(d) and (e). Each daughter cell now has the same number and identical chromosomes to those in the parent cell. The whole process can take place in six to twenty-four hours.

There is one type of cell in our bodies that is not produced by mitosis. These are the reproductive cells of a man and a woman. We have seen in Section 1.2 how we all develop from the fertilized cell formed when the nuclei of a sperm and ovum fuse together. If the sperm and ovum each contained forty-six chromosomes in their nuclei when they fused

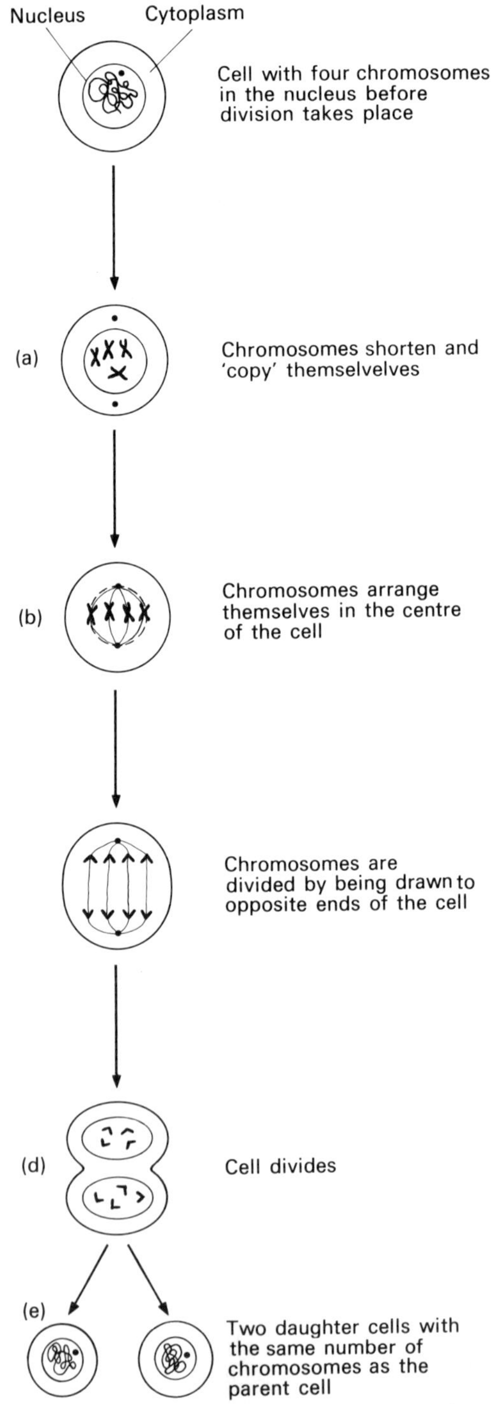

Fig. 1.14 *Mitosis*

together, the nucleus of the fertilized cell would have ninety-two chromosomes. This would mean that every new generation would have twice as many chromosomes as their parents! However, our human form depends upon our genetic material containing forty-six chromosomes. So sperm and ova must be produced with fewer chromosomes than their parent cells.

Question 1 How many chromosomes must a sperm and ovum each contain if the fertilized cell is to have forty-six chromosomes?

1.8 The formation of reproductive cells by meiosis

Meiosis is the name given to the cell division that takes place when reproductive cells are formed. In meiosis the number of chromosomes that the daughter cell contains is half that of the parent cell. This is a much more complicated form of division, but a simplified description is given here. The chromosomes arrange themselves into twenty-three pairs which match one another. As they lie close together they exchange ('swap') pieces of themselves. The pairs of chromosomes then separate and the partners from each pair move to opposite ends of the cell. The cell divides and forms two daughter cells. Each daughter cell now contains half the number of chromosomes of the parent cell (Fig. 1.15(a)).

Now when a sperm and ovum fuse the full number of forty-six chromosomes is restored within the fertilized cell. Half of these will have come from the mother and the other half from the father (Fig. 1.15(b)). A chromosome from the mother can be paired in each case with one of the same type from the father. The characteristic size and shape of the twenty-three pairs enables scientists to group them.

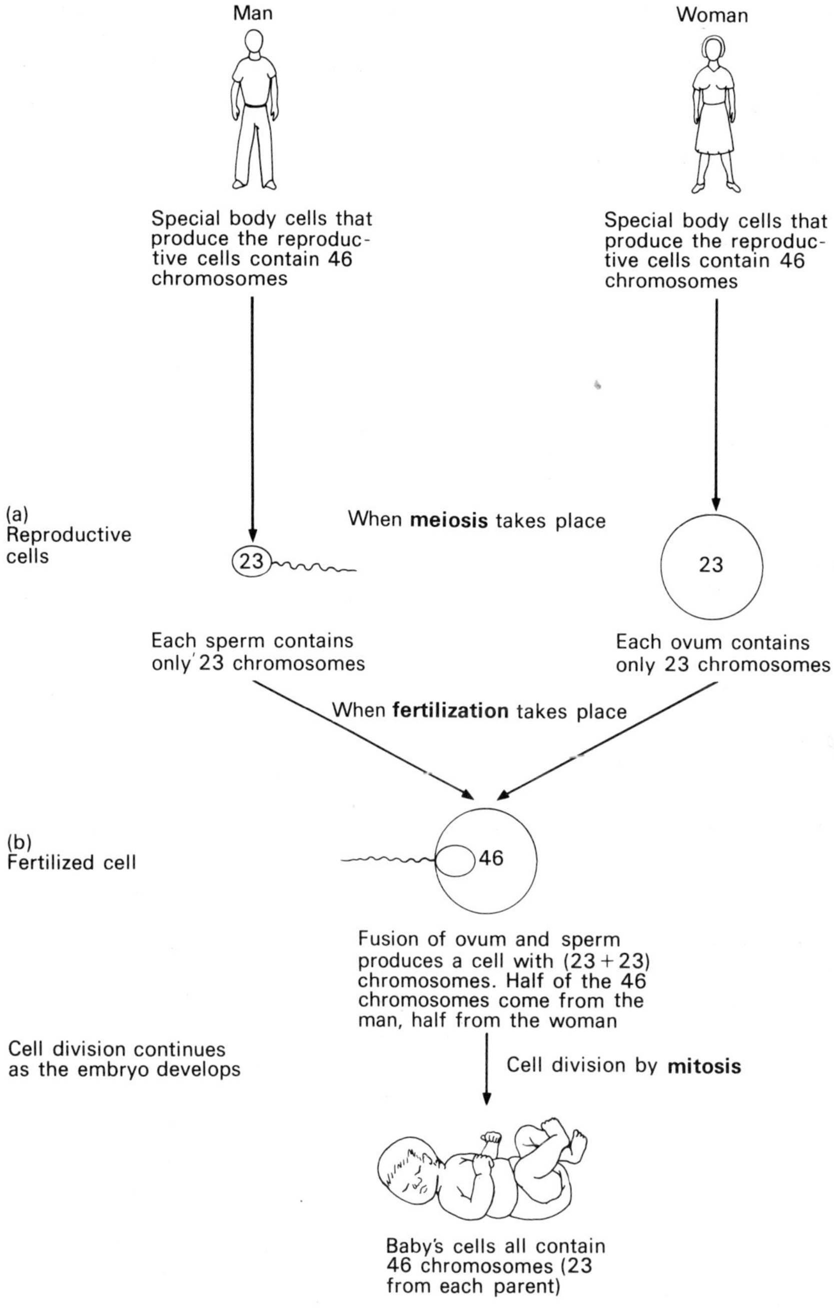

Fig. 1.15 *The formation of reproductive cells by meiosis*

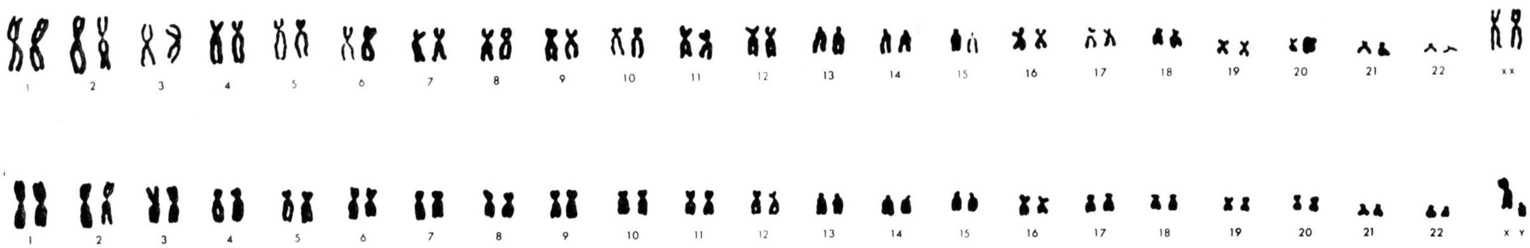

Fig. 1.16 *(a) Chromosomes from a woman's body cell. (b) Chromosomes from a man's body cell*

Question 2 What do you notice about the last pair of chromosomes in Fig. 1.16(a) compared with those in Fig. 1.16(b)?

1.9 How chromosomes determine sex

When chromosomes from a man's body cell are arranged in pairs we find, as shown in Fig. 1.16(b), that we are left with two chromosomes that do not match. These two chromosomes are called the *X-chromosome* and the *Y-chromosome*. It is only in cells from a man that these two unmatched chromosomes are found. In the cells of a woman there are instead two indentical X-chromosomes, as can be seen in Fig. 1.16(a). When a woman's reproductive cells are produced by meiosis, each ovum will carry an X-chromosome (Fig. 1.17). However, in the case of a man, half of the sperms that are produced at any one time will carry an X-chromosome, while the others will carry a Y-chromosome. It takes only one sperm to fertilize one ovum. If this is an X-carrying sperm then the baby will be a girl. If it is a Y-carrying sperm then the baby will be a boy. An X-carrying sperm and a Y-carrying sperm have approximately even chances of fertilizing an ovum. Recent figures show that on average, 106 boys are born for every 100 girls.

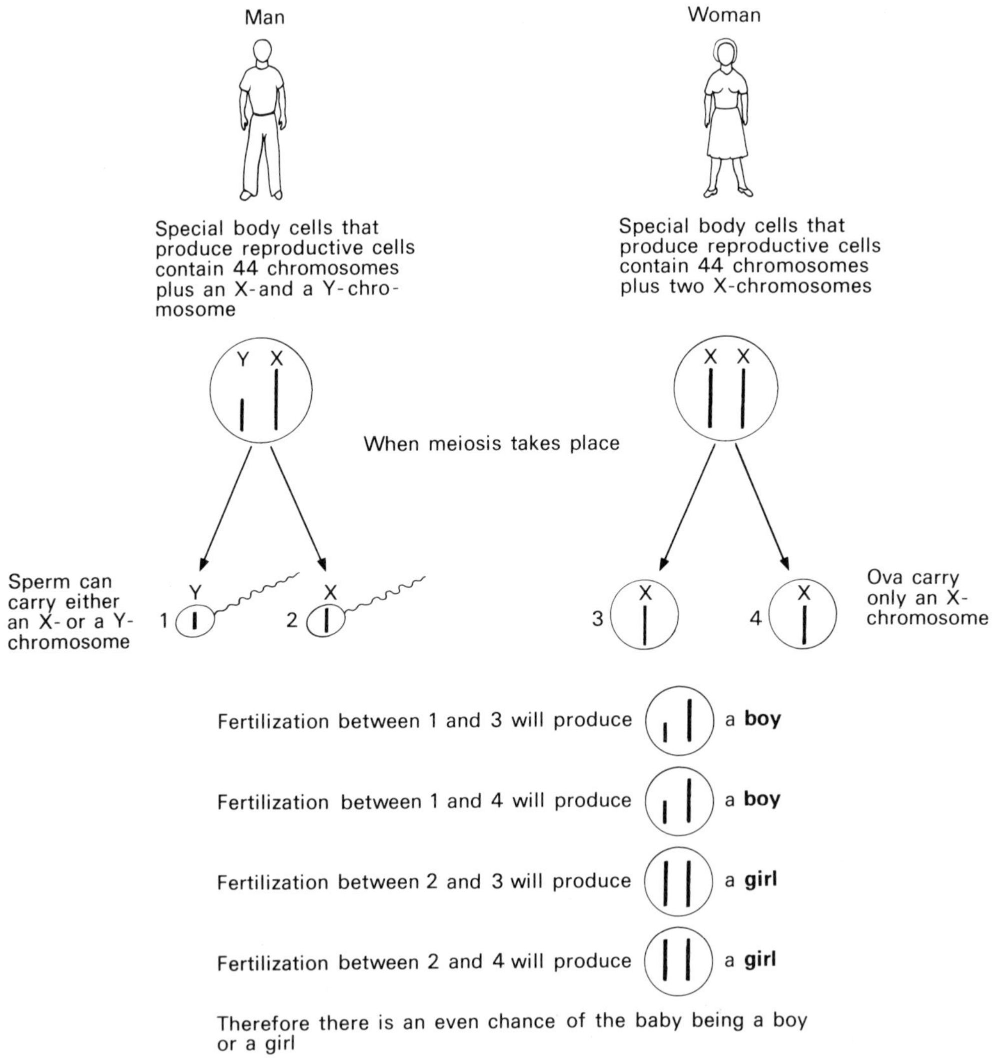

Fig. 1.17 *How the X- and Y-chromosomes decide the sex of the baby*

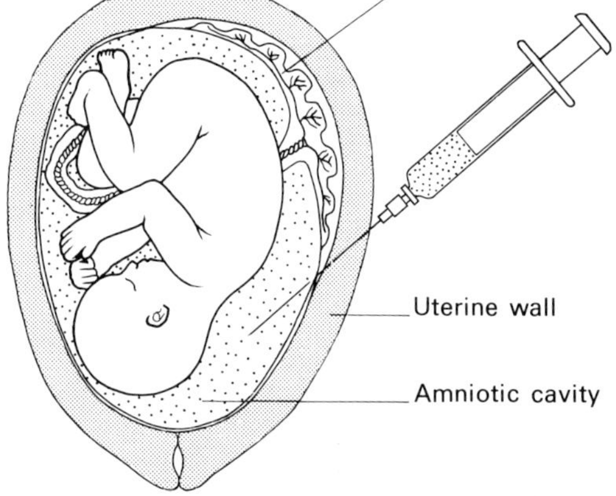

Fig. 1.18 *The process of amniocentesis*

The sex of a baby can be discovered before birth by collecting some of its cells and identifying the chromosomes. The baby's cells are shed into the fluid which surrounds it as it lies inside its mother's *womb* (page 243). The procedure in which some of this fluid is drawn off is termed *amniocentesis* (Fig. 1.18). The test may be performed in the early stages of pregnancy. However it is done only for special medical reasons, when the doctor knows there is a risk of the baby having a particular disorder. The disorder may be linked with the age of the mother (see Down's syndrome – page 14). Alternatively there may be a history of other disorders in the family. To understand this second example we need to know how we inherit our bodies' characteristics.

1.10 **Inheritance**

In Section 1.8 we saw how the nucleus of a fertilized egg ends up with twenty-three chromosomes from the mother and twenty-three chromosomes from the father. In this way chromosomes from each parent are handed down to a child. We say that the child has *inherited* its parents' genetic material. Each chromosome carries genes, which give the instructions for cellular and body structure. In a pair of chromosomes there are matching genes. For example, in Fig. 1.19 gene B on chromosome I instructs on eye colour and so does gene B at the same position on chromosome II. Therefore in every cell there are two genes that instruct on each characteristic. Which characteristic develops depends upon the combined effect of these genes. For example, let us imagine that gene B instructs that the eyes should be brown. The two genes BB will ensure that the person has brown eyes. However there may be another gene present in the pair of chromosomes, gene b, which instructs that the eyes should be blue. So a person with blue eyes will have on their chromosomes I and II two genes bb, instead of genes BB (Fig. 1.19).

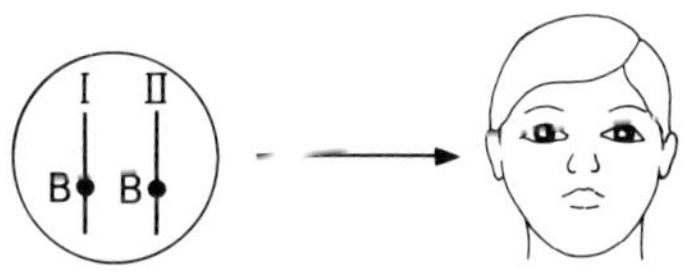

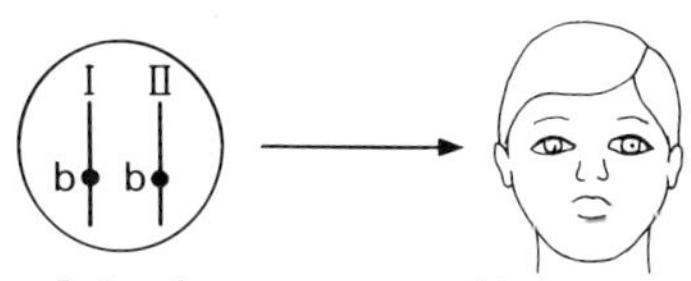

Fig. 1.19 *How our genes decide the characteristics of our bodies (example – eye colour)*

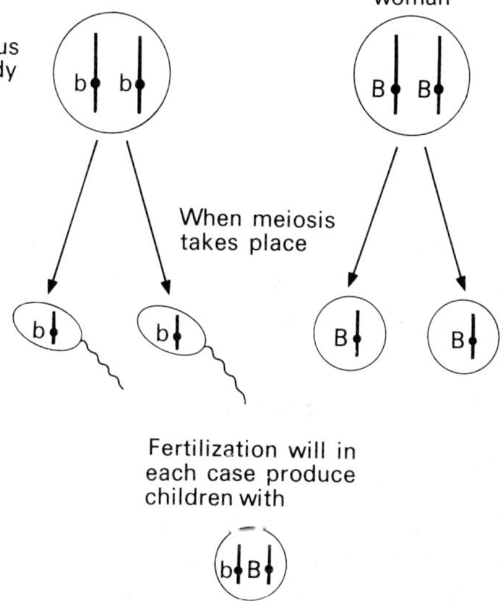

Fig. 1.20 *How we inherit our characteristics*

You may now be asking whether there could be a gene B on chromosome I and a gene b on chromosome II, and if so what colour the eyes would be. In fact, chromosomes I and II could well be Bb. Fig. 1.20 shows that this comes about through the separation of chromosomes I and II at meiosis. But what is the eye colour of the children? In fact it is brown, because gene B 'overpowers' gene b. We say that gene B is *dominant* while gene b is *recessive*. The eye colour of the children is therefore determined by the dominant gene. However the children's genetic material still contains the recessive gene and this can be handed on to future generations (Fig. 1.21). In this way our outward appearance (our *phenotype*) may not truly represent the genes that make up our genetic material (our *genotype*).

Question 3 Using Fig. 1.21, do you think it is possible to have brown-eyed children if both parents have blue eyes?

This description of the process of inheritance simplifies a complicated story. In fact there are usually several genes that instruct on an individual characteristic.

We can now see how sexual reproduction provides the opportunity for genetic material to be rearranged. At one stage in meiosis sections of the paired chromosomes are exchanged, so that a regrouping of genes takes place. Then at fertilization chromosomes from two different sources are mixed together. It is rather like pooling the letters from two or three words and seeing how many other words you can build from the assortment of letters. The rearrangement of genetic material leads to countless *variations*. Although we are more like our immediate family than anyone else, no two people are exactly alike. The genetic material in our cells is not exactly the same as our family's and this makes us *individuals*.

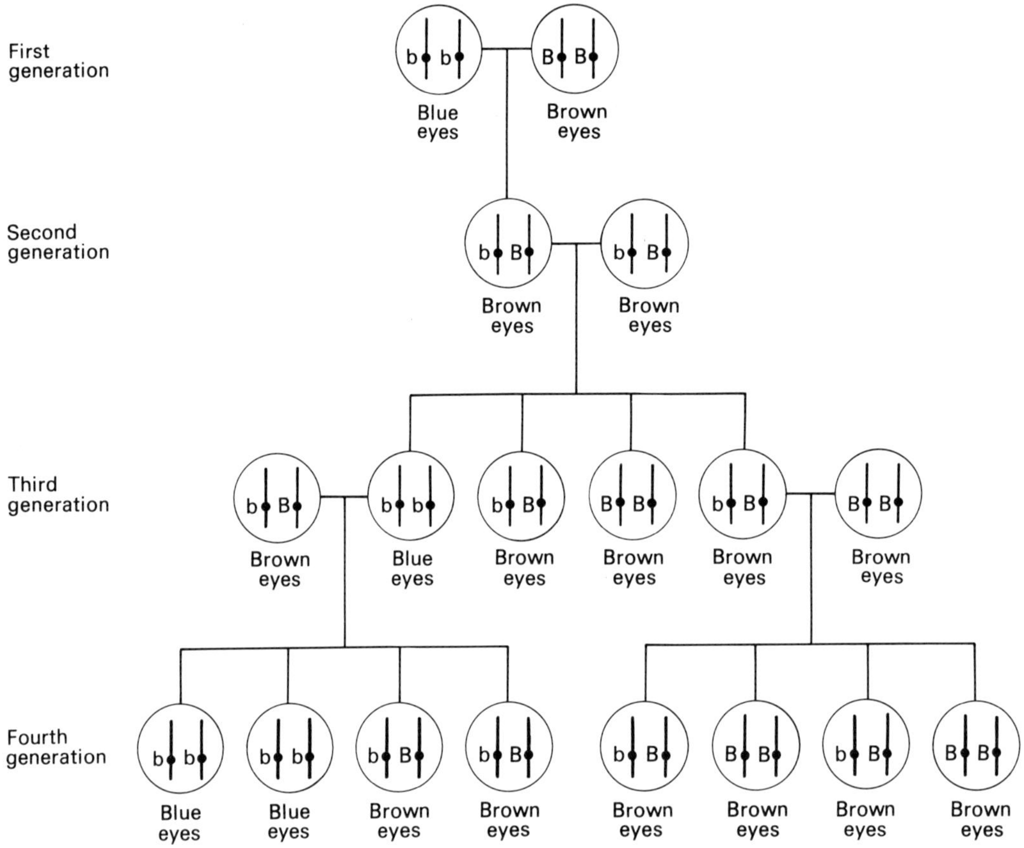

Fig. 1.21 *Eye colour through successive generations*

1.11 **Genetic disorders**

Some physical disorders within the body are the result of changes in the structure of the chromosomes or genes. *Down's syndrome* occurs when an extra chromosome is produced in an ovum. If this ovum is fertilized the developing child will have forty-seven chromosomes in its cells. When a chromosome or the genes change spontaneously like this the cell is said to have *mutated*. Mutations may cause a physical disorder so serious that the individual cannot survive.

Certain factors are known to increase the chances of mutation taking place. Examples of external factors are *radioactive rays* and *X-rays*. People working in environments where there is a high incidence of these rays are carefully protected and regularly screened (page 27).

In the previous section we saw how the characteristic of a recessive gene can appear when the gene is present on both members of a pair of chromosomes. Recessive genes can give rise to physical disorders such as *albinism*, when a person's skin and hair fail to manufacture their colouring pigments (page 194).

Red–green colour-blindness and *haemophilia* are caused by recessive genes which are linked to the chromosomes that determine sex. Although the X- and Y-chromosomes both carry genes that determine sex, the longer X-chromosome carries additional genes that instruct on other characteristics. Because the genes for these characteristics appear on the X-chromosome and are not paired on the Y-chromosomes, we say they are *sex-linked*. Fig. 1.22 shows how red–green colour-blindness is passed on through successive generations.

A *carrier* is a person who carries a recessive gene in their genetic material but does not show the characteristic. They can pass the recessive gene on to the next generation. Women act as carriers of haemophilia. Haemophilia is a disease in which the blood fails to clot (page 69). It is passed on in a family in the same way as red–green colour-blindness. When a woman whose family has a history of haemophilia becomes pregnant, she may wish to have only girls and not risk having a boy who could suffer from this very painful and disturbing disease. Amniocentesis will tell her the sex of her child. She and her family may wish to decide whether or not to continue with the pregnancy.

People who want to start a family and who are aware of genetic disorders within their family can get advice from local genetic clinics under the National Health Service (genetic counselling).

The state of our bodies at any one time is the result of our inherited make-up and outside influences. We have no control over our genetic inheritance but we may have some control over these other influences. This is important as it is these influences that will affect the well-being or health of our body.

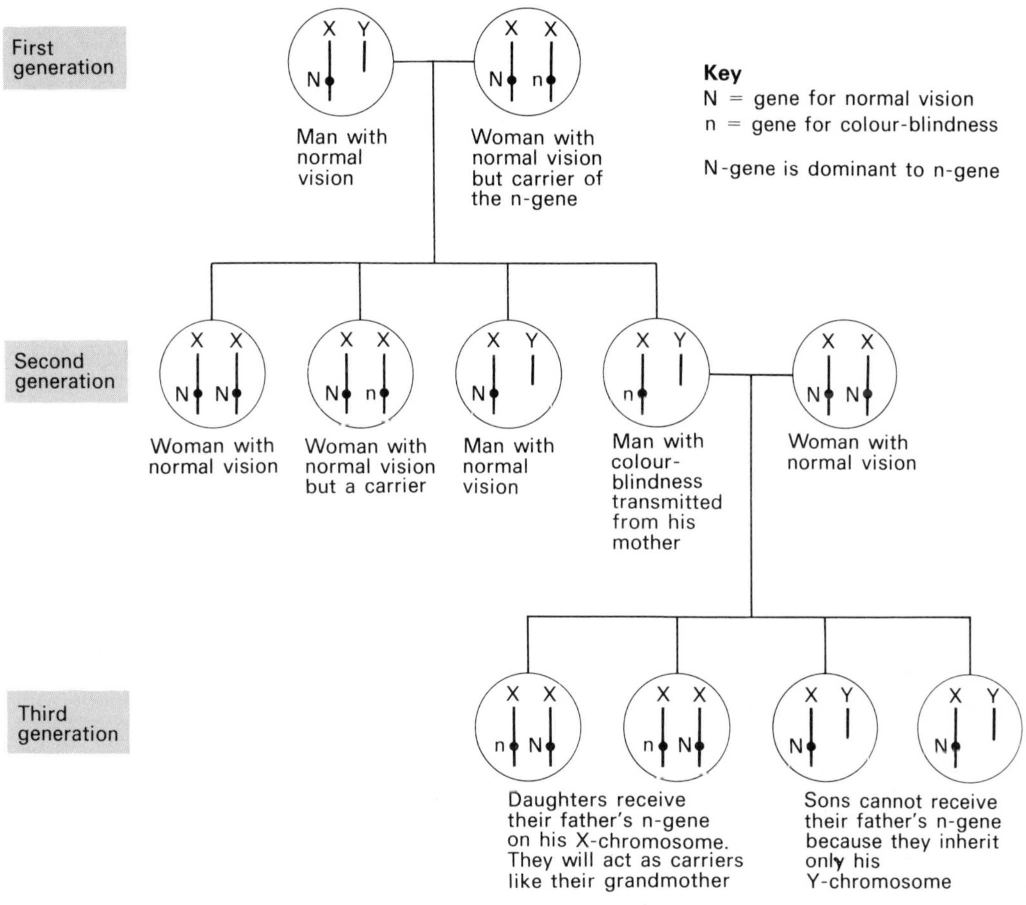

Fig. 1.22 *Red–green colour-blindness (a sex-linked characteristic): inability to tell the difference between red and green*

Answers to questions in the text

1 A sperm and ovum must each contain twenty-three chromosomes.

2 The last pair of chromosomes in Fig. 1.16(b) do not match: one chromosome is longer The last pair of chromosomes in Fig. 1.16(a) do match. They resemble the longer one of the pair in Fig. 1.16(b).

3 The gene b for blue eyes is recessive. Therefore the gene for brown eyes cannot be present in the parents' cells nor in the sperm or ovum. So there cannot be any brown-eyed children.

THINGS TO DO

1 Using thin card, draw and cut out an outline of the human figure. Draw and cut out different overlays which you can use on the figure to show the positions of the various organs and systems.

2 Obtain some thin pieces of card approximately 6 cm long and 3 cm wide. On one half of the cards draw a stage in the division of a cell by mitosis. Punch two holes in the opposite ends of the cards. Put a piece of string through the holes so that the cards are bound together. Now if you flick over the cards you should see a 'moving picture' of mitosis take place.

3 Collect information on the physical characteristics of as many members of your family as you can. Try to construct a 'pedigree chart' for one of these characteristics, for example eye colour.

4 Find out what is meant by genetic engineering. Select one crop or one animal and use it to illustrate how man's knowledge of genes has enabled him to improve the stock or fashion it to a particular environment.

THINGS TO WORK OUT

In Questions 1–4 write down the *one* answer which you think is correct.

1 A group of identical cells forms

A an organ
B an organism
C a system
D a tissue
E a membrane

2 A woman's skin cell has

A 23 chromosomes
B 48 chromosomes
C 24 chromosomes
D 23 pairs of chromosomes
E 46 identical chromosomes

3 The cytoplasm of a cell is

A the material inside the nucleus
B the contents of the cell which surrounds the nucleus
C the outer 'wall' of the cell
D all the cell's contents
E the outer 'wall' of the nucleus

4 A female brown mouse was mated with a male black mouse. The baby mice were all black. If the baby mice were interbred when they reached maturity the phenotypes of the next generation would be

A all black
B 3:1 black to brown
C 2:2 black to brown
D 1:3 black to brown
E all brown

5 Copy down the list of words. After each word write down the statement which you think describes the meaning of that word.

Words
respiration excretion reproduction
sensitivity growth

Statements
(a) removal of waste products
(b) response to changes in the environment
(c) breakdown of food to release energy
(d) increase in size or complexity
(e) parents producing offspring

6 (a) Give *one* function of the following regions of a cell: (i) nucleus, (ii) mitochondria, (iii) cell membrane.
(b) In which region of the cell is the genetic material located? Describe the appearance of this material before the cell divides.
(c) Name *three* types of tissue that are present in your ear. Draw a simple diagram to show the special features of *one* of these types.

7 Name an organ that is part of

(a) the respiratory system
(b) the circulatory system
(c) the digestive system
(d) the reproductive system
(e) the skeletal system

Give the approximate location in the human body of each of the organs that you have named (you could draw a simple diagram).

8 (a) Sperms may be grouped into two types according to the chromosomes which they contain. Explain this statement.
(b) Explain what is meant by dominant and recessive genes.
(c) How many chromosomes are present in a daughter cell after cell division by (i) mitosis, (ii) meiosis?
(d) What specialized cells are formed through cell division by meiosis? Give the name of this particular type of cell in a woman.

9 (a) Assume that hair colour is determined by a single pair of genes. Use the symbol B for black hair and b for fair hair. B is dominant to b. Write down the possible genotypes for a person with (i) fair hair, (ii) black hair.
(b) Work out all the possible genotypes of the children of a black-haired father and a fair-haired mother.
(c) The following are the genotypes of a husband and wife:

X X	X Y
H h	H

(i) What name is used to describe the inherited characteristics determined by genes such as H and h?
(ii) Write down the possible genotypes of the children.
(iii) Give an example of an inherited characteristic that could be determined by the genes H and h.
(iv) Describe the phenotypes of the children, using the characteristic you have named in (iii).

2 From the Beginning

2.1 **Care before birth**

Until now your well-being, like that of all other young people, has been the concern of others. This was true even before birth. In Fig. 2.1 the astronaut on his space walk depends upon the space capsule and his lifeline to it. As we saw in Fig. 1.10, the *foetus* inside the mother is dependent upon the mother and has a similar lifeline.

The cells of the foetus must be supplied with food and oxygen, and the waste products that they make must be removed. Both of these tasks are carried out by the blood system of the foetus. This blood system has to come into contact with the mother's blood system because it is her body which will supply the food and oxygen. It is the mother's body that will also dispose of the waste. The two blood systems intertwine within a special organ (Fig. 2.2) called the *placenta* (page 5). This is usually found on the rear wall of the *uterus* (womb). At first it is a mesh but quickly develops into a more solid, round organ (eventually 20 cm wide and 2.5 cm thick). On the side of the placenta which is next to the uterus wall there are *villi* (Fig. 2.3). Villi are like fingers, and they sink into the thick uterus wall. (Imagine pushing your fingers into a squashy balloon). The wall of the uterus, with its increased blood supply, surrounds the villi. Thus the mother's blood is separated from the foetus's blood by very thin walls. Food, oxygen and waste products can pass through these walls.

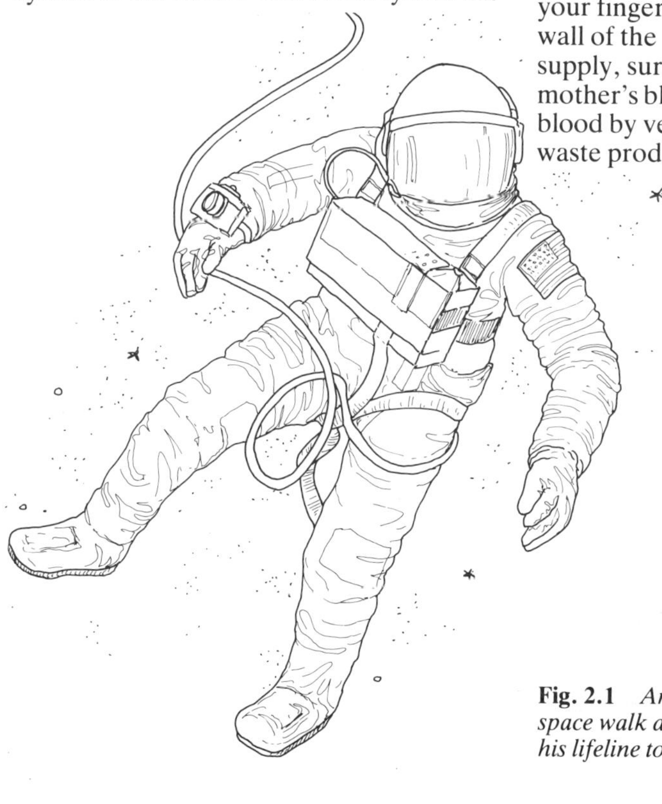

Fig. 2.1 *An astronaut on a space walk depends upon his lifeline to survive*

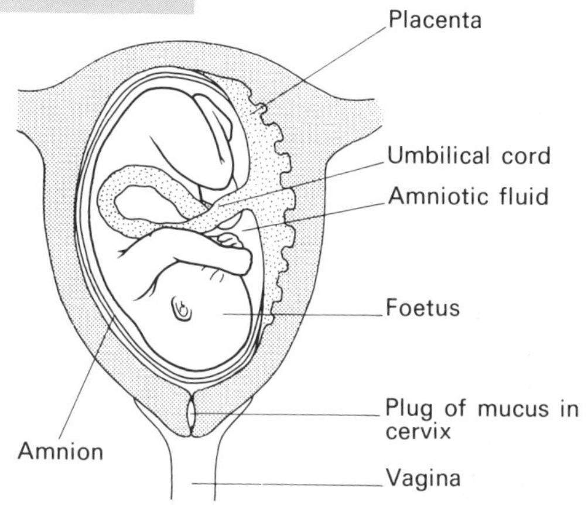

Fig. 2.2 *A foetus at full term*

Question 1 There is no direct connection between the blood system of the mother and the blood system of the foetus. Can you suggest why this is important for the foetus?

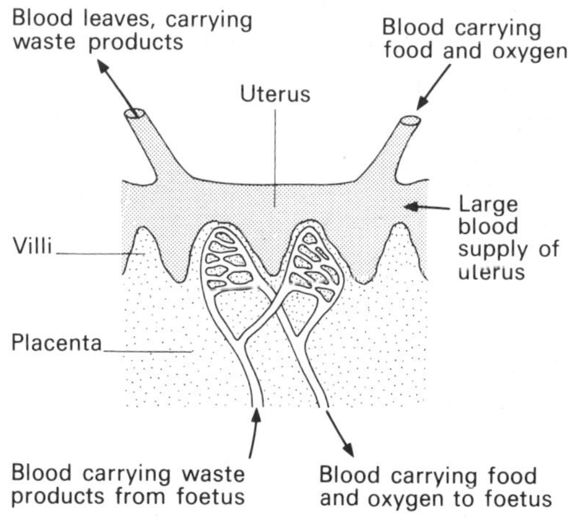

Fig. 2.3 *How the villi function*

The *umbilical cord* connects the foetus to the placenta. This cord grows from the stomach region of the foetus to a length of about 50 cm and gives it freedom to move. The umbilical cord contains three blood-vessels. One (a vein) carries oxygen and food to the foetus, the other two (arteries) carry waste products away to the placenta.

Question 2 What is your navel or 'tummy button'?

The developing foetus floats in fluid contained in a double-walled sac. The *amniotic fluid* and *amniotic sac* protect the foetus against extreme changes in temperature. They allow movement yet act as a cushion against the jolts and bangs that come as the mother moves around.

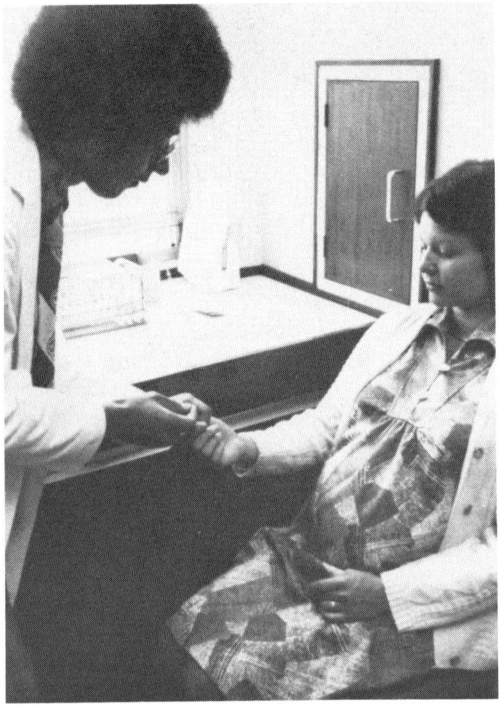

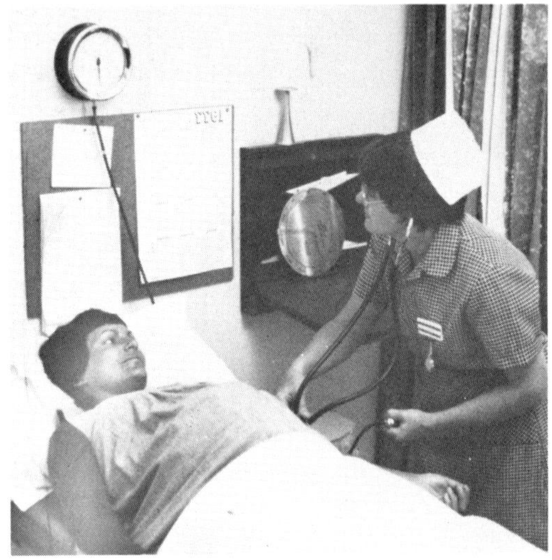

Fig. 2.4 *Some of the checks carried out on the expectant mother at an antenatal clinic. Left (a) A blood sample is taken. Above (b) Blood pressure is recorded. Below (c) From about the twenty-second week the heartbeat of the foetus can be heard.*

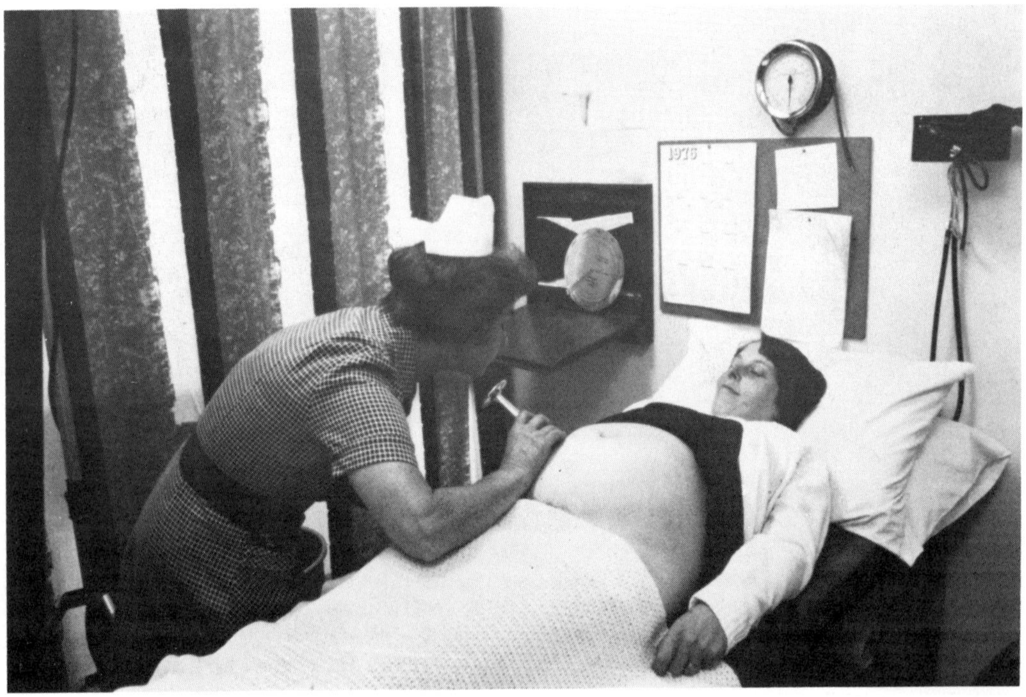

2.2 Care of the expectant mother

Generally an expectant mother can lead almost as active a life as she did before her pregnancy. However, as she is her child's life support system, she can be more certain about its well-being if she keeps her body in good health. Doctors, antenatal clinics and hospitals are able to provide the mother with all the advice and help she requires. However it is up to her to approach them, and it is best for her to do this in the very early stages by visiting her family doctor. A written record is kept of the mother's health. She is examined; her heart and lungs are listened to; her blood pressure (page 74) and a blood sample are taken (Fig. 2.4). The blood sample is tested to find the mother's blood group (page 70) and to check for infection (p. 248). Also the number of red blood cells present in the sample will show whether or not she is *anaemic* (page 68). This is a possibility because the foetus needs iron to make its own red blood cells. It takes this iron from the mother, in fact about one-third of her supply. The mother must replace the iron if she is to make enough red blood cells for herself. She may do this by eating food rich in

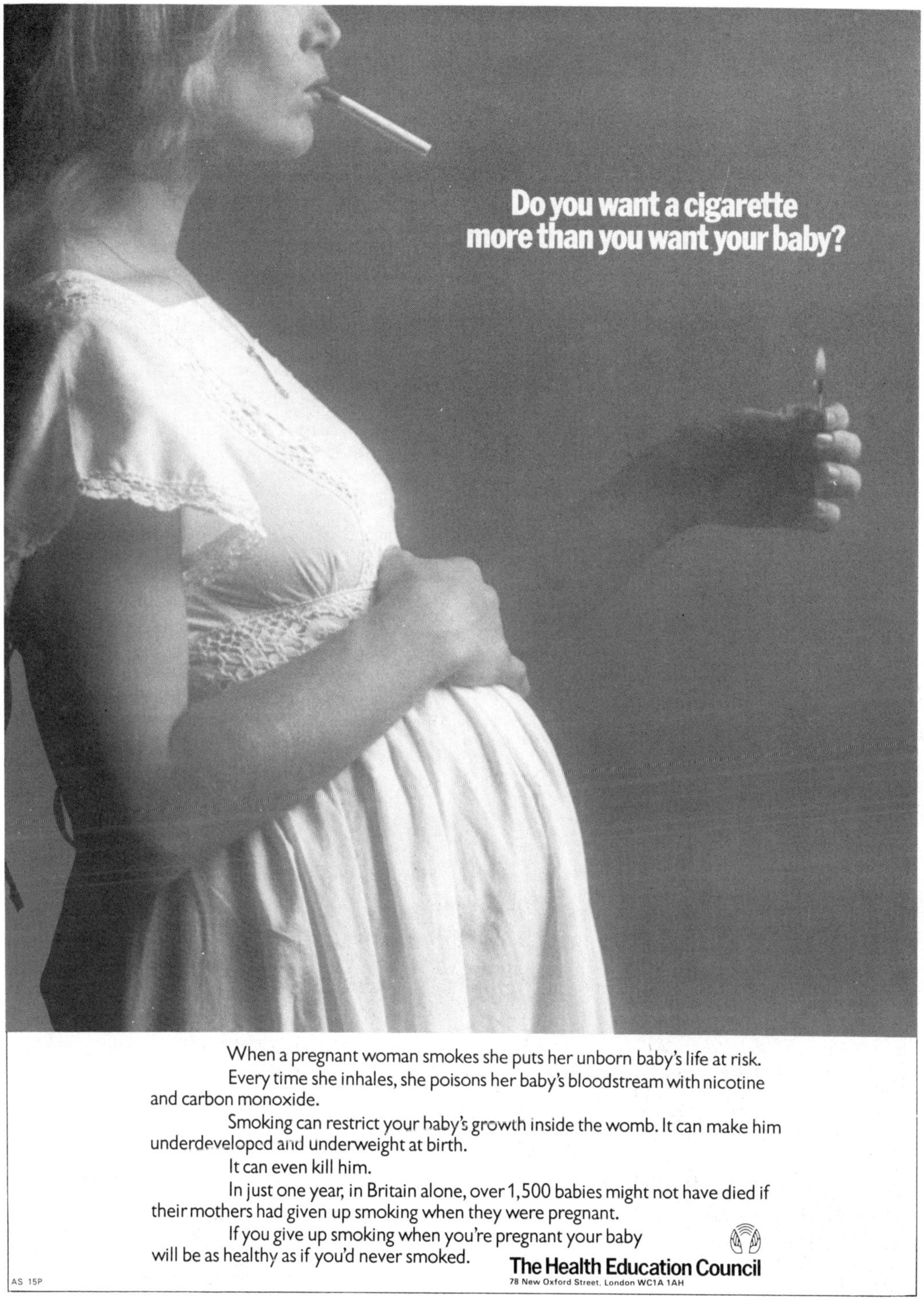

Fig. 2.5 *Health Education Council anti-smoking poster*

iron, such as liver, or the doctor may prescribe pills which contain iron and folic acid. Other vitamins may also be prescribed. The mother will be advised about her diet, and vouchers for a free pint of milk each day may be provided in some cases. She will also be given information about the money allowances that she can claim, and the free prescriptions and dental care that are available within the National Health Service. A urine sample is also taken and checked for diabetes (page 181) or kidney infection.

An expectant mother is advised not to take any drugs except under a doctor's supervision; also, if she smokes she may be advised to try and give it up. Recent figures indicate that, if an expectant mother smokes thirty cigarettes a day, her baby's birth weight could be reduced by 10 per cent (Fig. 2.5). It seems that smoking causes blood-vessels to narrow, allowing less blood to flow through them. This can result in less oxygen reaching the foetus, which may hinder its development.

The expectant mother will attend a clinic once a month for a routine check-up. In later months she will attend every fortnight, then once a week. It is this *antenatal care* (that is, care before birth) which has helped to make pregnancy and childbirth safer than in the past. It also helps to lay sound foundations for the child's good health in the nine months before birth.

2.3 **Care of the young child**

The baby's first medical check-up is carried out straight after birth. The doctor will examine the baby thoroughly, checking such things as heart rate, breathing rate, limbs, spine, and firmness of muscles, skin and responses. The baby's weight and sex will also be recorded.

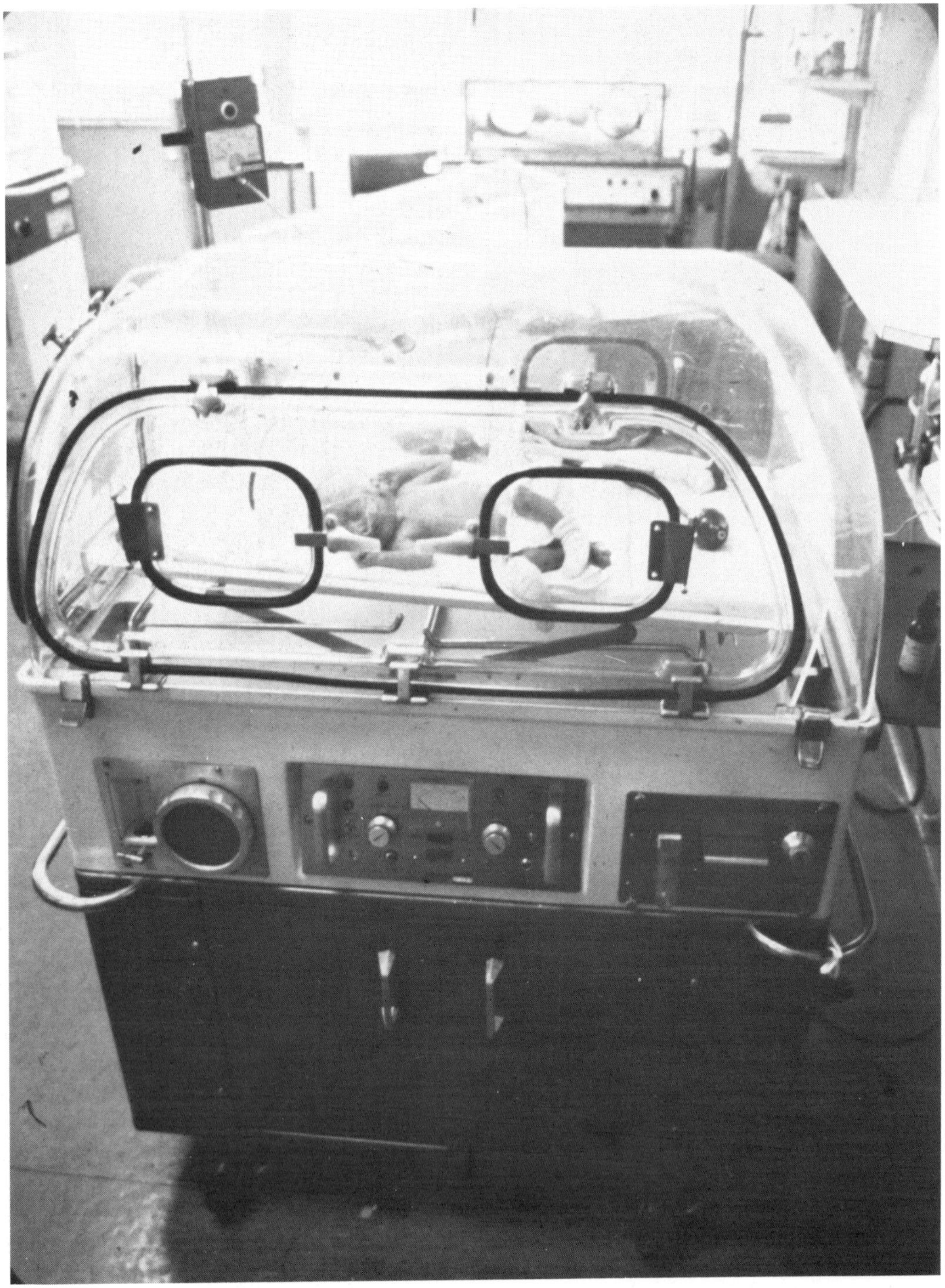

Fig. 2.6 *An incubator is used to control the heat and humidity of a pre-term baby's environment*

A baby whose birth weight is less than 2.5 kg may need special nursing. The baby may be placed in an incubator (Fig. 2.6) so that the temperature and humidity of its external environment can be controlled. This is necessary when the baby's own temperature control has not yet started to function (page 191). The baby may also need to be assisted in its breathing. If it cannot feed it must be tube-fed.

Once the mother and baby return home a midwife will check on their health and progress until the baby is ten days old. Soon after this the mother will be visited by her local health visitor. All health visitors are trained nurses. Their visits are made to give the mother information, advice and the support she may need in looking after the baby. The health visitor will tell the mother about the local child welfare centre and suggest that she goes there with her baby regularly. At the centre a record will be kept of the baby's development and weight, and any protection the baby is given against certain diseases by *immunization*.

Question 3 Can you name two diseases that the immunization programme can protect you against in early childhood?

There is little doubt that the services of child welfare centres have helped to obtain high standards in the health of children entering school. However the provisions do not end there. The Child Health Service, within the National Health Service, and organized at a local level, provides a range of services for children until they leave school. There are routine medical and dental inspections within schools.

The expert attention of general practitioners (G.P.s), hospital (Fig. 2.7) and specialist services are available to us all at any stage in our lives through the National Health Service.

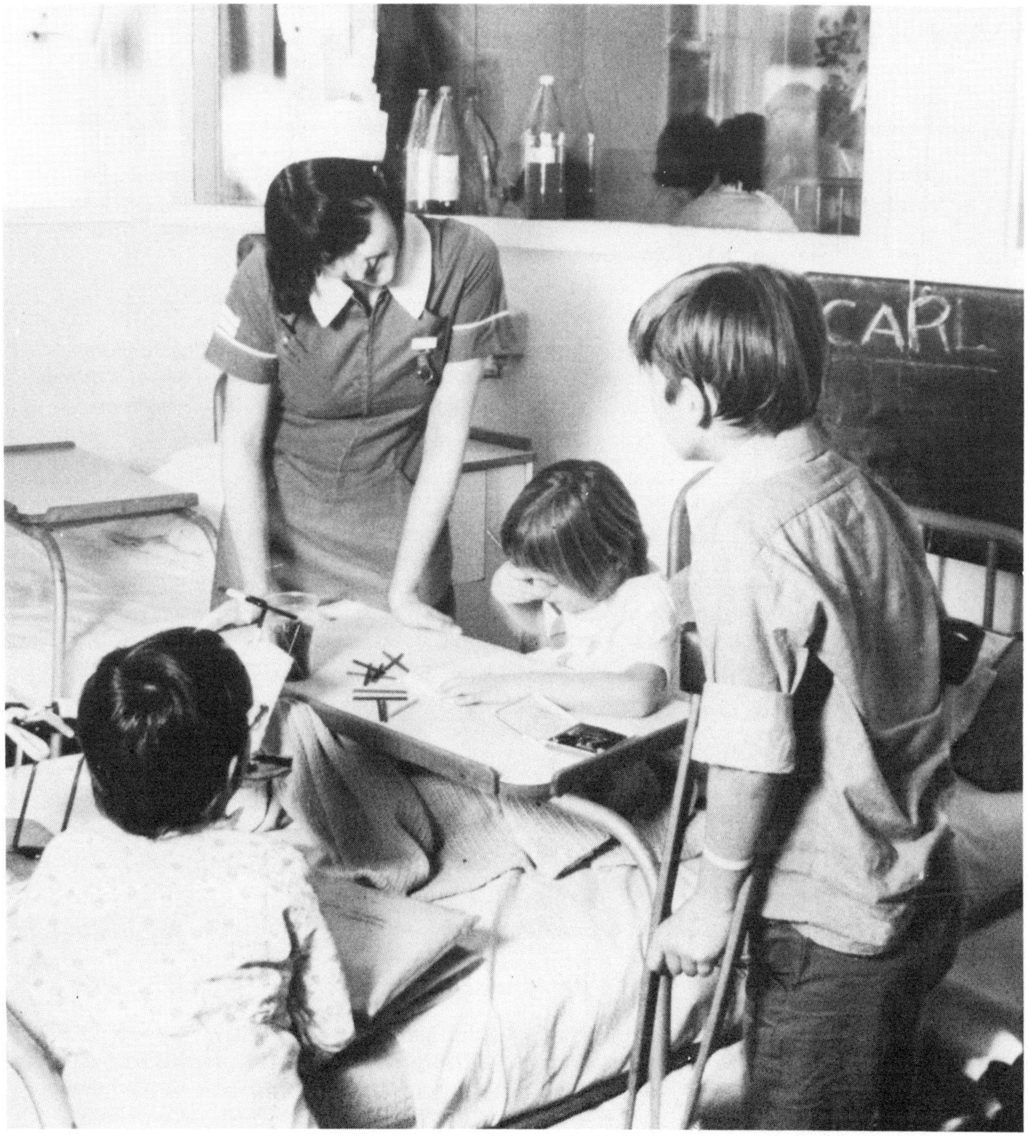

Fig. 2.7 *Hospital services are available to us all throughout our lives*

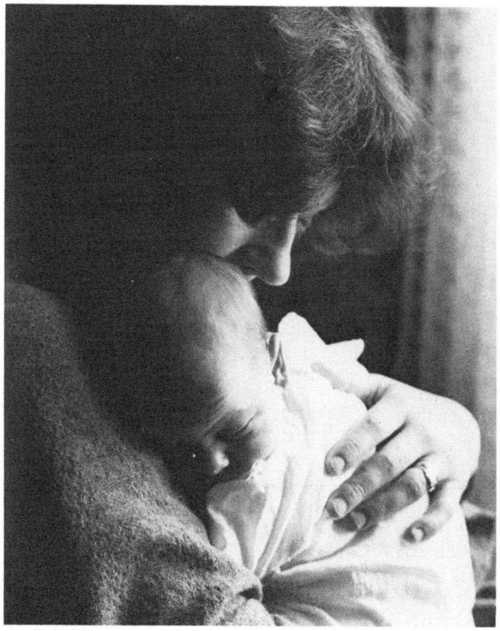

Fig. 2.8

So far we have seen how the young may be cared for by providing food, warmth, rest and protection but other things are also needed, for example attention and affection, security, guidance and opportunity for play (Fig. 2.8). All these factors make our development as individuals very complex.

Answers to questions in the text

1 The placenta can give the foetus protection against harmful pressure, organisms and some chemicals in the mother's blood. But this protection is not 100 per cent effective. Viruses and certain drugs can cross the placenta and some may damage the foetus. Thus a mother should protect herself against these harmful viruses and consult her doctor before taking any drugs.

2 The navel (umbilicus) or 'tummy button' is the scar left by the umbilical cord, which is tied and cut after birth.

3 The immunization programme offers protection against poliomyelitis, tetanus, whooping cough, measles and smallpox. The decision for the child to receive this protection is made by the parent.

Table 2.1. shows what we might expect in the development of a child up to eight years. You can see that behaviour patterns are considered, as well as stages in growth and physical maturity. But remember that we are not only individuals in size, weight and colour – we are also individuals in how we develop and in our behaviour.

TABLE 2.1 **SOME ASPECTS OF CHILD DEVELOPMENT**

Age	**Typical features**
Up to 8 weeks (2 months)	Sleeps and feeds a lot in the early weeks. Eyes begin to distinguish dark and light and by eight weeks they can usually follow objects that attract the baby's attention. Poor limb control. Head needs supporting. Obviously responds to attention from others
12 weeks (3 months)	Recognizes familiar faces. Possesses more limb movement but this is erratic. Can get hands into the mouth for sucking. Can support the head and turn it. Tries to reach out for objects. Can sit in a sloping position if supported correctly
16–18 weeks (4–7 months)	A period of rapid progress. Develops much more control over limb movement. Can roll over and sit up without support (Fig. 2.9). Can reach out and grasp things, first with two hands, eventually with one hand as head, eyes and hands develop co-ordination. Starts to make noises other than crying. Puts objects up to the mouth as part of general inspection. Curious and will spend time exploring. Begins to have more self-awareness. Identifies strangers and reacts
40 weeks–1 year (10 months–1 year)	Can crawl and use furniture to pull himself up on to his feet. May even be able to walk short distances with support. Hand control more advanced; shows great interest in picking up tiny objects. Tries to use a cup and spoon when feeding. Makes many more sounds and responds to people talking. Takes a more active part in family life and is more aware of people around him
1–1½ years	Stands and walks without support. Still needs help on steps and stairs. Can make use of large toys, pushing and pulling them. Can use building bricks and turn over pages of a book. Can recognize pictures. Can say a few simple words, usually naming objects. Still dependent upon those around him. Shows more concern with himself

Fig. 2.9 *Sitting without support and beginning to crawl*

2 years	More agile, can run but has difficulty in stopping or changing direction. Can tackle steps and stairs more easily. Hand control improves, so a higher tower of bricks can be built and crayons used for scribbling. Likes helping adults and copies them. Dresses up. Still plays on own even when other children are around. Starts to use simple sentences
3 years	Becoming less dependent, for example can wash, dress and feed without help. Confident in walking and running. Much more agile and generally well co-ordinated. Can use climbing apparatus, pedal cars, etc. Can do simple drawings using curved and straight lines. Likes arranging things, showing an idea of order. Speaks in simple sentences and can repeat rhymes. Asks such questions as 'What?', 'Where?' and 'Who?' Can express ideas. Much more sociable. Prepared to share activities and play with others. More aware of others and relationships
4–5 years	Much more independent. Can play for long periods alone or with others. Chooses friends and enjoys their company. Games are more complex, can involve making up stories and acting. Can construct things from simple kits or rubbish (Fig. 2.10). Eats with a knife and fork. Copes easily with washing and dressing. Asks such questions as 'When?', 'Why?' and 'How?' Can hold a conversation. Much more aware of responsibilities and the needs of others
6–8 years	The stage when all the development that has taken place seems to become united. Rapid physical growth and development of physical strength. Friendships very important. More independent of adults, yet knows how to get on with them. Development in speech and writing allows thoughts and ideas to be stated. Likes collecting information. May have definite interests which develop into hobbies. Team spirit high

Fig. 2.10 *Finer co-ordination of the fingers enables models to be constructed*

THINGS TO DO

Your teacher may arrange for you to visit an antenatal clinic and a child welfare clinic. Also, a health visitor may visit you to talk about her work.

1 Find out about the development of the National Health Service. What are its provisions at a local level?

2 Draw a map of your area. Indicate on the map the location of any provisions for community health, for example G.P.s' surgeries, health centres and hospitals.

3 Construct a wall chart (or frieze) which illustrates some of the main stages of development from babyhood up to the age of eight years.

4 Construct a toy which you think would be suitable for a very young child.

5 If possible record the activities over twenty-four hours of

(a) a baby
(b) a pre-school child
(c) a secondary-school child
(d) a middle-aged person
(e) an elderly person

Compare your results.

6 Construct a chart which illustrates how we use our bodies to communicate with one another, for example facial expressions, eye movements, hand gestures, general posture of the body, tone of voice. Indicate how these forms of communication help us to convey feelings such as anger, fright or depression.

7 Plan the birthday party of a child who is at infant school.

THINGS TO WORK OUT

In questions 1–3 write down the *one* answer which you think is correct.

1 A mother can seek advice about caring for her young baby from the

A Medical Officer of Health
B local health visitor
C probation officer
D district nurse
E home help

2 At an antenatal clinic blood tests are

A never taken
B carried out when the expectant mother requests it
C part of a regular routine
D done only when there is a family history of a genetic disorder
E done only in the early stages to confirm the pregnancy

3 The foetus is enclosed in the uterus, which contains

A amniotic fluid
B water
C protoplasm
D fatty tissue
E synovial fluid

4 The immunization programme in early childhood offers protection against certain diseases. Which *one* of the following diseases is not included in this programme?

A whooping cough
B tetanus
C poliomyelitis
D measles
E mumps

5 The sex of a baby is determined at fertilization. Describe how this happens.

6 (a) State *two* functions of the placenta.
(b) State *two* functions of the amniotic fluid.
(c) Why is it dangerous for an expectant mother to contract German measles in the early months of pregnancy?
(d) Why should an expectant mother be encouraged to give up smoking?

7 What work is done by antenatal and child welfare clinics that helps to ensure the health of the community?

8 (a) State which government department is responsible for the National Health Service (N.H.S.).
(b) Where does the money come from to pay for the N.H .S.?
(c) Name *four* local health services.
(d) State *two* of the responsibilities of a regional health authority.
(e) Describe *three* different types of medical inspection that are carried out in school. In each case state the advantage of these inspections to personal health and, if any, to communal health.

3 Ready for Action

3.1 The framework of the body

The children in Fig. 3.1(a) are trying to build a tent, using an old sheet. They quickly discover that the sheet has to be supported on a frame made from sticks or their mother's clothes-horse. Similarly, a frame is needed to support the human body. The frame of the human body is its *skeleton*. Without our skeletons we would collapse in a shapeless heap (Fig. 3.1(b)).

The skeleton supports internal organs and keeps them in position. It protects delicate and vital organs of the body, such as the brain, heart and lungs. It also makes it possible for the body to move.

When a manufacturer makes a tent, he knows he had to provide a frame which is strong yet light enough for people to carry around. The human skeleton has similar qualities. It is made up of *bones* (206 in all), some of which are hollow tubes filled with a tissue known as *bone marrow* (Fig. 3.2). This makes them light; the total weight of the adult skeleton is about 9 kg. The tissue from which bones are made gives them strength under all sorts of forces.

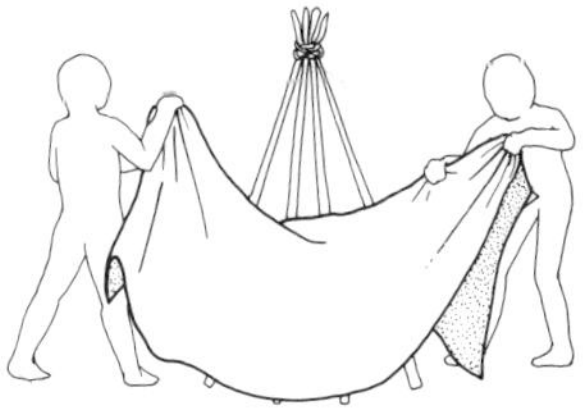

(a)

(b)

Fig. 3.1 *As a framework is needed to support a tent, so the skeleton is the framework which supports the body*

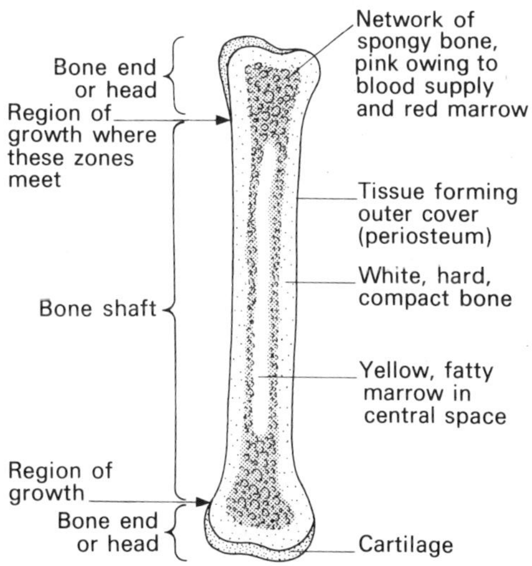

(a)

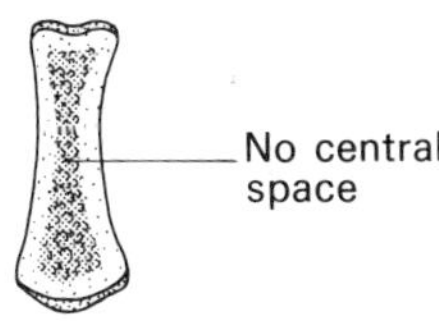

(b)

Fig. 3.2 *Longitudinal sections of bone. (a) The long bone of the upper arm (humerus). (b) A short bone of the hand (not to scale)*

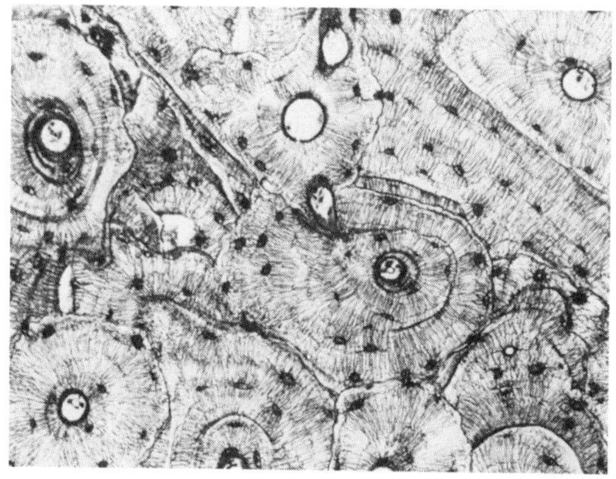

Fig. 3.3 *Transverse section of compact bone*

3.2 'Oh, them bones, them bones, them dry bones'

Bone tissue is a mixture of living and non-living material. The living material is a fibrous tissue consisting of fibres of *collagen*, which form glue when bones are boiled. This flexible living material is strengthened by a deposit of non-living material, consisting mainly of calcium phosphate and calcium carbonate. Such a mixture makes an excellent building material which is strong yet flexible.

Bone cells are found in small cavities in solid bone tissue. They develop the bone by depositing the non-living material around themselves. However, as you will see in Fig. 3.3, the cells are not cut off from their supplies of food and oxygen, because they are arranged around canals which contain small blood-vessels. The cells are connected to each other and to the canals by even smaller channels.

Bone cells begin to deposit calcium compounds as the foetus develops in the womb (Colour plate 1). The calcium is obtained from the expectant mother, so her diet must be rich in calcium. If her diet lacks calcium then it is taken from her own bones and teeth (page 121).

At birth and through early childhood the bones contain more fibrous tissue than solid calcium compounds, so a child's bones are softer and more flexible and they are less likely to break. However they can splinter easily, a condition known as a *greenstick fracture*.

3.3 **Broken bones**

Accidents, however, occur everywhere and every day. They may be minor or serious. Among the accident victims which come to the casualty department of our hospitals there are always some suffering from broken bones. A vital tool which will be used to help these people will be an *X-ray machine*. This machine depends upon two properties of X-rays:

(a) X-rays can pass easily through some substances but not through others, just as light can pass through glass but not through brick.

(b) When X-rays hit a film or photographic plate they cause a change in the chemicals. When the film is developed the areas where the X-rays hit the film are black, while the areas where the X-rays failed to pass through are white.

Question 1 The bones of an elderly person are very brittle and more likely to break. What do you think has taken place within the bone?

If there is a shortage of calcium or vitamin D (page 120) early in a child's life then the bones remain soft and may become deformed. For example, leg bones may become bent under the weight of the body. The child is said to have *rickets*. Nowadays a good diet, generally better standards of living, antenatal care and more child welfare facilities (page 20) have made this disease less common. It may still occur where there are poor social conditions. Bones continue to grow in width and length up to early adulthood. Growth in width takes place under the *periosteum*, the layer of tissue which covers all bone surfaces. Growth in length takes place in the regions indicated in Fig. 3.2.

Bone cells continuously remove and replace the calcium compounds throughout life. Research has shown that more calcium compounds are laid down if stress on the bone is increased, so exercise naturally strengthens bones. If a person has to stay in bed for a long time then the bones become weakened as the calcium salts are removed. This also happens to the bones of astronauts, who experience zero gravity in outer space. In both cases some form of exercise is essential to stop too great a loss of bone mass.

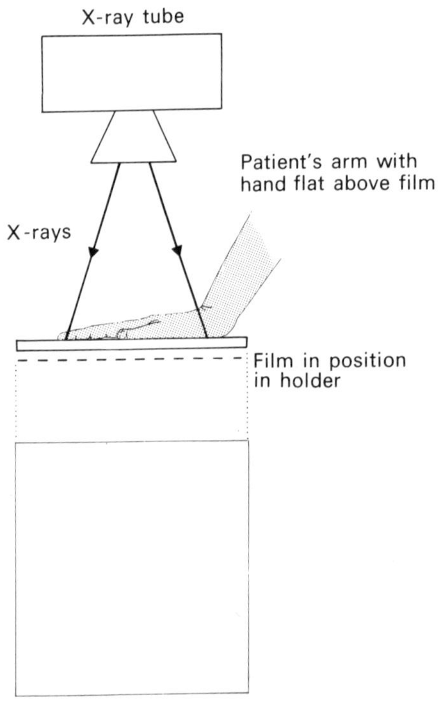

Fig. 3.4 *How an X-ray photograph is taken*

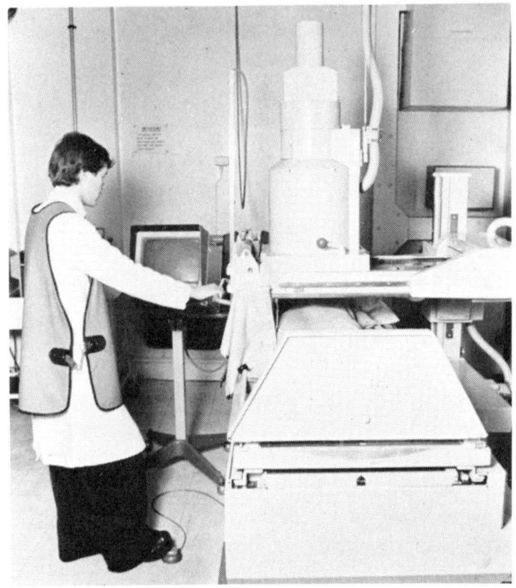

Fig. 3.5 *A person must be protected against overexposure to radioactive rays. A radiographer is protected by a lead shield or 'apron'*

Question 2 Fig. 3.4 shows how an X-ray photograph of the hand is taken, and the result. The bones in the hand are clearly seen as white regions. Are the X-rays passing through the bones? If a bone in the finger was cracked across what would you expect to see on the X-ray plate?

X-rays allow us to 'see' into the body. A radiographer (Fig. 3.5) is an expert who will take the photographs needed by the doctor. You can also see in Fig. 3.5 that the radiographer is protected by a lead apron. This is because high doses of X-rays can destroy tissue and produce changes in living cells (page 14). A person's exposure to X-rays must be recorded.

If a bone is broken the doctor will use the X-ray photograph to see how to place the pieces together again. The break will then be left to mend by the natural activities of the bone cells. They will remove the damaged tissue and rebuild it. However this does take time, so that areas of the body will have to be strapped up or set in plaster to stop any movement around the break.

3.4 **Cartilage**

An accident which often happens to footballers is a torn *cartilage*. You can feel a piece of cartilage if you give the end of your nose a twist! This will show you that cartilage is quite flexible. It is also strong and has a smooth surface. The arrangement of the cells within cartilage is shown in Fig. 3.6; you can compare this with Fig. 3.3. In fact most bones in the human body start as areas of cartilage. Bone is laid down, transforming the matrix from a soft to a hard material. This process is called *ossification* and will continue to occur in regions of bone growth.

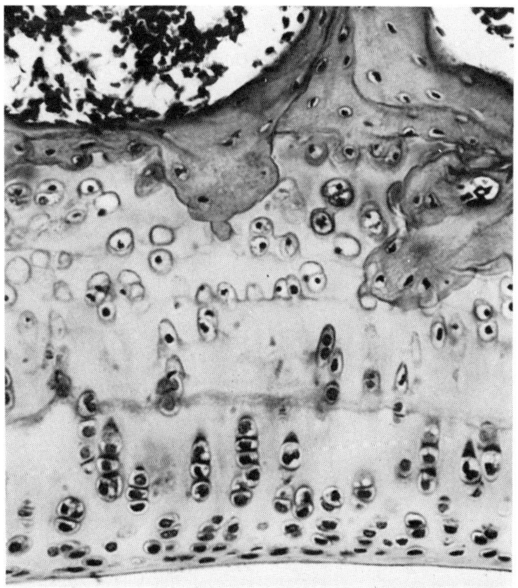

Fig. 3.6 *Transverse section of cartilage*

Where bones meet at *joints* there are areas of cartilage. Its smoothness allows the surfaces of the bones to move across one another with very little friction.

The flexibility of cartilage makes it act as an excellent cushion against jolts to the bones. However it can be damaged by a sharp blow such as a kick. If this happens to the cartilage at the knee-joint then a piece may break off and become trapped between the bones, preventing them from moving. This is a torn cartilage.

3.5 **Accidents** (Fig. 3.7; Table 3.1 and 3.2) overleaf

An accident is not caused by bad luck. There is a reason behind every accident, whether it be mechanical failure or human error. Because this is so it is the responsibility of individuals and the community to take action to try and reduce accidents.

You might like to discuss how helpful the following could be in a campaign aimed at reducing accidents in your area:

Fig. 3.7 *We think of home as being somewhere safe but if care is not taken it can be a dangerous place. Thousands are killed on the roads each year, but until recently many more were killed in the home. This surprises most people.*

(a) providing information about the body, its strengths and weaknesses, so that people can protect themselves against injury
(b) providing instruction so that people can acquire and develop a skill (Fig. 3.8)
(c) making and enforcing laws that will protect people against accidents (Fig. 3.9)

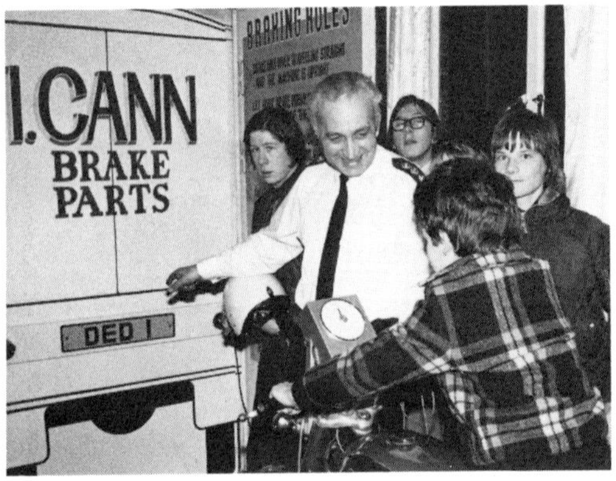

Fig. 3.8 *A group of youngsters learning how to handle a motorcycle*

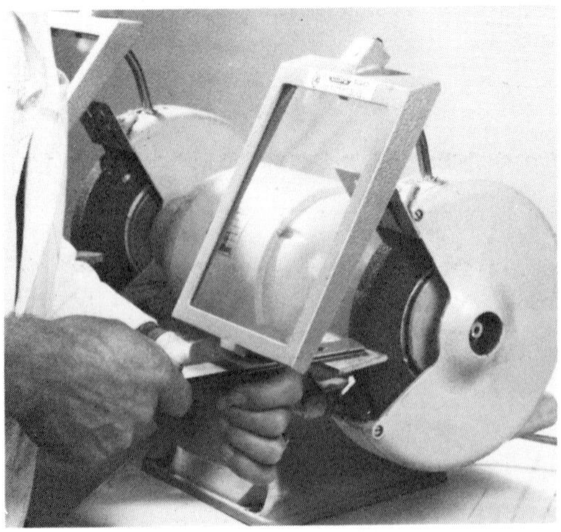

Fig. 3.9 *'Safety at work' regulations require certain machines to have guards*

TABLE 3.1 **ROAD-ACCIDENT STATISTICS FROM A METROPOLITAN COUNTY [FIRST QUARTER 1981]**

Class of casualty	Age	Number of casualties		
		Fatal	Serious	Slight
Pedestrians	Under 15 15–59 60 and over	4 2 10	83 57 34	375 297 68
Pedal cyclists	Under 15 15–59 60 and over	– 1 –	14 96 4	53 118 8
Motor-cycle riders	Under 60 60 and over	4 1	96 11	340 3
Motor-cycle passengers	Under 15 15–59 60 and over	– – –	– 6 –	3 28 –
Car/taxi drivers	Under 60 60 and over	3 1	71 9	571 55
Other drivers	Under 60 60 and over	2 –	9 –	70 –9
P.S.V. passengers	Under 15 15–59 60 and over	– – 1	2 – 7	5 48 44
Other passengers	Under 15 15–59 60 and over	– 4 –	5 61 9	73 366 46
Total		33	486	2571

1

In an emergency, keep calm

2

Find out how to get in touch by telephone with the Emergency Services in your area BEFORE an emergency. Dialling '999' is probably correct but this can depend on where you are and what telephone you use.
A quick way to find out is to read the dial label of the telephone. Emergency call instructions are displayed in telephone kiosks

3

Although a telephone in a kiosk may APPEAR to be 'out of order', you may still be able make an emergency call. Lift the handset. If you hear 'dialling tone' (a continuous purring sound) you should be able to make your call

4

After dialling, when the telephone operator answers, LISTEN CAREFULLY to what she asks. Tell her what emergency service you require. (If you do not know which Emergency Service you need, or if you require more than one, ask for 'Police'.) She will want to know the number of the telephone YOU are using. Give her the number clearly and accurately

5

When she has connected you to the Emergency Service, do LISTEN CAREFULLY to the person speaking to you. He will ask what the trouble is, where it is, and from where you are telephoning. Answer clearly and correctly

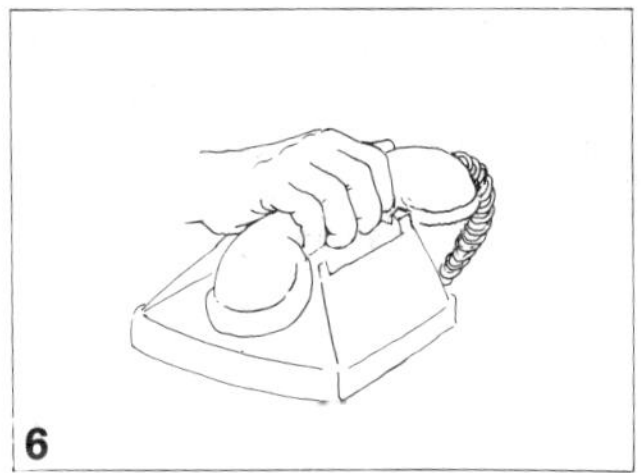

6

When you have finished the call, remember to replace the handset on the telephone rest

Fig. 3.10 *How to make an emergency telephone call*

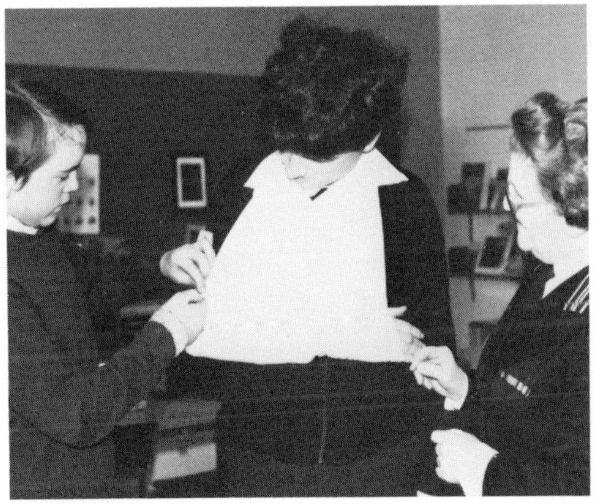

Fig. 3.11 *A first-aid class at a school*

When an accident has happened prompt action often saves lives. It is important to know how to call for expert help (Fig. 3.10) and how to act before assistance arrives. When helping anyone with injuries, it is essential to know exactly what to do, because the wrong action can make matters worse. First-aid courses provide this information and they are available to anyone at many centres all over the country (Fig. 3.11).

TABLE 3.2 **FATAL ACCIDENTS IN THE HOME 1978**

Age	Poisoning	Falls	Fire, flames etc.	Suffocation and choking	Other accidents	Total
0–4	16	24	74	157	54	325
5–14	5	2	44	21	14	86
15–44	281	98	101	137	55	672
45–64	233	320	140	129	64	886
Over 65	148	3302	461	167	273	4351
Total	683	3746	820	611	460	**6329**

3.6 The limb bones of the human skeleton

Fig. 3.12 shows a human skeleton. Some of the bones have been labelled with their common names (which you may know already) and with their medical names. By the side of the photograph is a simplified diagram showing the general plan of the skeleton. If you draw plans of the skeletons of other vertebrates (animals with backbones) you will see that they are very similar. The only difference is in the sizes and shapes of the bones within the plan. This is because the bones have become adapted to suit the different animals' ways of life.

All vertebrates have the five-fingered (*pentadactyl*) limb. But unlike most other vertebrates man can use his hands with great skill: he can grasp and pick things up; he can use tools and write. These actions are possible because the *thumb* can be moved so that its front surface touches the *palm* or the front surface of the other fingers. The thumb is said to be *opposable* to the other fingers.

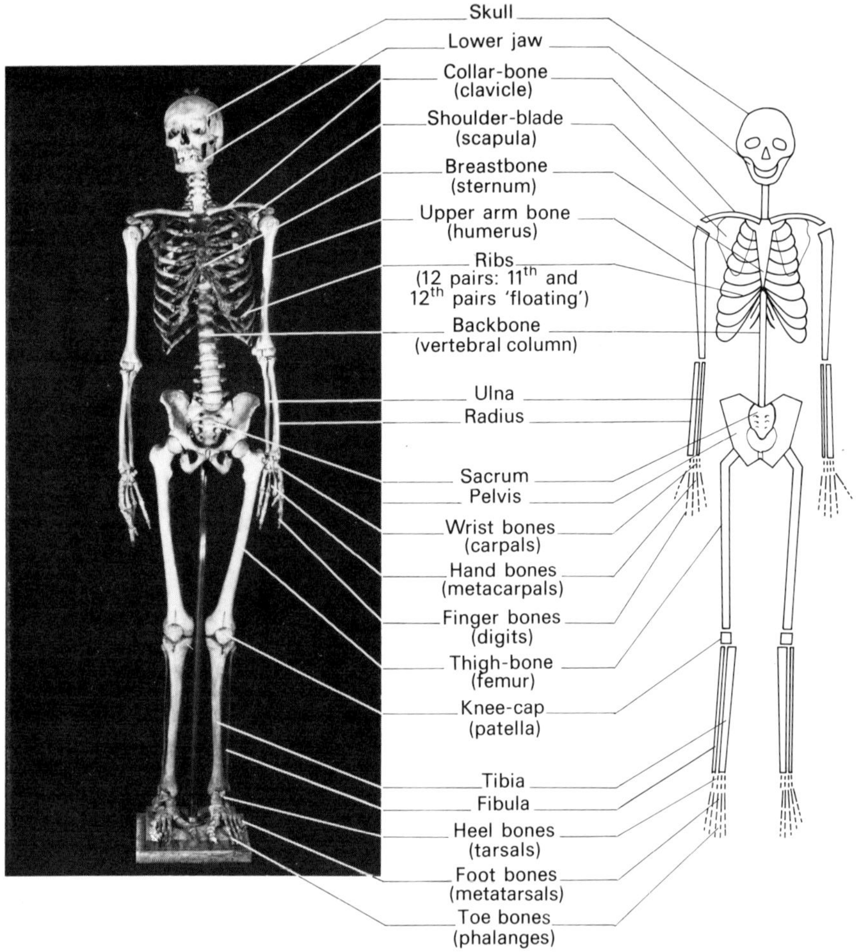

Fig. 3.12 *Human skeleton*

Question 3 Can you name another vertebrate that has an opposable thumb?

Man's foot has adapted to his upright position. The big *toe* is not in opposition to the other toes: instead, the *foot* bones are at right angles (90°) to the lower bones of the leg. The bones of the foot are so arranged that they add a 'spring' to our walk (see Section 3.12).

The limbs (arms and legs) are able to move easily outside the line of the body. This is possible because the limbs are supported by *girdles* which are attached to the *vertebral column*.

The *collar-bones* (*clavicles*) and the *shoulder-blades* (*scapulae*) make up the *shoulder* (*pectoral*) *girdle*, from which the arms hang.

The girdle for the legs is the *hip* or *pelvic girdle*. This is composed of the *pelvis*, which has two halves. Each half is made up of three bones fused together. You can feel the top of these bones quite easily if you put your hands on your hips. The region of the vertebral column called the *sacrum* makes up the back part of the pelvic girdle. The sacrum is shaped like a wedge so that, as it takes the weight of the body, it is driven deeply between the pelvic bones. This wedge shape and the fibres that link the bones together produce a very strong joint. The fibres joining a bone to another bone are known as *ligaments*.

The pelvis protects organs such as the bladder and, in a woman, the womb. A baby at birth has to pass through the pelvic girdle. If the girdle is small and the baby large then there could be some difficulty. However, nature helps as much as possible by allowing the ligaments to soften, so that the joints at the back of the pelvis can open up a little to give extra width. This softening of the ligaments can lead to backache in pregnancy, and at this time it is even more important that the woman knows how to lift weights correctly (see Section 3.10). If the baby still cannot pass through the pelvic girdle then a Caesarean section will be carried out. This involves making a cut in the body wall and womb and removing the baby through it.

When the pelvic girdle and the legs take the weight of the body they act like an arch that takes the weight of a building.

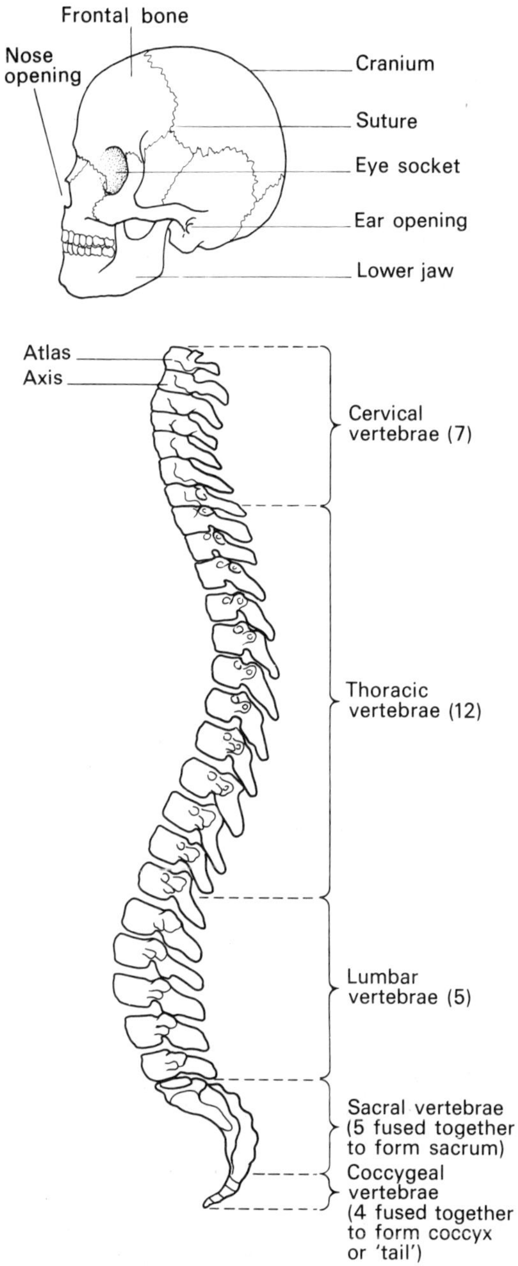

Fig. 3.13 *Skull and vertebral column, viewed from the side*

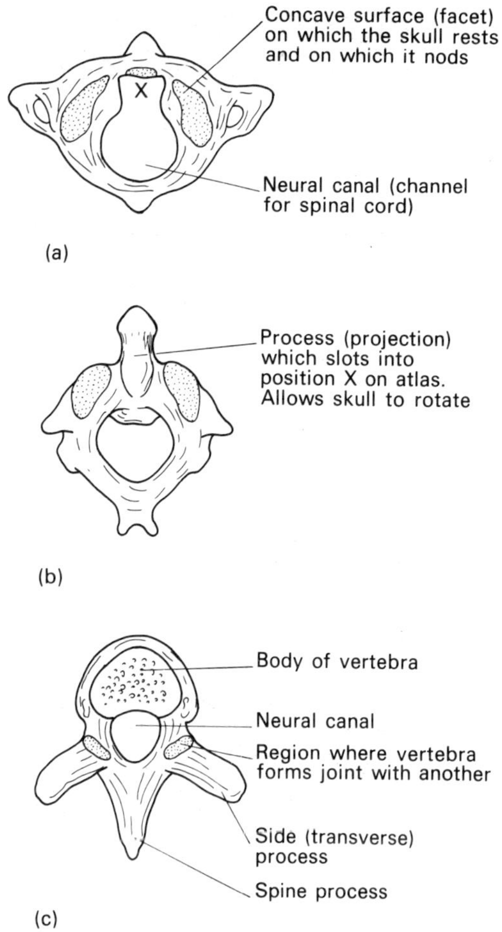

Fig. 3.14 *(a) Atlas viewed from above. (b) Axis viewed from above. (c) A dorsal vertebra viewed from above to show the general plan (processes provide attachment points for muscles and ligaments).*

3.7 The skull and vertebral column (spine)

The *skull* appears to have hair-line cracks all over it (Fig. 3.13). These are *sutures*, the joints where the wide skull bones meet. During birth the skull bones may overlap one another slightly. This allows the head to pass more easily along the birth canal. *Soft spots* (*fontanelles*) where the bones are not joined may still be felt up to one year after birth. As a child grows the skull bones because firmly fixed together and protect the brain. However it is important to remember that the skull may itself need protection in some circumstances.

The skull is supported on the top of the *vertebral column* by a small bone called the *atlas* (Fig. 3.14(a)). This is specially shaped so that we can nod our heads. The second bone at the top of the column is the *axis* (Fig. 3.14(b)). This has a finger-like projection which slots into the atlas. It allows the skull and atlas to swivel round on the axis, so that we can turn our heads.

Altogether thirty-three small bones, known individually as *vertebrae*, form the vertebral column. In an adult nine of these vertebrae are fused together, in one group of five and another of four. The general plan of a vertebra can be seen in Fig. 3.14(c).

If the vertebral column is viewed sideways, as in Fig. 3.13, several curves can be seen. These produce the action of a spring to cushion and protect the head from jolts. The vertebral column also forms a bony tube to protect the main nerve 'cable', the *spinal cord* (page 226).

3.8 Joints

As we have already seen, joints are formed where bones meet. At some joints there may be no movement of the bones. Such *immovable joints* can be found in the skull at the sutures.

At other joints the bones may move slightly against one another. Such joints can be found between the bodies of two vertebrae (Fig. 3.14(c)). Discs of cartilage separate the vertebrae at this point, and act as cushions between the bones. A person who suffers from a slipped disc has a cartilage disc out of place.

Synovial joints (Fig. 3.15) are found where the bones move freely. At these joints, such as the knee, *synovial fluid* separates the bones and allows movement with very little friction. This is very like oil in an engine which keeps moving pieces of metal apart. The synovial fluid is kept in place around the joint by a form of sleeve, the *synovial membrane*. Surrounding this membrane are the ligaments which hold the bones together. The cells of the membrane produce, i.e. *secrete*, the synovial fluid and under certain conditions may produce too much. This produces a swollen condition of the knee known as water on the knee.

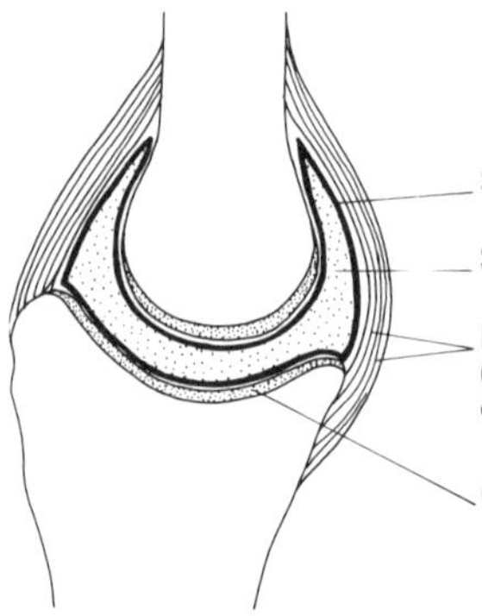

Fig. 3.15 *A typical synovial joint*

Synovial joints may be grouped into four main types: *ball-and-socket, hinge, pivot,* and *gliding* joints. These have different structures which gives variety in both type and size of movement. They are illustrated in Table 3.3.

Question 4 Move these joints in your body and try to work out what type they are: (a) hip-joint; (b) finger joint; (c) knee joint; (d) jaw; (e) wrist.

Regular exercise, particularly as we grow older, helps to keep joints working smoothly with no creaks! Arthritis and rheumatism are two diseases of joints where there is inflammation and pain. In an arthritic joint the cartilage has become broken down so much that the end of the bone cannot move freely. It is now possible to repair such a diseased joint so that there is free movement once more (Fig. 3.16) and no pain.

A physiotherapist's job is concerned with helping people to keep mobile. The movement of bones at joints is brought about by the action of *muscles*.

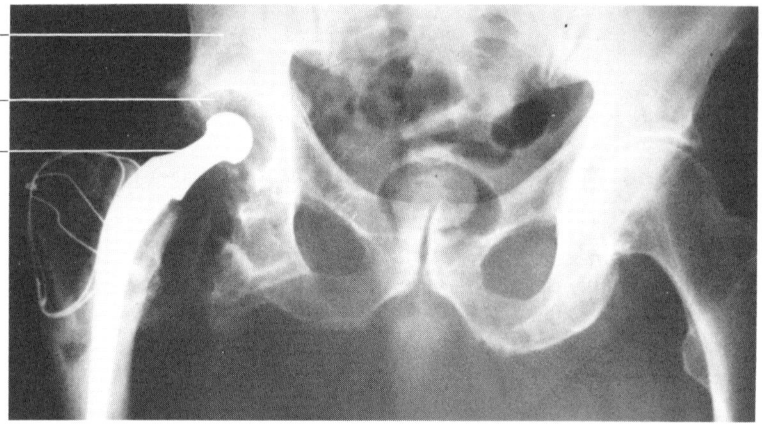

Fig. 3.16 *X-ray of a hip joint which has been totally replaced*

3.9 Muscles and how they act

Muscles are made up of cells which are drawn out to form long fibres (Fig. 3.17). These fibres are bundled together within a protective sleeve of tissue. The greater part of the human body is made up of such bundles. When we eat red meat we are eating mostly muscle.

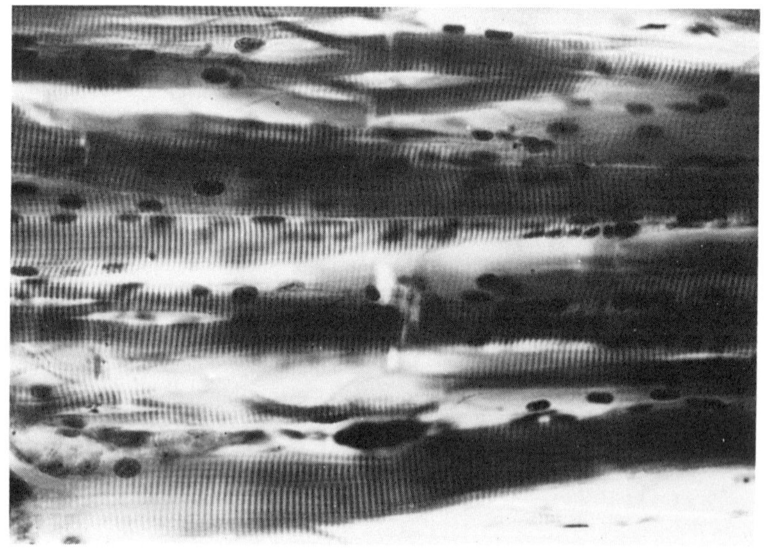

Fig. 3.17 *Muscle fibres (longitudinal section of striated muscle)*

TABLE 3.3 **FOUR TYPES OF SYNOVIAL JOINT**

Type of joint	Movement of bones	Examples
Hinge joint	Like a hinge on a door. Allows bones to move closer together or further apart. However there is a set distance beyond which the bones cannot move. If they are forced to do so the 'hinge' or joint will be damaged	Elbow Ankle Toes Nodding of head 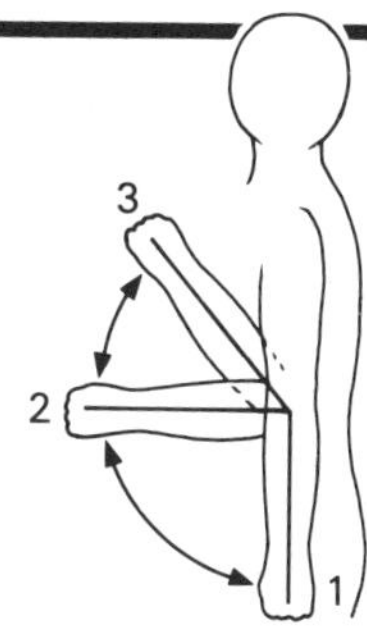
Ball-and -socket joint	Rounded end (ball) of one bone fits into a hollow (socket) in another bone. Allows movement in three planes. Also allows rotation	Shoulder 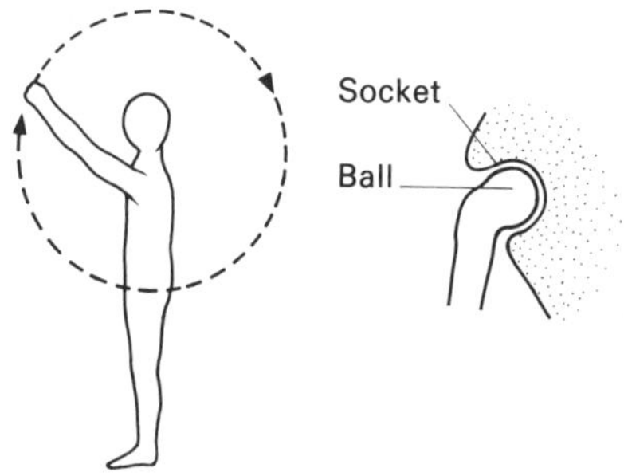
Pivot joint	One bone moves round another bone, allowing turning. Only allows rotation	Lower arm Rotation of head (between 1st and 2nd vertebrae) 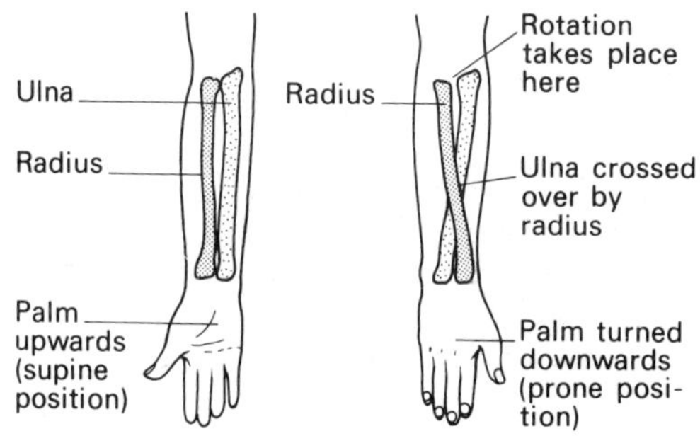
Gliding joint	One bone slides over another	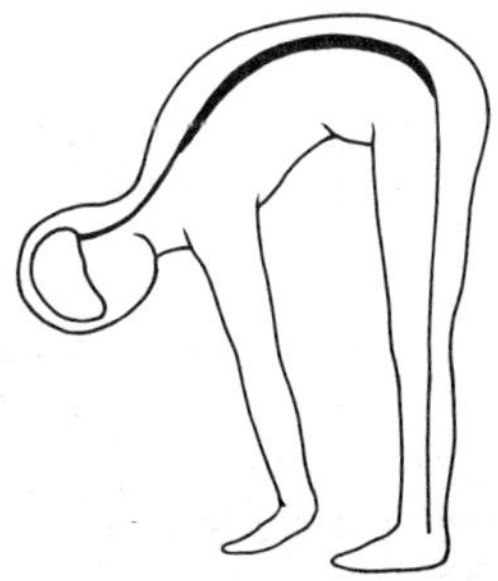 Small bones of wrist and ankle. Gives great variety of movement but not a strong joint Between vertebrae Between thumb (metacarpal) and carpal (Section 3.6)

A muscle fibre can get fatter and shorter by *contracting*. It can then return to its original state by *relaxing*. A fibre is either contracted or relaxed: there is no condition in between. However, it takes many fibres to make a muscle, and some may be relaxed while others are contracted. Therefore a muscle contraction can be weak or strong according to how many fibres are involved in the contraction. Try this by squeezing your friend's hand!

In other science lessons you may have learned that anything which can cause movement is called a *force*, and that a force can be either a push or a pull. A muscle exerts a force on a bone, but it can only pull the bone and cannot push it.

To pull a bone the muscle must be attached to it. This is done by means of a *tendon*, which is made up of long parallel fibres. The tendon is attached over a small area of bone called the *insertion* of the muscle. Tendons may operate over long distances, like control wires. You can see this in Fig. 3.18, which shows that the tendons which control the finger bones run from muscles in the lower arm. Injuries to the wrist may cut through these tendons.

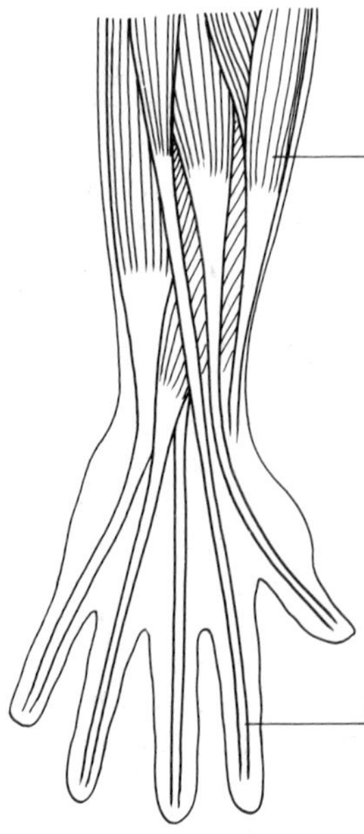

Fig. 3.18 *The tendons controlling the fingers*

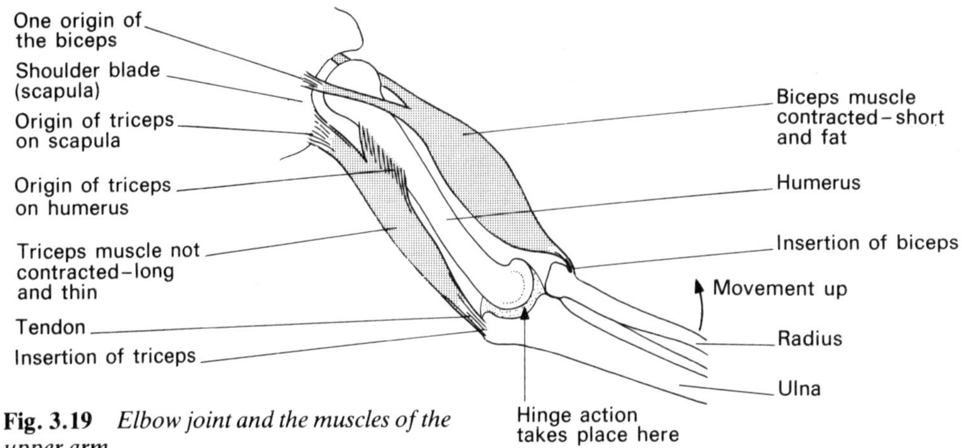

Fig. 3.19 *Elbow joint and the muscles of the upper arm*

Question 5 Do you know where your Achilles tendon is?

If the muscle is to pull the bone its other end must be firmly anchored. At the opposite end to the insertion the muscle is usually attached to a bone which remains still. At this end, the *origin* of the muscle, the fibres are attached over a large area.

To produce a movement which is controlled, muscles have to work together, that is *co-ordinate*. This can be seen at its simplest when muscles work as a pair. Such muscles are found at hinge joints, like the elbow shown in Fig. 3.19. The two muscles cause movement in opposite directions, and are said to be *antagonistic*.

The muscle which pulls one bone closer to the other bone is called the *flexor* muscle. The muscle which pulls one bone away from the other bone and straightens the limb is called the *extensor* muscle. As one muscle contracts so the other relaxes.

Question 6 In the arm, which is the flexor muscle and which is the extensor muscle?

3.10 Levers of the human body

If you have studied physics you may have already realized that the movement of bones by muscles is like the movement of *levers*. A lever is a machine which exerts two forces (a load and an effort) acting in opposite directions around a pivot (fulcrum), as seen in Fig. 3.20. The turning action of the forces is called their *moment*, and this is measured by multiplying the force by its distance from the fulcrum. Thus in Fig. 3.20

the moment of $L = L \times x$
the moment of $E = E \times y$

If the lever is not moving then the moments of the forces are in equilibrium and are equal:

$$L \times x = E \times y$$

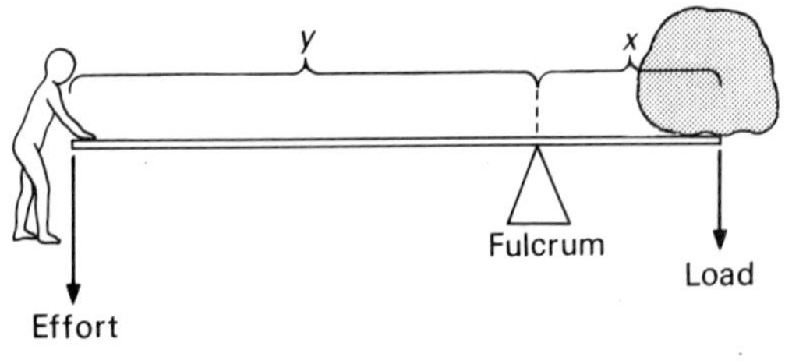

Fig. 3.20 *A simple lever (crowbar) which enables us to use less effort*

So if L is large, x is small and y is large, then E can be small. In this way a large load placed near the fulcrum can be lifted by using a small effort exerted at a great distance from the fulcrum. A machine with such a *mechanical advantage* is the crowbar.

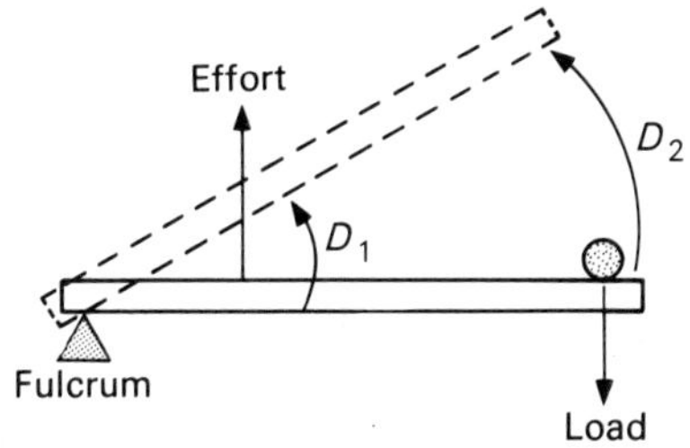

Fig. 3.21 *A type of lever which increases the amount and speed of movement*

However, if the load and effort are in different positions the effect changes. For example, with the lever in Fig. 3.21 there is no way that E can be smaller than L. So what is the advantage? Consider the distances that the load and effort move through. As the effort moves through D_1 the load moves through the greater distance D_2. These movements will occur within the same space of time. In this way the lever has increased the amount and speed of the movement. If you look at Fig. 3.22 you can see that, when the flexor muscle contracts, the arm is this type of lever. As a result, we can move our hands very quickly and through large distances by using short muscle contractions. We can be quick enough to swat a fly, but what force is needed from the muscle? Imagine that the force (load) needed to squash the fly is 10 g. Using the distances given in Fig. 3.22 we can calculate the force (effort) needed from the muscle:

$$L \times x = E \times y$$
$$10 \times 45 = E \times 2$$
$$E = 225 \text{ g}$$

This illustrates how strong muscles have to be!

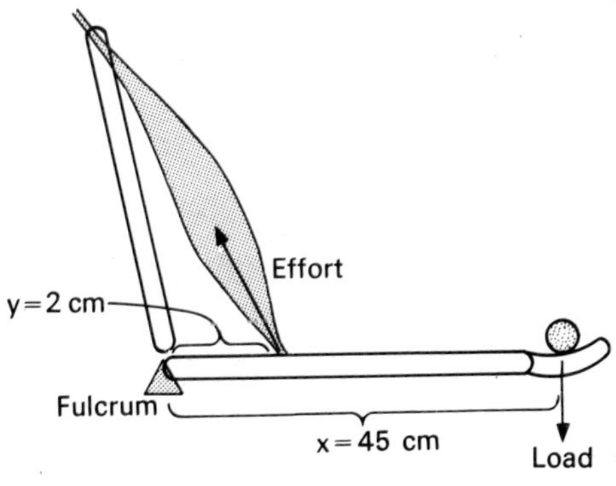

Fig. 3.22 *When a flexor muscle contracts, the amount and speed of movement are increased*

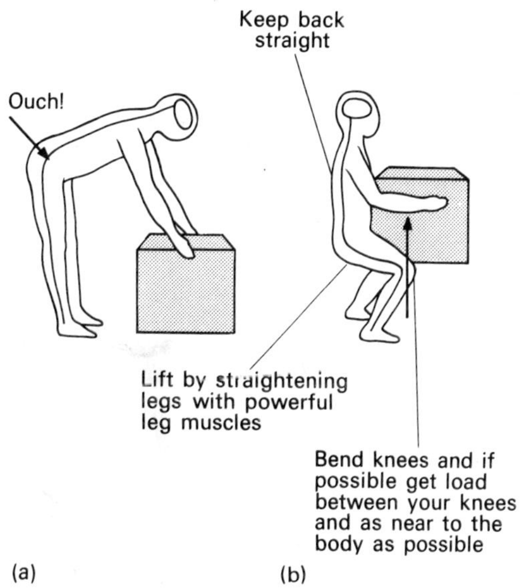

Fig. 3.23 *(a) Watch out! You could strain your back by lifting this way. (b) How to lift without strain – make the weight act along the length of your backbone,* not *at an angle to it.*

Yet, like most building materials, there is a limit to the force a muscle can stand. If this force is exceeded then the muscle is stressed and may be damaged. In such instances we feel pain and talk of a strained muscle. So it is important to understand that muscles have these limits and why we should not add further stress by using them incorrectly. This is especially true in regions where the body is mechanically not very sound, for example, along the curve of the lower spine. Many strains in this region could be avoided if lifting were not done as in Fig. 3.23(a) but as in Fig. 3.23(b). By bending at the knees, the weight is taken by the shoulders and arms through the line of the body and not just by the muscles pulling at the base of the spine. Back problems account for the loss of thirteen million working days a year. So workers in factories may find the firms' industrial nurses advising them on how they can avoid such injuries.

3.11 Posture

In Section 3.1 we discussed how the frame of a tent provides support. Yet to keep the tent rigid and upright we also need guy ropes. Each guy rope must be under tension. If a guy rope is slack then that area of the tent sags, and if all the guy ropes give way the tent collapses in a heap.

Muscles are the guy ropes of the body. Even when the body is resting, some muscles are contracted, and this is known as *muscle tone*. This tone of the muscles maintains the posture of our bodies and we have conscious control over it (page 234). When a person is unconcious or drugged, this control over the muscles is lost, so tone and posture are no longer maintained, causing the body to collapse.

Muscle which is controlled by our wills is called *voluntary muscle*, and its fibres appear striped. You can see this *striated muscle* in Fig. 3.17. *Involuntary muscle* is not striped nor is it under the control of the will. You would find such muscle, for example, in the digestive system. A special form of muscle, *cardiac muscle*, is found in the heart. This is not under the control of the will, and it continues to contract rhythmically throughout life. If it stops because of an illness or accident it must be restarted very quickly if the person is to survive (page 74).

Fig 3.24 illustrates good body posture when standing and lying. With good posture the muscles are doing the minimum work necessary, giving less chance of muscle fatigue, strain and aches.

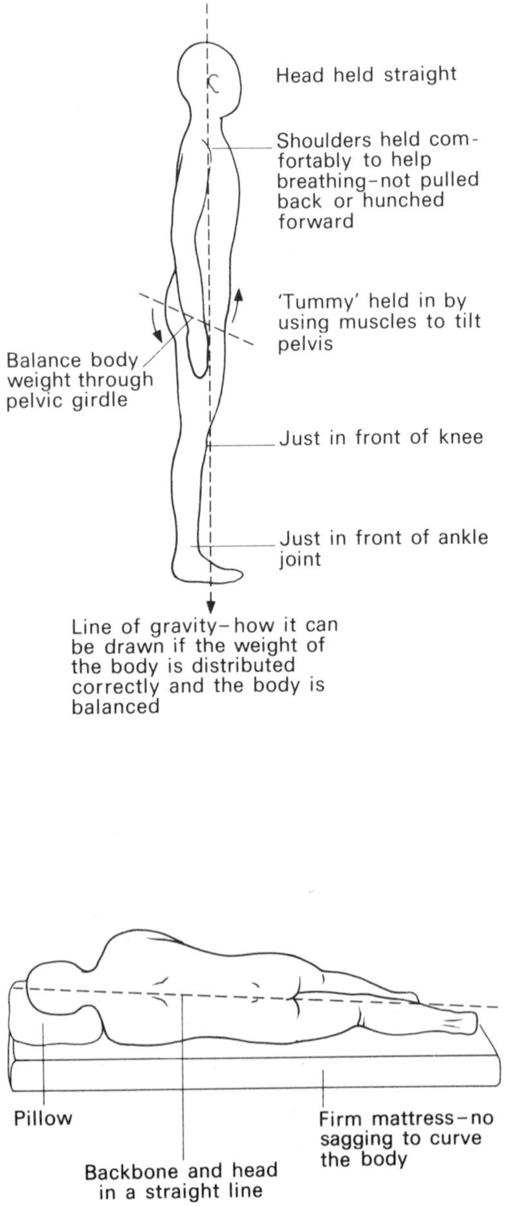

Fig. 3.24 *What is good posture?*

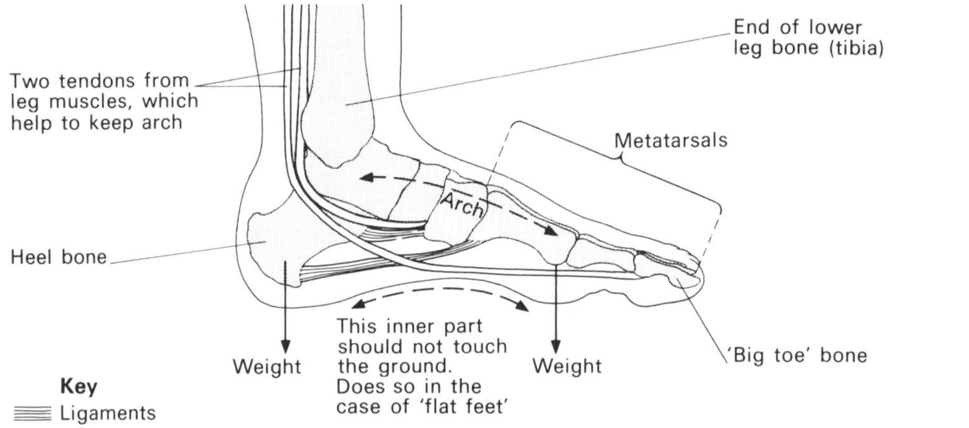

Fig. 3.25 *Bones and ligaments in the foot provide a foot arch*

3.12 **Feet**

Muscle tone is essential in the feet (Fig. 3.25). Here muscles and ligaments work to support the small bones and to maintain the *foot arches*. These arches are:

(a) from heel to big toe on the inner side of the foot
(b) from heel to little toe on the outer side of the foot
(c) half of an arch across the middle of the foot

When we put weight on our feet the arches become flattened by the movement between the bones. This ability of the arches to flatten or curve upward gives an action like a spring to the foot and prevents jarring of the legs. Exercise helps to maintain these arches and prevent flat feet.

Any deformity or pain in the feet will affect posture and our ability to move around. Foot strength, like all other features of the human body, varies from person to person, but exercise and care of the feet can help in all cases. The need to protect our feet in everyday life by covering them brings its own danger. The covering of socks and shoes may restrict the foot and even change its shape by buckling the toes (Fig. 3.26). When choosing shoes the following questions should be asked:

(a) Is the length correct when standing? In children's shoes, is there room to grow? The recommended space is 18 mm from the big toe to the end of the shoe.
(b) Is the width correct, so that toes are not cramped?
(c) Are the heels and insteps gripped so that the foot cannot slide forwards to cramp the toes?
(d) Is the shoe flexible enough to allow a natural springy step.

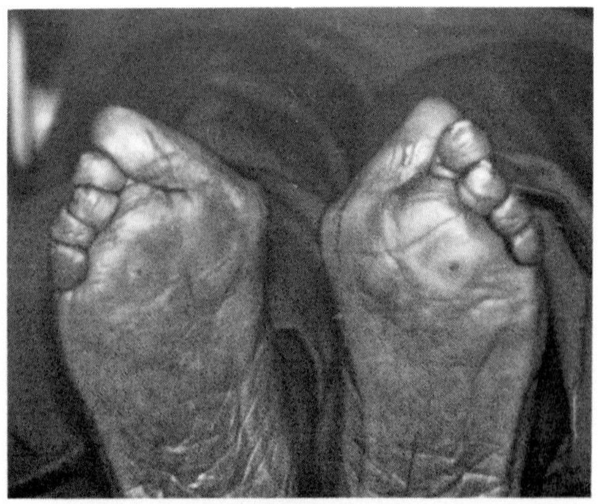

Fig. 3.26 *'Buckling' of the toes*

You may have realized that we need to choose our footwear to suit what we are doing at a particular time. if this is done our feet are comfortable at all times and will remain so throughout life. However, we often disregard what we know is suitable and choose our footwear for the sake of style and appearance. It is well to remember that we have a lifetime of appearances to make, and how active we are at sixty could be decided by what we wear on our feet at fifteen.

Feet need to be kept clean, especially if they sweat a lot. Daily washing, proper drying and changing of socks will keep feet fresh and feeling comfortable. Toe-nails will also need cutting, and they should be trimmed straight across.

A chiropodist can help with all foot care, particularly with discomfort and infections. If buckling occurs, *corns* may develop because of the pressure. Swollen and mis-shapen joints will give rise to *bunions* (Fig. 3.27). *Athlete's foot* is a fungus infection and appears as a spongy, itchy area, sometimes between the toes. *Verrucae* (Fig. 3.28) are thought to be the result of a virus (page 80) infection. They are spread from an infected person in communal places where people go barefoot, such as swimming-baths. As a pre-caution, in these places floor areas are regularly disinfected and foot-baths are provided. It is best to use them for your own protection.

3.13 Exercise

There is no doubt that the human body deteriorates physically without use. Muscles need to be used if they are to stay firm, and muscle activity can improve muscle strength. This does not mean training to become Mr Universe or a champion weight-lifter. It means we should take regular periods of *exercise*. During the period of exercise our muscles, heart and lungs should work harder than they do normally. Running, cycling and swimming are examples of good dynamic exercise. The reasonable demands that are put on muscles, heart and lungs over a period of planned regular exercise will gradually improve their performance.

Physical activity also encourages the body to rest. When muscles are very active, a sub-stance known as *lactic acid* (page 41) builds up in the cells and causes *muscle fatigue*. A period of rest allows the body to see to the removal of this lactic acid. A person who has difficulty in sleeping may find that brisk exercise, such as a walk, before bedtime will help him to sleep.

From early childhood through to old age we make different demands on our bodies. How our bodies cope with these demands will depend on their condition, readiness and ability to go into action. Coping or not coping with a situation may also affect the way we feel, our sense of well-being. So how we care for our bodies can affect our quality of life.

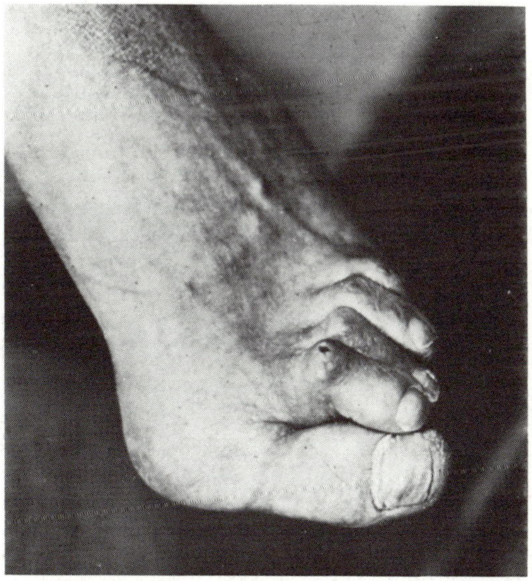

Fig. 3.27 *Corns are areas of thick, hard skin caused by pressure. A bunion is a cushion of bone and fluid formed as protection against pressure on a joint that is already out of line.*

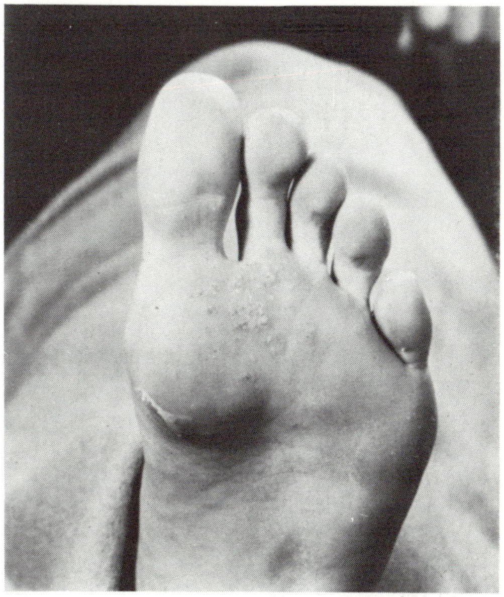

Fig. 3.28 *Verrucae infection of the foot*

Answers to questions in the text

1 As we grow older more calcium com-pounds are deposited by the bone cells to replace fibrous tissue. Because of this the bones become less flexible and more brittle.

2 No: X-rays pass through flesh but not bone. A crack across the bone would appear as a black line across it.

3 Apes and monkeys are the most common. Such a thumb allows these creatures to grasp and climb.

4 (a) Ball-and-socket joint; (b) hinge joint; (c) hinge joint; (d) hinge joint; (e) pivot joint.

5 At your heel.

6 The flexor muscle is the *biceps*; the extensor muscle is the *triceps*.

THINGS TO DO

1 Simple observations of bones.

(a) Ask your local butcher to saw the leg bone of a sheep in half lengthwise so that you can examine its internal structure.
(b) Leave a chicken bone in vinegar for 3–4 days. Notice how the bone becomes more flexible as the acid dissolves away the hard calcium salts.
(c) Carefully place a chicken's leg bone on the hot ashes of a fire and allow the protein (living material) to burn away (you will be able to smell it). Notice how you are left with a very light but brittle substance which will easily crumble. This is the remains of the mineral content of the bone.
(d) Note the lightness of bone. To test its strength, hang up the leg bone of a sheep so that it is horizontal. Then suspend a number of weights from the centre of the bone. See how many weights you can add before the bone breaks. Repeat this exercise with bones from different regions of an animal's skeleton.

2 Next time your family has a chicken for dinner strip the skin from one of its legs and study the muscle. (Note that this muscle is the meat we eat.) A good cookery book will show you how to remove the leg from the body of the chicken. This will give you the chance to observe the 'hip' joint. If you carefully use a knife to free the muscles from the bones you will be able to compare the 'knee' joint with the 'hip' joint. Also try to trace the tendons of the muscles and the bones they act upon.

3 You don't have to skin yourself to trace most of the bones in your body! All you have to do is to feel for them. By moving your bones you can study the function of joints and the action of muscles. For example, feel the action of the muscles of your upper arm as you bend and straighten your arm. Feel for the tendon of the biceps muscle in the crook of your arm when your arm is bent.

4 Try to construct a simple working model of a limb, or any other region of the skeleton, to show the movement of bones and the action of muscles. Use stiff card as bones and rubber bands as muscles.

5 Support a plank of wood horizontally so that it forms a shelf 1 m above the floor. Place ten 1 kg bags of sugar on the floor. Ask someone to use a watch with a second-hand to time how long it takes you to lift all the bags, one by one, from the floor on to the shelf.

Your body does work when it lifts things. To calculate work scientists use the simple formula:

$$\text{work} = \text{force} \times \text{distance moved in the direction of the force}$$

If the force exerted is measured in newtons (N) and the distance moved is measured in metres (m) then the units of the work done will be *joules* (J). A joule is also a unit of energy (see p. 120).

The downward force exerted by the earth's gravity on a mass of 1 kg is nearly 10 N. Therefore to lift a 1 kg bag of sugar your body must pull upwards with a force of 10 N. So:

work done by body in lifting one bag on to shelf = force (N) × distance (m)

$$= 10 \times 1\ \text{J}$$
$$= 10\ \text{J}$$

Thus:

work done in lifting all the bags

$$= 10 \times 10 = 100\ \text{J}$$

The amount of work which must be done to lift the ten bags of sugar from the floor on to the shelf will be the same whoever does the lifting and however fast they do it. But the *power* used to do the lifting will be greater if it is done quickly. Power is measured in watts (W) and is calculated using the formula

$$\text{power (W)} = \frac{\text{work done (J)}}{\text{time taken (s)}}$$

So if a man could lift the bags on to the shelf twice as fast as his son he would have twice the muscle power.

The work done in lifting all the sugar has already been calculated, so you can find your power output by dividing by the time taken (in seconds).

6 Stand at the bottom of a flight of stairs. Ask someone to time how many seconds it takes you to run up to the top stair. Work out the total height of the flight of stairs by measuring the vertical height of each stair. Next, find out how much you weigh in kg. Now you can work out your power output:

power output (W)

$$= \frac{10 \times \text{mass of your body (kg)} \times \text{height of flight of stairs (m)}}{\text{time taken (s)}}$$

(Remember that the force exerted by gravity on *each* kg of your body mass is about 10 N.)

7 Stand in front of a wooden box, or stool, about 50 cm high. Check that you can comfortable step up on to the top surface and that there is no chance of the box tipping over. If you find it a strain to step up on to the box, replace it with one that is 10 to 15 cm lower. Step up on to the box with one foot, follow it with the other. Straighten both legs. Now step down with one foot, then with the other. Repeat this stepping up and down and try to get into a rhythm. With another person watching

the time, count how many 'step-ups' you can do in 30 seconds. If you know how much you weigh in kg you can work out your power output:

$$\text{power output (W)} = \frac{10 \times \text{mass of your body (kg)} \times \text{height of box (m)} \times \text{number of step-ups}}{\text{time taken (s)}}$$

(Note that you must use the height of the box in *metres* in this formula.)

8 Try repeating one of the previous exercises several times, each time under different conditions. For example, do the exercise several times without resting. Then try resting between each attempt. In this way study what conditions might cause your power output to vary. Also try to find out if you can improve upon your power output or performance by doing an exercise regularly every day for a period of several weeks.

9 Measure the strength of your muscles using bathroom scales. Here are two methods:
(a) Finger muscles. Hold the scales in your hands with your fingers underneath and just your thumbs on the top surface. Squeeze the scales and note the reading.
(b) Chest muscles. Turn the scales on to one edge. Then pick them up between the palms of your hands and hold them in front of your chest as if you were praying. Push your hands together as hard as you can and note the reading.
Now work out other ways of using the scales to measure the strength of your arm and leg muscles.

THINGS TO WORK OUT

In Questions 1–5 write down *one* answer which you think is correct.

1 Muscle fatigue develops when there is an accumulation in the tissue of

A blood
B oxygen
C salt
D hormones
E lactic acid

2 Muscles are arranged in antagonistic pairs because

A muscle fibres act by contracting
B muscles can push but they cannot pull
C they require a good supply of blood
D they are controlled at will
E a bone must have an equal weight of muscle on either side of it

3 Muscles in the human body are attached to bones by

A ligaments
B cartilage
C sutures
D tendons
E synovial capsules

4 Synovial fluid is present in the human body. Which of the following gives its correct function?

A It assists digestion in the small intestine.
B It acts as a lubricating fluid in joints.
C It circulates in the semicircular canals of the inner ear and assists body balance.
D It is present as insulation in the body cavity.
E It acts as a shock absorber to protect the foetus in the uterus.

5 The bones which meet at our elbow are

A humerus, radius, ulna
B femur, fibia, radius
C humerus, radius, fibula
D femur, fibia, axis
E tibia, fibula, patella

6 Select the bones from the list below that are found in the regions of the body known as (a) knee, (b) lower leg, (c) toes, (d) palm, (e) head, (f) neck, (g) chest, (h) wrist, (i) hips.

Bones ribs, humerus, metacarpals, carpals, patella, tibia, skull, pelvis, tarsals, metatarsals, fibula, phalanges, vertebrae

7 With the aid of a labelled diagram explain

(a) how the bones of the upper and lower arm are held together at the elbow joint
(b) how the bones at the elbow joint are able to move with the minimum of friction

What type of joint is the elbow joint?

8 Using very simple diagrams compare the structure and function of (a) a hinge joint, (b) a ball-and-socket joint. In each case name *one* place in the body where you would find each joint.

9 (a) The diagrams show three types of lever:

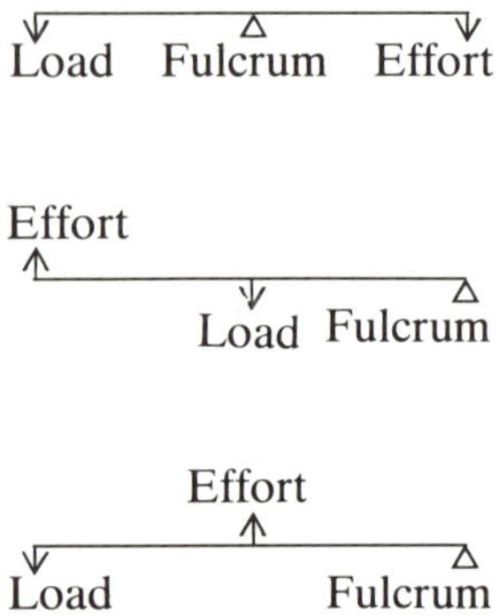

Draw simple diagrams to show *two* of these types of lever in operation within a particular region of the body. In each case indicate on the bones the position of the fulcrum, load and effort.
(b) If the hand lifts a weight of 10 N, is the muscular effort (i) 10 N, (ii) less than 10 N, (iii) more than 10 N?

10 Draw a diagram of the human arm when it is extended (straight) to show the position of the bones and the state of the muscles. Label
(a) the bones
(b) the muscles
(c) a muscle origin
(d) a muscle insertion

Give *one* reason why we have two bones in the lower arm.

11 Describe the action of the muscles and the movement of bones in the left leg when the left foot is raised and placed on the bottom rung of a ladder. Continue the description to cover what happens as the left leg is straightened to bring the right foot on to the rung.

12 (a) Name *three* different types of muscle tissue that are found in the body. Name one place in the body where each type is found.
(b) Draw labelled diagrams to show the structure of *two* types of muscle. State what control the body has over these types of tissue. Give an advantage of each method of control.
(c) What is meant by antagonistic muscles? How do these muscles help to maintain body posture? Give *two* conditions under which body posture is lost.

13 (a) Draw *two* simple labelled diagrams that would help a person identify bone and cartilage by their microscopic structure.
(b) State *one* property of cartilage that is different from that of bone, and give two functions of cartilage within the skeleton.
(c) What is the process of ossification?
(d) In what way are the bones of an elderly person different from those of a young child? How are these differences noticeable in the number and types of injuries resulting from falls that are recorded within these age groups?
(e) Why are blows 'below the belt' dangerous? State *two* other functions of the skeleton.

4 Breathing

4.1 Exercise and breathing rate

Most of you know how the man in Fig. 4.1 is feeling as he runs along. He feels his straining leg muscles, his heart beating quickly and his chest heaving as he takes in air. *Breathing*, or *external respiration*, is the taking in and giving out of air by the body. A person at rest will take on average sixteen breaths per minute.

When muscles are active the energy they need is taken from the energy store of the cell. The cell's energy store is a substance known as *adenosine triphosphate*, called *ATP* or *APPP* for short. You can see in Fig. 4.1 that ATP contains three phosphate (P) groups. When a molecule bond that holds a phosphate group on to the molecule is broken, energy is released. This is a 'one-step' reaction, so energy is available immediately. However, there is only enough ATP stored within the cell to supply a few seconds of vigorous activity. The energy from ATP is the only form of energy that the cell's activities will run on, so the store of ATP must be renewed. The energy used to rebuild ATP is obtained from the cell's respiration of food (page 186). Under normal circumstances respiration of food is carried out using oxygen (*aerobic respiration*). Therefore a good supply of oxygen is essential for muscular activity. In fact fifty times more oxygen is needed by our muscles when they are active than when they are at rest. To meet this extra demand there is an increase in our breathing rate and depth of breathing.

There is a limit to the amount of oxygen that can be supplied in a given time by deeper and faster breathing. If demand exceeds supply then the cells become starved of oxygen. However, the muscles still manage to carry one working for some time. They do this because the cell turns to another form of respiration, the *anaerobic respiration* of food. In this process the food is chemically broken down without the use of oxygen. Anaerobic respiration is not so efficient as aerobic respiration and it produces a substance known as *lactic acid*. The accumulation of too much lactic acid in the cells causes muscle fatigue and muscle cramps, which prevent the muscle from working. In these conditions our bodies are said to have an *oxygen debt*. This debt has to be repaid at the end of the activity, so when the man in Fig. 4.1 stops and sits down he continues to breathe at a faster and deeper rate. Eventually his breathing returns to its normal rate. The length of time that this takes is known as the *recovery period*.

A person's recovery period after exercise can vary. It will depend upon how hard the body has to work, i.e. the nature of the physical activity. On page 38 there are some suggestions on how you can measure the work done by your body during various physical activities.) The other fact that determines the recovery period is the *physical fitness* of the body. Physical fitness is regarded generally as a person's ability to enjoy a moderate physical activity such as walking, cycling or jogging (suitable to their age) without discomfort either during or after the exercise. We discussed in Chapter 3 the fact that muscles need regular exercise if their capacity and power are not to deteriorate. This can be applied to the body as a whole. A planned daily exercise which produces *mild* breathlessness is recommended as a way of promoting physical fitness and good health. Regular exercise that makes the body work harder than at other times during the day stretches and helps to extend the capacity of muscles, heart (which is itself a muscle, page 35) and lungs. This can give us a sense of well-being: we feel that our bodies can cope with the many demands of life without physical strain or stress.

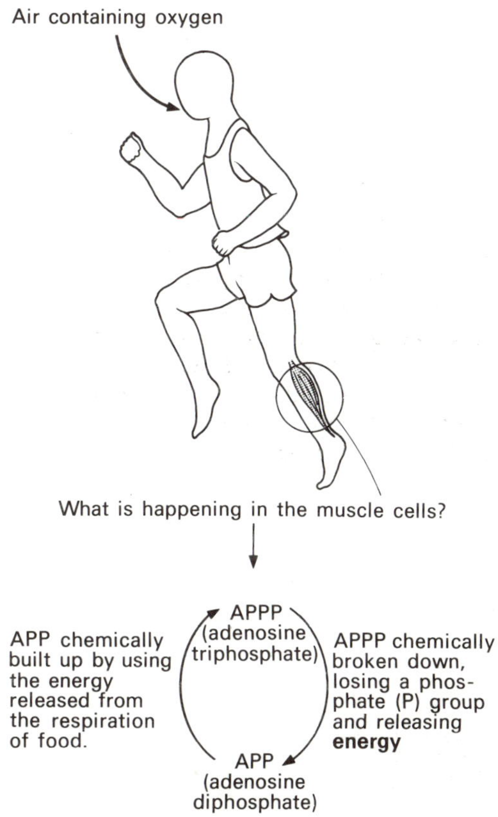

Fig. 4.1 *Exercise and the energy store of the muscle cells*

4.2 **Oxygen in the air**

Oxygen is essential for life and is present in the surrounding air. From other science lessons you may already know that air is a mixture of gases. The main ones are oxygen (21 per cent), nitrogen (78 per cent), argon and carbon dioxide. It is the percentage of oxygen present in the air which decides whether human beings can survive or not. A level of only 12 per cent would be dangerous to life, and a level of 7 per cent would make survival impossible. Where necessary artificial life support systems are needed, for example in space travel and emergencies (Fig. 4.2).

Nature provides us with clean air, but it does not always remain so. Industrial growth in countries all over the world produces the same result: massive air pollution, which before long comes to the notice of everyone because of its effect on health and the quality of life.

Fig. 4.3 *(a) Smog hangs over London (pre-1956). (b) London is now generally clear of smog*

Fig. 4.2 (Below) *In dense smoke, respirators are needed*

Fig. 4.4 *Tokyo's Pollution Control Centre. The degree of air pollution is indicated on the city map from measuring devices at various places.*

4.3 **Air pollution**

Man-made pollution of the air comes mostly from the burning of fuels which release waste matter into the air. Fig. 4.3(a) is a photograph of London in the 1950s which shows clearly the result of pollutants being released into the air. When a fog occurred in the winter the resulting mixture was smog. There is no doubt that unclean air injures health. People living in such conditions are more likely to suffer from diseases of their

TABLE 4.1 **SOME IMPORTANT AIR POLLUTANTS**

Pollutant	Composition	Fuel burning	Example of source	Harmful effects
Smoke	Small particles of unburnt soot Tar Hydrocarbons	Incomplete burning of natural coal and oil	Low household chimneys	Affects the tissue of the lungs. Increases incidence of bronchitis. Blackens vegetation and buildings
Sulphur dioxide	Gas formed when sulphur burns	Most coal and oil contain a certain amount of sulphur	Power-stations Oil-fuelled boilers	Dissolves in rain, producing an acid solution. This harms plant and animal tissues. Corrodes metal. Eats away stonework
Exhaust fumes from vehicles	Carbon monoxide Sulphur dioxide Hydrocarbons Small amounts of lead compounds Oxides of nitrogen	Petrol	Motor vehicles (diesel engines can be cleaner). Legislation has been introduced controlling oxides of nitrogen, carbon monoxide and hydrocarbon emissions, also the quantity of lead in petrol	Carbon monoxide is poisonous. In a very large city traffic-jam fumes can affect breathing. Lead absorbed into the body
Dust and grit	Solid particles	Solid fuel or produced from a manufacturing process	Iron and steel works Cement and brick works	Creates dirt and dust layer over surroundings. Respiratory system irritated by particles that are inhaled

breathing systems, and naturally the very young and the old are most at risk. In London in 1952, 4000 deaths were attributed to smog. Clearly action was needed on a national scale and in 1956 the Clean Air Act was passed. The aim was to stop pollutants reaching the air that people were breathing. The most common pollutants are described in Table 4.1.

Fig. 4.3(b) shows how the Clean Air Act (1956), the Control of Pollution Act (1974), and the Health and Safety at Work Act (1974) have helped London. Fig. 4.4 shows one of the methods that is being used in Tokyo, Japan, to fight their air-pollution problem. You may think that the anti-pollution drive requires too many regulations, yet consider the alternative in terms of our general health and the state of our environment (see Table 4.1). Nor is the fight against air pollution at an end. There may be less smoke and no smog in this country now but our air still needs to be cleaner. You might like to discuss what other sources of energy could be used in the future that would not pollute the atmosphere.

Fig. 4.5 *Team work. (Please don't try this yourself!)*

4.4 Man's interdependence

Look at the motor-cyclists in Fig. 4.5. They keep the pyramid shape because they support and are linked to one another. You can imagine very clearly what would happen if the centre person slipped. All forms of life on earth are linked to one another in a similar complex pattern. Man is part of this pattern and relies upon his links with other life forms for his survival. This reliance on others is termed *interdependence*. Fig. 4.6 shows how this interdependence works for the supply of oxygen and how oxygen is recycled in nature.

Question 1 In the redevelopment of towns, parks and green open spaces are regarded as essential. Why are they called the 'breathing spaces' of the town?

Living things can respond to changes, but if these take place too quickly or are too drastic then the normal balance of nature can be upset. This is seen most clearly when the survival of an animal is threatened as a result of the destruction of its natural habitat by man. Less obvious but just as important is the estimate that in America the total demand for oxygen is greater than the rate of supply through photosynthesis. Such demands are leading to an increase in carbon dioxide in the atmosphere, assisted by the destruction of forests and natural vegetation as man seeks more land for agriculture and building. This in turn may be causing a slight rise in the temperature of our atmosphere. Therefore it is important that, both as individuals and as a community, we should understand how to live without upsetting the delicate balance of nature.

4.5 The body's air-filter system, the nose

The body does have some defence against dirty air and this is found in the 'air inlet tube' of the body, the *nose* (Fig. 4.7). The nose is made up of two *nasal passages*, which at the inner end open out into the *nasal cavity*. Cartilage at the end of the nose gives flexibility yet prevents the passages from collapsing. Air enters the nasal passages at the *nostrils*. Here there are stiff hairs which act

Fig. 4.6 *Natural recycling of oxygen and carbon*

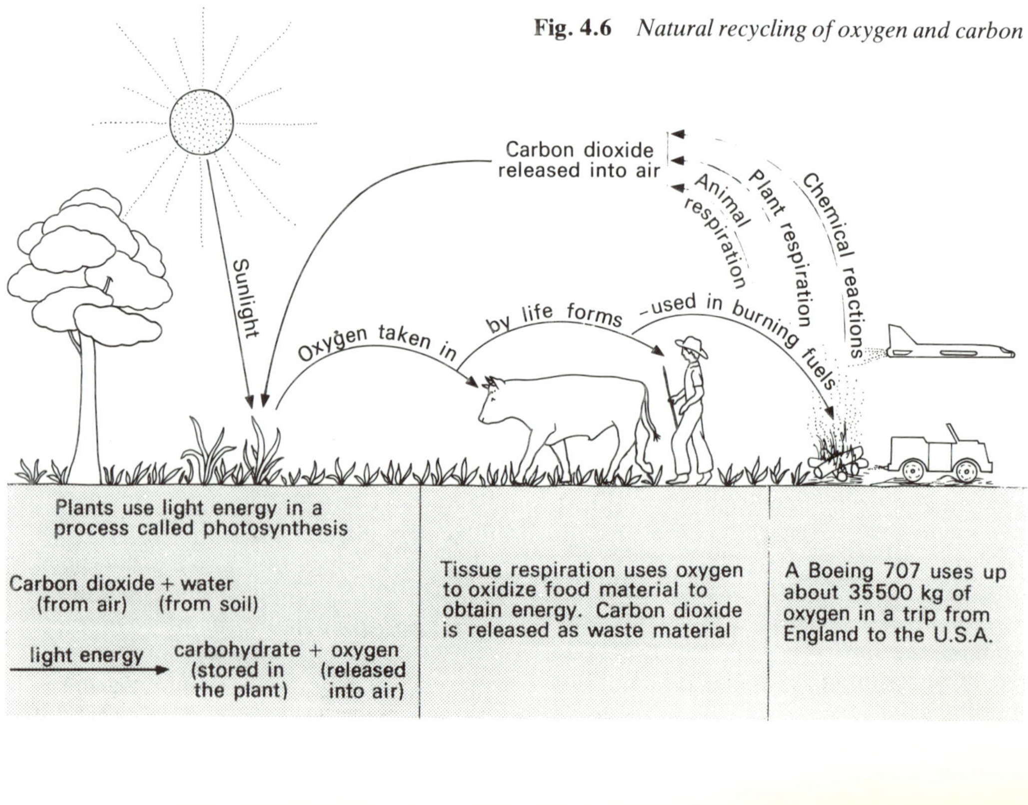

like a net and trap large particles of dirt. The side walls of the nasal cavity form a very large surface area within a small space, because the tissue lies in folds. The membrane covering this area is called a *mucous membrane*. This is because it contains many cells which produce a thick fluid called mucus, and germs get stuck or trapped in this mucus. By this means the air is cleaned and moistened. An increase in the supply of this mucus is one of the first signs of a cold!

There is another special type of cell in the lining of the nose. This is a cell which has fine bristles on its upper surface. The bristles are called *cilia* and the cell is a *ciliated cell*. The cilia are able to move and push any mucus and dirt towards the nostrils and out of the nose.

Question 2 What action can you take to help the removal of dirt and mucus?

If we blow our noses after being in a very dusty atmosphere we can see how much dirt has been collected. This shows how important it is that as children we develop the habit of breathing in through our noses and not through our mouths.

There is another good reason for this nasal breathing. The lining of the nose contains a large network of small blood-vessels (capillaries). The blood in these capillaries is at body temperature, so it warms the air as it is drawn through the nasal passage. This stops air passages and the lungs being damaged by excessively cold air.

The quantity of blood that is supplied to the nose becomes obvious during a nose-bleed. Such bleeding can result from a bump on the nose, picking of the nose or fierce blowing. To stop the bleeding the person should sit with his head forward and pinch just below the hard bridge of the nose to apply pressure to the blood-vessels. After ten minutes the bleeding should have stopped and the pressure may be released. It is not advisable to tip the head backwards, and you can see the reason for this if you look again at Fig. 4.7. With the head held backwards the blood would flow into the cavity (pharynx) at the back of the mouth and nose. From here it can enter the *oesophagus*, which is the tube leading to the stomach. Trying to swallow with the head back is very uncomfortable and can make a person choke.

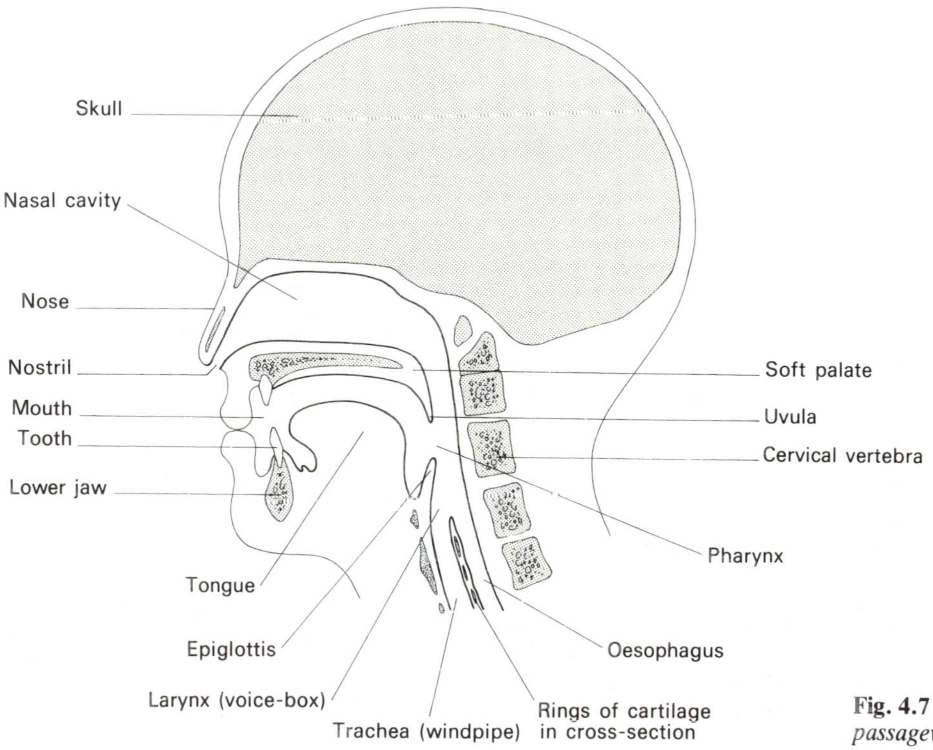

Fig. 4.7 *Cross-section of head showing passageways of nose, mouth and throat.*

4.6 The passageways from nose to lungs

In Fig. 4.7 you will see that there is a second tube leading downwards from the pharynx. This is the *trachea* (windpipe) and it is positioned in front of the *oesophagus*. The trachea is the passageway for the air, whereas the oesophagus is the passageway for food. Food and air must be directed into the correct passageway and this is the job of the flap of skin called the *epiglottis*.

When food is swallowed the epiglottis is dropped over the opening of the trachea to close it. If food does 'go down the wrong way' the reaction of the body is *automatic*, that is it responds without the person thinking about it (page 232). In this case an attack of coughing will bring the food out of the trachea. Patting the sufferer on the back can help in such an emergency and small children may even have to be turned upside-down to clear the blockage. Also, during swallowing the *soft palate* moves upwards, closing off the nasal cavity. You can see the *uvula*, the lower end of this soft palate, hanging down at the back of your mouth.

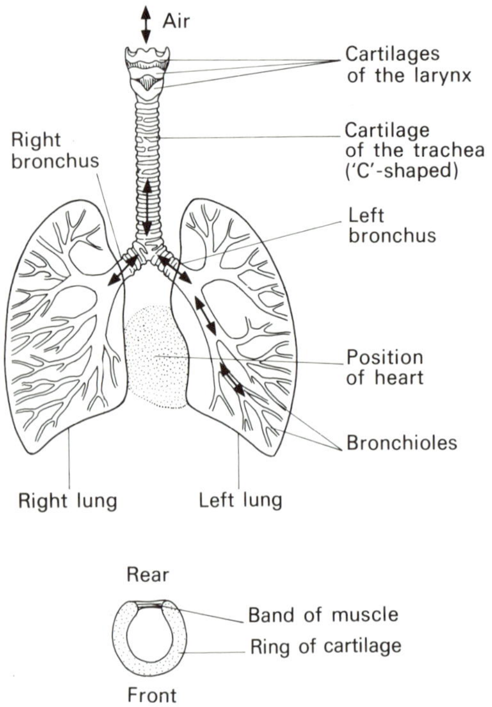

Fig. 4.8 *The passage of air into the lungs*

When air is breathed in, the soft palate drops down, closing off the mouth cavity. At the same time the epiglottis is raised so that air can enter the trachea.

The top end of the trachea is much wider than the rest of the tube (Fig. 4.8). This wide section is the *voice-box* or *larynx*. Its shape is kept by rings of cartilage. You can feel the largest of these if you tilt your head backwards and run your fingers along your throat. The largest cartilage is sometimes visible as your Adam's apple.

Two membranes are stretched across the floor of the larynx. These *vocal cords* are shown in Fig. 4.9(a). Most of you will have at some time stretched a blade of grass between your thumbs and blown through them to make a whistle (Fig. 4.9(b)). You may have noticed that if the grass is slack you get a low-pitched whistle, while if the grass is taut you get a high-pitched whistle. The sound is produced because the grass *vibrates* in the stream of air.

In our voice-box, the vocal cords vibrate as air passes between them. Muscles can tighten or relax the cords to give high- or low-pitched sounds. The length and thickness of the cords also alters the sound of the voice. In boys the thickness of the vocal cords increases in adolescence and the voice becomes deeper. We say that the voice *breaks*.

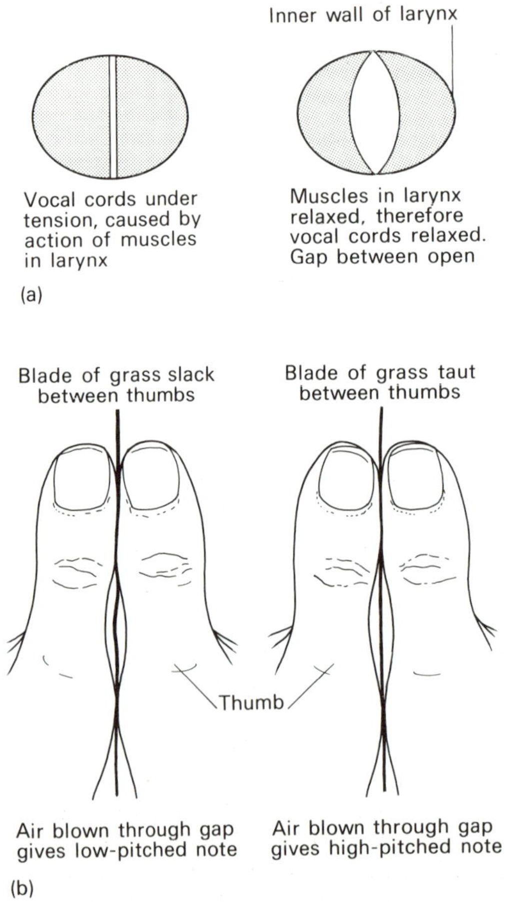

Fig. 4.9 *Looking down the larynx to view the vocal cords*

From our earliest days we learn to use tongue, lips and teeth to form sounds into syllables and words. When we speak the soft palate drops down so that the sounds can travel round the cavities of the nose. These cavities work like the sounding-box of a guitar, reinforcing the sound and giving it depth or *resonance*.

Question 3 If you pinch your nose or have a cold what differences does this make to your voice?

The rings of cartilage present in the larynx are also found in the trachea. They allow the tube to be flexible, yet stop its fibrous tissue from collapsing inwards. These rings of cartilage have a small gap in them. Stretched across the gap is a band of muscular tissue (Fig. 4.8), enabling the trachea to constrict and relax. This framework of cartilage continues when the trachea splits into two, forming the left and right *bronchus* (Fig. 4.8 and Colour Plate 2). The framework of cartilage also continues into the *bronchioles*, which are the smaller tubes branching out from the bronchi as they enter the *lungs*.

If the muscular tissue in the cartilage of the bronchioles begins to constrict in spasms it can make breathing quite difficult. This reaction may be seen in someone who suffers from *asthma*. An asthmatic attack can be brought on by an infection, an emotional factor, or as an allergic response.

A person is said to have an *allergy* if their body visibly over-reacts to a substance which is not harmful, and which does not affect most other people. For example, a person who suffers from hay fever is allergic to pollen (Fig. 4.10). Skin rashes, such as *nettle-rash*, are allergic responses to contact with a particular substance or the eating of a certain food. Many allergic responses are triggered off by the substance *histamine*, which is always present in the body. In an allergic reaction there is a large increase in the amount of histamine at the point of response. *Antihistamine* drugs can lower the body's

Fig. 4.10 *Hay fever is an allergic response to pollen. During summer months a pollen count is recorded every day.*

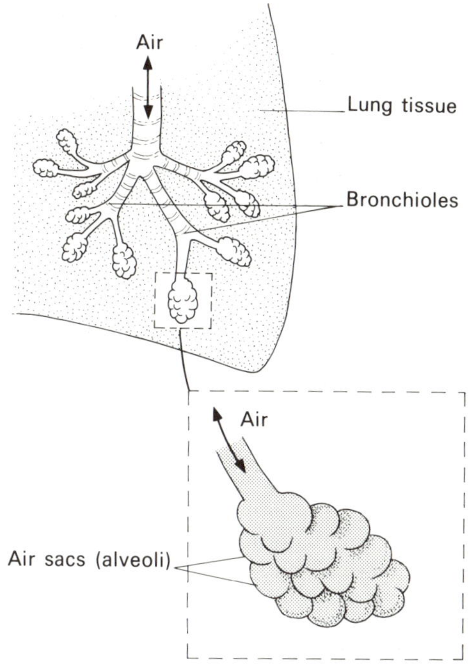

Fig. 4.12 *Alveoli*

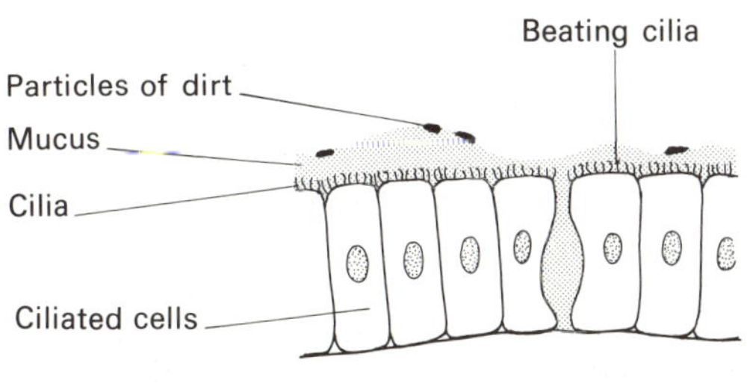

Fig. 4.11 *Movement of mucus and dirt by the cilia in the respiratory passageways*

production of histamine and so control the allergy.

The trachea, bronchi and bronchioles are lined with a mucous membrane and ciliated cells. Dirt is trapped and carried upwards by the movement of cilia to the mouth cavity (Fig. 4.11) and swallowed. This ensures that dirt does not reach the small fragile sacs, the *alveoli*, which are at the ends of the bronchioles within the lungs (Fig. 4.12).

4.7 What happens in the alveoli?

The alveoli are very small, about the size of the fine point of a needle. They have elastic walls and, like balloons, they can expand a little when air enters them. In every person there are approximately 700 000 000 alveoli and all are surrounded by a network of capillaries (Fig. 4.13). These capillaries contain blood that has been pumped by the heart to the lungs after it has travelled around the body. Thus this blood carries a high concentration of carbon dioxide, a waste product of tissue respiration.

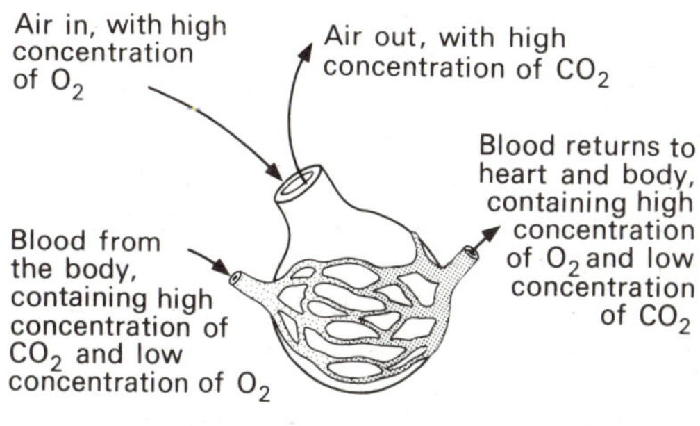

Fig. 4.13 *Exchange of gases in the alveoli*

The alveoli, however, contain a low concentration of carbon dioxide but a high concentration of oxygen. These are exactly the right conditions for the movement of molecules by a process known as *diffusion*. Substances diffuse when their molecules move from a region where they are highly concentrated to a region of low concentration (Fig. 4.14). Thus oxygen can diffuse across the tissue walls, out of the alveoli and into the blood; at the same time carbon dioxide moves in the opposite direction. The blood now carries a high concentration of oxygen as it flows back to the heart. The heart pumps it round the body to supply the cells. The air that is breathed out of the lungs consists of 16 per cent oxygen, 4 per cent carbon dioxide and 79 per cent nitrogen. It is also saturated with water vapour taken from the moist linings of the alveoli. You only have to sit in a car on a cold day to see evidence of this water vapour as it condenses on the windows.

To obtain the best exchange of gases there must be a large moist surface area at which air and blood can meet. This vast area is provided by the total surface area of the alveoli. It is estimated that if the surfaces of the alveoli were spread out flat they would cover a tennis-court (about 80 m^2). The membrane is moistened by tissue fluid (page 62).

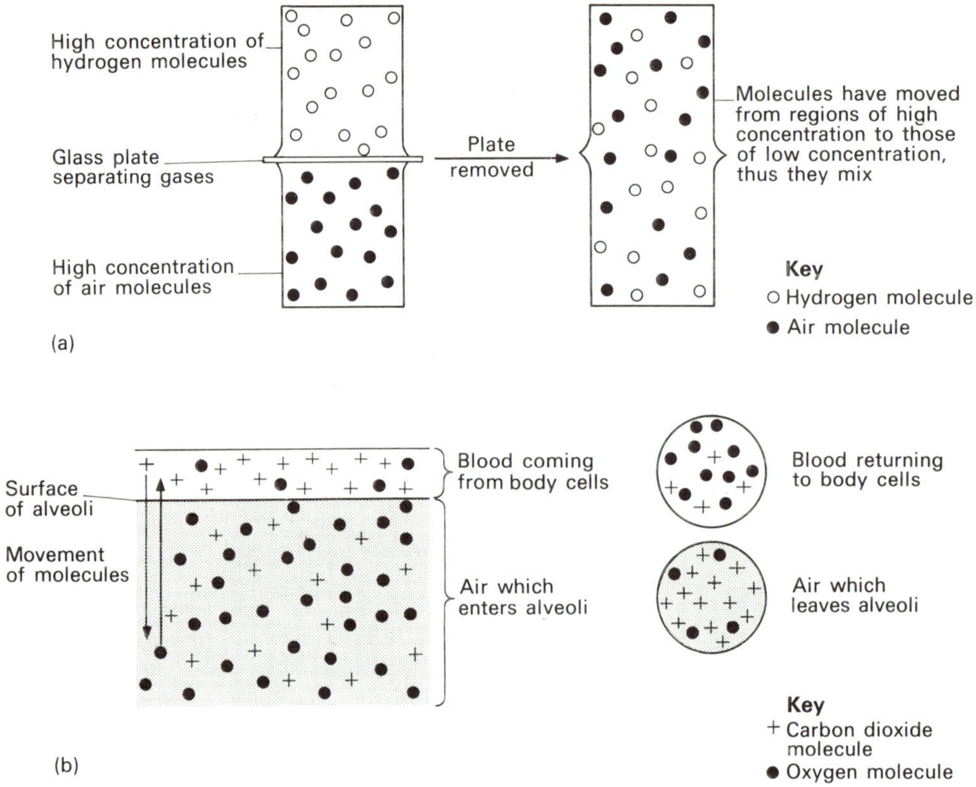

Fig. 4.14 *Diffusion. (a) In gases. (b) In the alveoli*

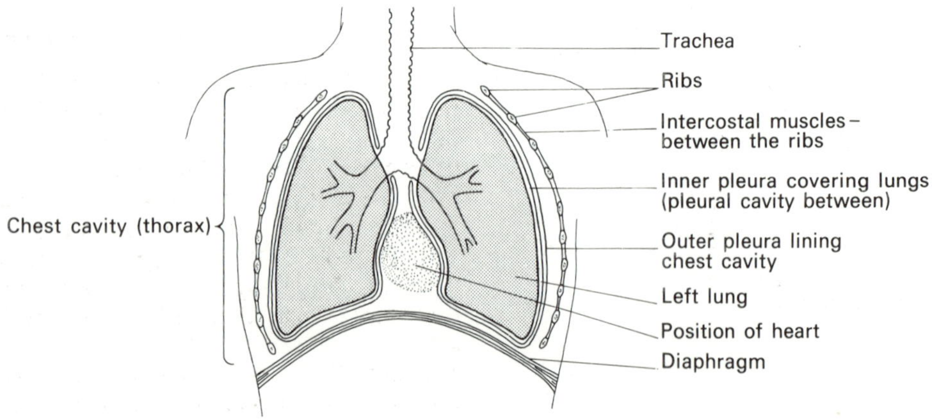

Fig. 4.15 *The lungs as they appear within the chest cavity*

It is this complex structure of bronchioles, alveoli, air spaces and blood-vessels that makes up the two spongy structures known as *lungs*. The outer surface of the lungs is covered by a shiny membrane, the *pleura* (Fig. 4.15). Each lung is split into *lobes*. There are three lobes in the right lung and two lobes in the left lung.

We have considered what happens to air when it is inside the lungs – now let us consider how it gets inside them. It cannot just drift in, for the lungs, like balloons, cannot inflate themselves. The air must be forced in.

4.8 How air enters and leaves the lungs

The lungs are suspended inside the chest cavity, or *thorax* of the body. The sides of this cavity are formed by bones (the *ribs*) and by the strong *intercostal muscles* that link the ribs together (Fig. 4.15). The floor of the cavity is made up of a sheet of tissue, the *diaphragm*. Attached to the diaphragm are muscles which control its movement. In an attack of hiccups you will have felt the muscular spasms of your diaphragm.

The air around us exerts a pressure of 100 000 N on every square metre. If there are areas of different pressure, air will move from where the pressure is high to where it is low in an attempt to make the pressures equal. The necessary changes in pressure inside the lungs are brought about by movement of the ribs and diaphragm.

Breathing in – inhaling (Fig. 4.16(a))

(a) The intercostal muscles contract, lifting the ribs upwards and outwards.

(b) The diaphragm muscles contract and the diaphragm is lowered.

Both of these movements cause the volume of the chest cavity to increase. The chest cavity is completely sealed, so the pressure on the sides of the lungs is less. The elastic tissue of the lungs allows them to expand, thus increasing their volume and reducing the air pressure inside them. The air pressure is greater outside the body than inside the lungs so air is forced through the nose, down the trachea and into the lungs, thereby inflating them.

Breathing out – exhaling (Fig. 4.16(b))

(a) The intercostal muscles relax, letting the ribs drop downwards and inwards.

(b) The diaphragm muscles relax and the diaphragm is raised by the pressure of the organs underneath.

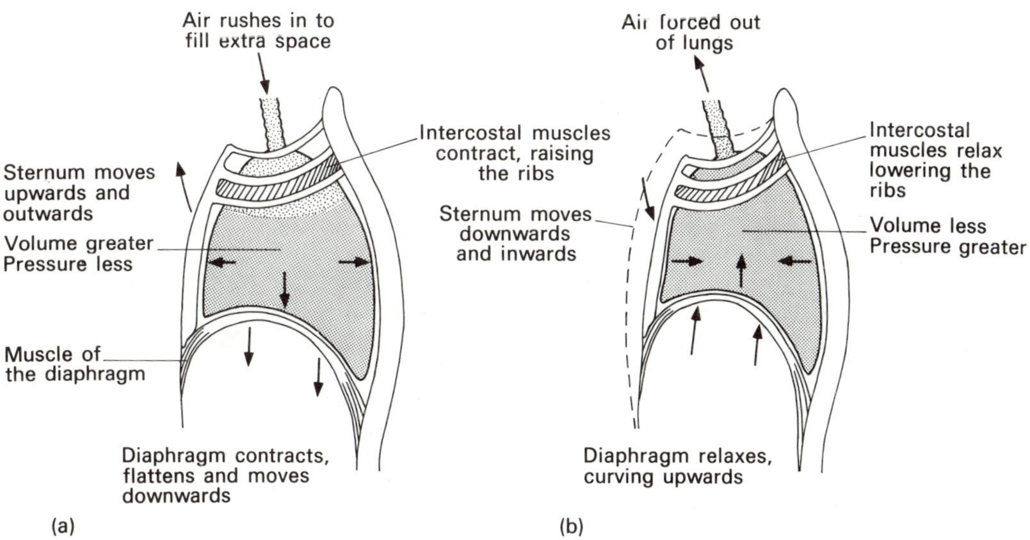

Fig. 4.16 *Movement of ribs and diaphragm in breathing. (a) Breathing in – inhaling. (b) Breathing out – exhaling*

Both these movements cause the volume of the chest cavity to decrease. This exerts pressure on the lungs so that their elastic walls are squeezed, increasing their inner pressure and pushing out the air.

The pleural membrane of the lungs and a similar membrane lining the chest cavity ensures that there is no friction between the moving surfaces. Normally all these breathing movements are automatic (page 232). If we exercise they increase automatically. This action is triggered off by the high concentration of waste carbon dioxide in the blood. Coughing is an automatic response to particles or large quantities of mucus in the air passages. Air is exhaled with great force in an attempt to move the material. However there are also times when we use our minds to control our breathing (page 234).

Question 4 Try to write down three occasions when you might consciously control your breathing.

We can normally hold our breath for only a short time before the automatic reaction to breathe takes over.

4.9 How much air can be held in the lungs?

If an adult inhales and holds his breath then the lungs will contain approximately 5 litres of air. Taking a deep breath will absorb 1½ litres of this 5 litres and this is called *complemental air*. The *vital capacity* of our lungs is the maximum amount of air that can be breathed out after taking a deep breath.

Normal breathing moves only about ½ litre of air in and out of the lungs. This is *tidal air*.

A further 1½ litres, *supplemental air*, can be removed if a person forcibly exhales. However there will always remain about 1½ litres, *residual air*, which cannot be removed.

This residual air does not become stale, because it exchanges gases by diffusion with the incoming air.

A simple way of finding the vital capacity of your lungs is described on page 00.

Question 5 Will a measurement of the volume of the air you exhale give you your total lung capacity?

Individual lungs have different vital capacities. A person who plays a brass instrument regularly may find that the vital capacity of his lungs is much larger than that of a non-player. People with asthma, bronchitis and other chest infections find that they become breathless because the efficiency of their lungs is reduced. To monitor this change in efficiency the person may be asked to record the *peak expiratory flow* of his lungs, that is the maximum rate at which he can breathe out air. Fig. 4.17 shows someone recording her peak flow with a Wrights peak flow meter.

Fig. 4.17 *Measuring a person's peak expiratory flow (P.E.F.) of air can indicate the efficiency of the lungs. A person with asthma will find that their P.E.F. value is decreased.*

4.10 Diseases of the lungs

Every square centimetre of lung area is needed by the body if it is to work efficiently. Damage to lung tissue can produce breathlessness and make lung infections more likely. *Bronchitis* is inflammation of the bronchial membrane resulting from an infection (page 00). *Pneumonia* is inflammation of lung tissue and can occur as a complication to another illness.

Tuberculosis (T.B.) of the lung is one of the oldest known diseases (page 101) and until the 1920s and 1930s it caused many deaths. In Britain T.B. is now under control. The change has been brought about mainly by better living conditions, better diet and a higher standard of health for most of the population. T.B. is a disease that is closely linked with social conditions because of the way in which it is spread (page 82). It will always spread where there is overcrowding, poor food and poor health. Mass radiography (introduced in 1943), treatment by drugs and the B.C.G. vaccine (page 93) have all helped to overcome T.B. Mass radiography can be carried out using mobile X-ray units, which visit an area and take chest X-rays of a large number of people. Any infection of the lungs by T.B. will show up on an X-ray. By this method it is easy to detect T.B. in its early stages before the person is aware of being infected. Carrying out medical checks on groups of people in this way is known as *screening*. Nowadays various screening techniques are proving invaluable in the fight against other diseases. The fact that T.B. is under control does not mean that screening to detect the disease should stop. T.B. and other diseases that have declined will stay at these low levels only if methods of controlling them are continued.

Other methods of controlling disease involve protecting a person against any conditions that might cause the disease. This applies to diseases which develop because a person has been doing a particular job; these are termed *occupational diseases*. For example, people who work in conditions where there are large amounts of industrial dust are likely to inhale very small particles which can damage their lung tissue. Miners can develop *pneumoconiosis* through inhaling coal-dust. Particles of a substance known as blue asbestos can cause *asbestosis*. The Health and Safety at Work Act requires factory inspectors to check that industrial dust is kept below certain levels. Industrial regulations may require workers to wear masks that filter the air that they breathe. Dangerous substances, like blue asbestos, may be banned from use altogether.

It is unfortunate that at present there is no decrease in the number of cases of *lung cancer*. In the U.K., in 1900 about 500 people died of lung cancer. In 1968 approximately 32 000 deaths were reported and in 1977 38 000, of which 8000 were women. The rapid rise in deaths from lung cancer has led to many investigations all over the world into its causes. Changes in the environment have been studied, such as pollution and industrial developments, but the closest link is the increase in cigarette smoking.

4.11 What is cancer?

In Chapter 1 we saw that among the activities of a cell were growth and division. Sometimes the body's control over this growth and division breaks down. More cells are then produced than are worn away, and this leads to a small lump of cells. This is how warts and cysts are formed, but they are *not* cancer for they are made up of normal body cells (Fig. 4.18(a)).

A cancerous growth is a lump of abnormal cells called 'rebel' or cancer cells. They divide rather like body cells but they grow by taking their 'food' from the protoplasm of the normal body cells that surround them. Because of this action some surrounding cells are destroyed by the cancer cells

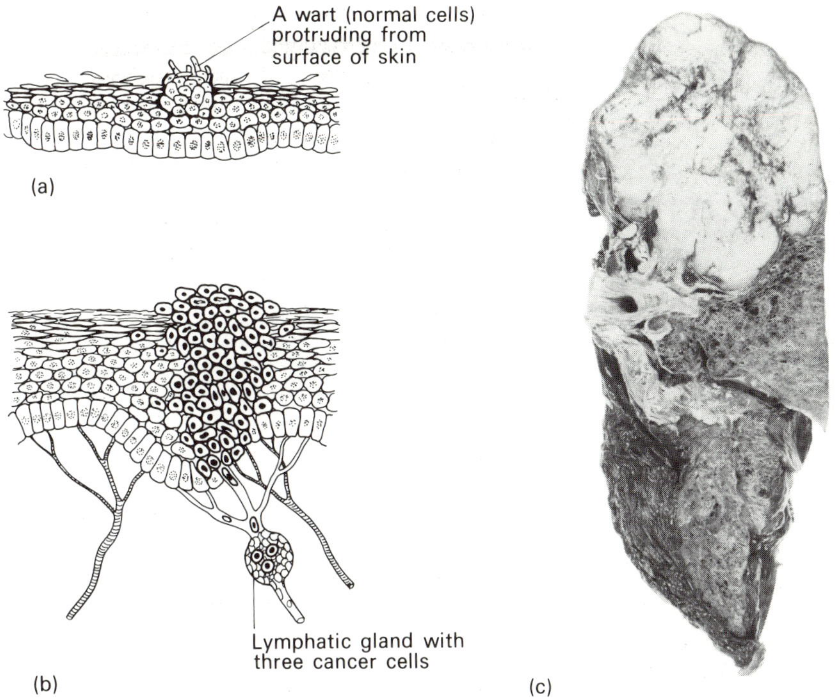

Fig. 4.18 *Abnormal cell growth. (a) Innocent. (b) Malignant. (c) Cancer of the lung*

(Figs 4.18(b) and (c)). Cancer cells are abnormal in another way: they can split off from the primary growth and enter the bloodstream. In this way they can move to other areas of the body and start new colonies of cancerous cells (secondary growths). Therefore cancer is said to be malignant. It is this that makes it very important that cancer should be detected and treated at the earliest possible stage. Already screening techniques are available to do this for some forms of cancer (page 247). It is up to the individual to make use of these facilities. There is one universal screening technique – our own observations on the health of our bodies. This makes it essential that everyone understands the facts about cancer. Every year in Britain more than 30 000 people who have been treated for cancer are cured. Treatment at present is of three types: surgery, radiotherapy and chemotherapy. Radiotherapy is based on the fact that cancer cells are more easily destroyed by radioactive rays (X-rays), from substances such as radium, than normal cells. Chemotherapy involves threatment with drugs.

4.12 Smoking and cancer of the lung

Cigarette smoke contains gases and very small drops of tar. The tar collects at the unlit end of the cigarette, so more is breathed in as the cigarette burns low. You can see this tar collect in the smoking-machine described on page 54.

Irritant substances in cigarette smoke cause more mucus to be made by the mucous membranes. They also stop the movement of the cilia in the ciliated cells. This means that the cilia can no longer move dirt, germs or mucus away from the lungs. The lung tissue therefore gets very dirty. This dirt contains substances sixteen of which are known *carcinogens*, that is cancer-forming chemicals. The more cigarettes that are smoked the dirtier the lungs become and the higher the risk of disease. You may have noticed that many addicts of cigarettes have a 'smoker's cough'. Coughing is the only way of moving the mucus once the cilia are out of action. However, if it is fierce this coughing can burst the alveoli, leaving less lung tissue for the exchange of gases. A person is then more likely to develop complications such as bronchitis when suffering from a cold or flu. In fact chronic bronchitis kills approximately 30 000 people a year. When a large amount of lung tissue is destroyed the person becomes constantly breathless, a condition known as *emphysema*.

If a person gives up smoking then the cilia gradually begin to function again, but a return to normal action could take several years – doctors reckon about ten – for a person who has smoked heavily for some time. However surveys have revealed that giving up smoking at any stage results in an improvement of health and reduces the risk of developing lung cancer. It also lessens the risk of developing other diseases of the respiratory and circulatory system (page 76).

When people know these facts they often resolve to give up smoking, but there is no doubt that they find this difficult. The reasons for this are complex but one common factor can be considered. *Nicotine* is present in cigarette smoke. The effect of this varies from person to person but generally it may calm or excite by acting on the body's nervous system (page 222). Because of this people become dependent upon its 'pleasure effect'. This dependence may cause a light smoker to become a heavy smoker, inhaling deeply. Nicotine may make a first-time smoker carry on experimenting even when the first experience was unpleasant. On average most smokers begin to smoke before they are twenty.

(**a**) All my friends smoke

Is it true to say all?
Will you have the same friends ten years from now?
Should a person copy everything his friends do?
Should other people encourage others to risk their health?
Could smoking upset a relationship between two people?

(**b**) I was told not to

Why should anyone bother to tell people not to smoke?
Is there anything wrong in taking advice from others?
Does smoking cause fires?

(**c**) I like it

Smokers may, but do others?
Should smokers be more considerate towards non-smokers in their homes and public places?
Does anyone have the right to 'pollute' someone else's atmosphere?

(**d**) I feel more grown up

Does smoking make people look older, or is it to do with confidence and being admired by others?
Are there other ways of impressing people?

(**e**) My family smoke and they are alright

They are alright now, but what about later on in life?
Even if parents smoke, do they like to see their children doing it?
Can children help parents to give up smoking?
When you are a parent how will you feel?

(**f**) It makes me feel good

Is this the 'drug' effect?
When a person smokes do they look or smell attractive?
Are there other ways of improving your appearance?
Could the money be spent on other things – new clothes, a new bicycle?

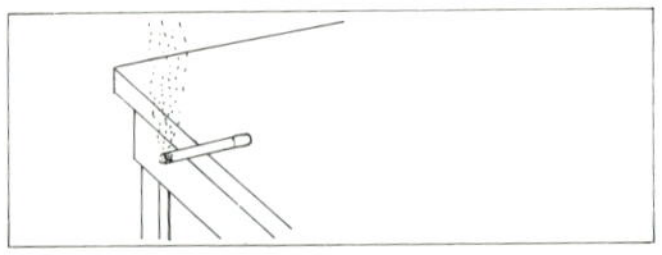

(**g**) I only smoke occasionally

Will this continue?
Are there any reasons why a person might start to smoke more?

(**h**) You have to die sometime

True, but can our actions increase or decrease our life span?

Fig. 4.19 *Reasons people give for smoking*

(**i**) Smoking does not harm your health

Is this like the ostrich with its head in the sand?
Why is there a government warning on every cigarette packet?

(**j**) The facts in black and white

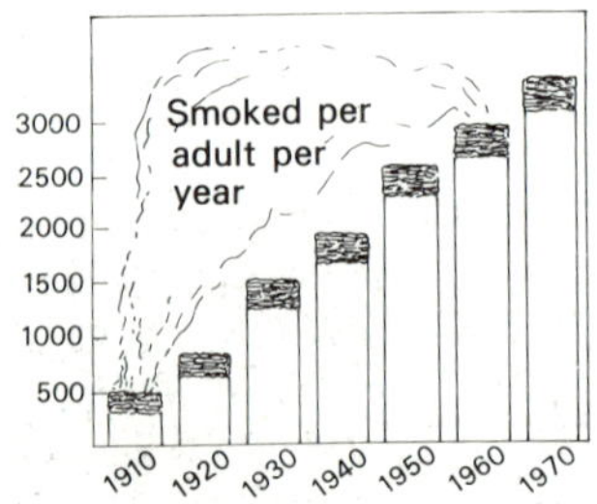

Increase of cigarette consumption since the beginning of the century (U.K.)

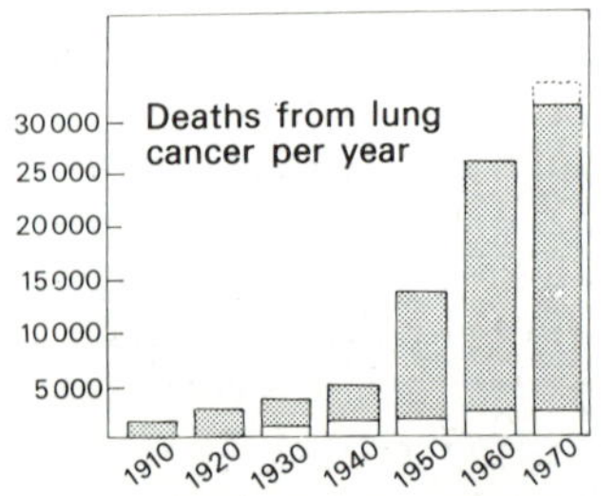

Corresponding increase in lung cancer deaths. Small area in white at foot of each column indicates approx. number of lung cancer deaths from other causes i.e. non-smokers

Fig. 4.19 illustrates some of the answers that people give when asked why they first started smoking. These answers, and the reasons why people continue to smoke, help to show that smoking becomes part of a person's pattern of behaviour. You may like to discuss the illustrations and comments.

4.13 **When a person stops breathing**

It has already been said that coughing helps to remove material from the air passages. An unconscious person cannot cough, so one result of an accident may be *blocked breathing*, where blood, vomit or other debris may stop air passing along the trachea. There is also the danger in a state of unconsciousness of the tongue falling backwards. This can be prevented by putting the head in the position shown in Fig. 4.20.

A person's breathing may stop because of an accident or illness. If the person is to live, breathing must be started again within 4–6 minutes. *Artificial respiration* is one way of trying to do this. The aim is to breathe air directly into the unconscious person's lungs, through either the nose or the mouth. It is essential to know how to do this properly and any first-aid course or lecture will provide the necessary training (Fig. 4.21). Without such training a life may be lost which could have been saved, or damage done to the casualty by doing the wrong thing.

Another body system which must be restarted rapidly is the circulatory system. In fact, if a person's heart and breathing have both stopped, artificial respiration and attempts to restart the heart must be carried out together. We will be looking at the action of the heart in the next chapter.

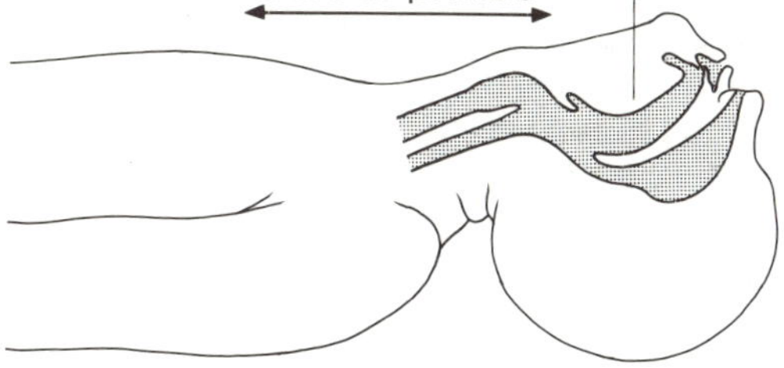

Fig. 4.20 *How to position the head of an unconscious person*

Fig. 4.21 (Below) *A St John Ambulance instructor teaching artificial respiration*

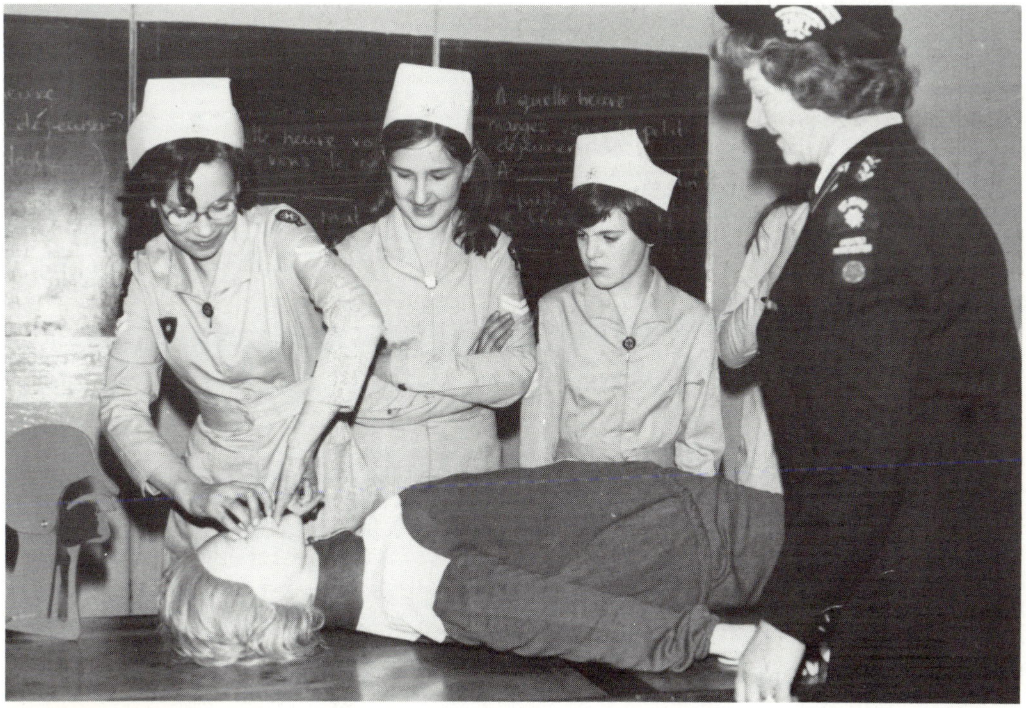

Answers to questions in the text

1 Plants and trees use up carbon dioxide from the air and release oxygen into it. By this process they freshen up a town's air supply, as well as looking pleasant and providing space for open-air activity.

2 By closing the mouth, air can be blown forcibly through the nose, carrying with it mucus and dirt.

3 The voice will sound flat. The sound cannot travel round the blocked nasal cavities, so it has no variety of tone.

4 These are a few examples:
   (a) when diving and swimming
   (b) when playing a wind instrument
   (c) when doing physical exercise such as yoga
   (d) when a doctor is listening to your chest!

5 No. Even when you exhale forcibly residual air is left in the lungs. This residual air accounts for about one-third of the total lung capacity (T.L.C.). In order to calculate the lung capacity this one-third must be worked out and added to the experimental reading of vital capacity.

THINGS TO DO

1 (a) Record the breathing rate of people from as many different age groups as possible. Carry out your recording under the same conditions.
(b) Select one person and record their breathing rate after three minutes of
(i) sitting still
(ii) steady walking
(iii) jogging
(iv) a moderate gym exercise
(v) a 'step-up, step-down' exercise (page 38)
(c) After the exercises record the person's breathing at intervals of half a minute. Find out how long it takes the breathing rate to return to normal, that is, the recovery period.

2 Carry out a set exercise regularly each day for a period of several weeks (page 38). Monitor your performance (breathing rate and recovery period). Consider whether regular exercise encourages the fitness of your body. Tabulate your results in the form of a graph. If you can persuade other people to take part you can compare your results.

3 If you have a large enough container you can measure the vital capacity of your lungs. The container needs to have a volume of 5 litres (washing-up liquid bought in bulk comes in such a container). Completely fill the container with water. Turn it upside-down in a sink half full of water. The neck of the container must remain below the water-level in the sink. Insert one end of a piece of rubber tubing into the container. Take a deep breath, then with the other end of the rubber tubing in your mouth breathe out as much air as you can in one go. Pull the tube out of the container. Screw the plastic cap on to the neck of the container and then remove it from the water. Take off the cap. Use a measuring jug to find out how much water must be added to the container to refill it.

volume of water added
= volume of air breathed out
= vital capacity of lungs

4 Study air pollution in your area. Here are a few examples of what you might do:
(a) Smear a thin film of vaseline on a tin lid. Leave it outside for a set period of time and estimate the quantity of solid particles that settle on it.
(b) Use a jam-jar to collect rain-water. Examine it for any solid particles.
(c) Plants called lichens are good pollution indicators.
(i) Heavily polluted area – no lichens. Single-cells plant *Pleurococcus* present. Looks like green dust on walls, trees, etc.
(ii) Less polluted areas – greyish-green crusty lichens present. Look like green crazy paving.
(iii) Very little pollution – 'fleshy', branching lichens present.

5 Make your own smoking-machine as shown in Fig. 4.22.

6 Organize a market research programme on smoking within your school. You can find out how many people smoke, when they started, why they began smoking, if they have ever attempted to give it up, and if there is any evidence of ill health due to smoking. You could consider running an anti-smoking campaign and survey the results at regular intervals.

7 Ask your local butcher to provide you with the lungs of a sheep or bullock (they are sold as 'lights' for pet food). Place a thick plastic tube into the windpipe and blow through it to inflate the lungs. Cut up the lung tissue and examine its structure. Why does a piece of this tissue float when put into water?

8 Collect illustrations to show the use of breathing equipment in environments where man cannot breathe naturally.

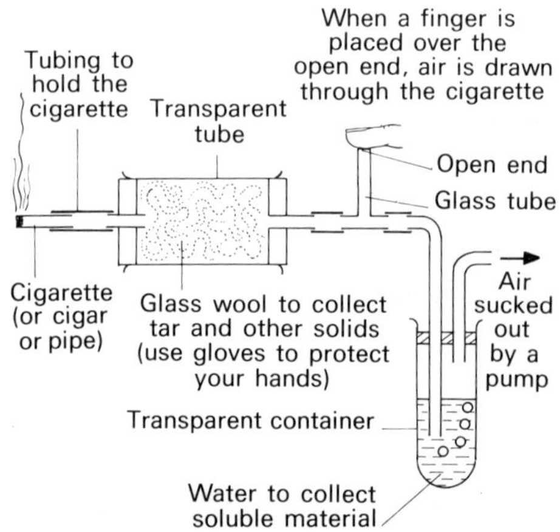

Fig. 4.22 *Smoking-machine*

THINGS TO WORK OUT

In questions 1–3 write down the *one* answer which you think is correct.

1 The air that we breathe out is different from the air that we breathe in. It has
- **A** more oxygen, less carbon dioxide, and is cooler and damper
- **B** less oxygen, more carbon dioxide, and is warmer and drier
- **C** more oxygen, less carbon dioxide and is warmer and damper
- **D** less oxygen, the same amount of carbon dioxide and is warmer and drier
- **E** less oxygen, more carbon dioxide, and is warmer and damper

2 A miner is more likely to contract the occupational disease of
- **A** silicosis
- **B** tuberculosis
- **C** pneumoconiosis
- **D** asbestosis
- **E** lung cancer

3 The molecules of gases and liquids tend to move at random until they are evenly distributed throughout the volume. This process is called
- **A** boiling
- **B** osmosis
- **C** diffusion
- **D** evaporation
- **E** settlement

4 Copy the following, putting in each space *one* word which you think is correct:

Breathing in takes place because of the . . . . . of the diaphragm muscle and the . . . . . of the intercostal muscles. In this way the volume of the chest cavity is . . . . . . The lungs expand so that the pressure of the air inside them becomes . . . . . than that of the air outside. Therefore air . . . . . the lungs.

5 (a) Name *two* pollutants of air. Give *two* harmful effects of *each* of the pollutants you have named.
(b) Why do more people suffer from chronic bronchitis in an industrial area than in a rural area?
(c) Describe *two* ways in which a community can control the amount of pollutants that enter the air.

6 Describe the function of the following:
(a) the epiglottis
(b) the mucous membrane in the nose
(c) the larynx
(d) the rings of cartilage in the bronchial passageways
(e) the ciliated cells lining the bronchial passageways

7 Study Fig. 4.23 then answer the following questions:
(a) What form of energy is *a*?
(b) What type of organism is *A*?
(c) What is the name of the process which takes place within *A* and what other substance is needed?
(d) What are the products obtained from *A* by man?
(e) Describe the process which takes place within man using *B* and *C* and releasing carbon dioxide?

8 (a) What is meant by the term diffusion?
(b) Describe the diffusion which takes place across the membrane of the alveoli.
(c) State *two* other characteristics of the lungs which you think help to achieve the best exchange of gases.

9 (a) Describe, using a labelled diagram, an experiment to show that a person breathes out carbon dioxide. If the person has just sprinted round the playground, how would this affect the results?
(b) Describe briefly the method of resuscitating a person by the mouth-to-mouth method of artificial respiration. How may the greater quantitiy of carbon dioxide in our breath help to revive the person?

10 (a) What are the differences between anaerobic respiration and aerobic respiration within a cell?
(b) Which of these two types of tissue respiration would produce more energy from the same amount of glucose?
(c) How does the process of anaerobic respiration affect muscular tissue?
(d) Describe the function of ATP and how levels of ATP are maintained within a cell.
(e) Why is oxygen needed even when the body is at rest?

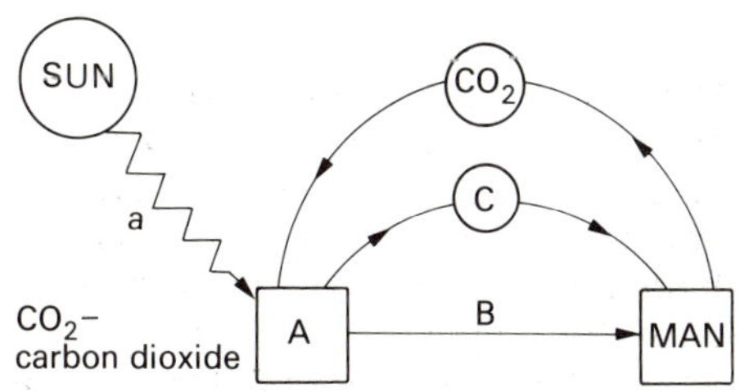

Fig. 4.23

5 The Heart and Circulation

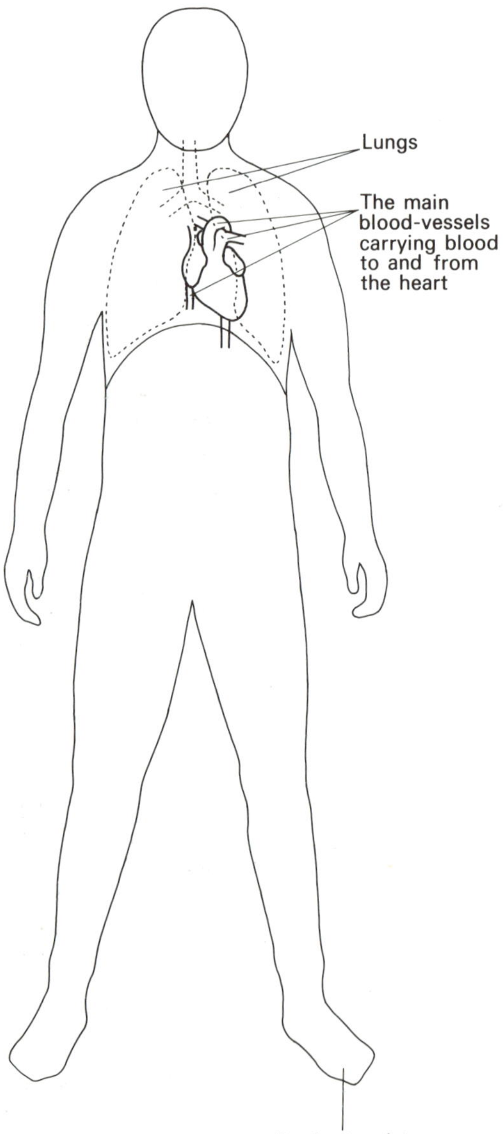

5.1 Circulation

Most of you will have at some time used a bicycle-pump to push air into a tyre. A pump is a machine used when we want to move air or liquid from one place to another. In a central-heating system the pump forces the water to flow round the system from the boiler, through the pipes and radiators and back to the boiler again. This flow of water round a circuit is called *circulation*. The process of circulation allows materials to be carried from a point of supply and distributed – the central-heating system distributes heat from the boiler to various places around the house.

The human body has a circulatory system for distributing materials such as food and oxygen to the cells. The same system transports away the waste products. The body's circulatory system is made up of blood flowing in a continuous network of blood-vessels. The pump in the system is the *heart* (Fig. 5.1).

5.2 The human heart

It is very easy to detect whether or not a pump is working, as you can often hear it or feel it. The human heart is no exception and a stethoscope is designed to pick up the sounds of the heart working, i.e. the *heartbeat*. On page 65 you can see how to construct your own stethoscope.

Fig. 5.1 *Position of the heart within the body. The heart is the pump in the body's circulatory system*

Question 1 Where on your chest would you place your fingers to feel your heartbeat?

Even a valentine card is able to show us the general shape of the heart; it is wider at the top than at the bottom and about the size of a clenched fist. It is a muscular organ with walls made of special *cardiac muscle* (page 35). Internally it is divided into four compartments or chambers. The two chambers at the top of the heart are called *auricles* (or *atria*) and the two at the bottom are called *ventricles*. You can see these in Fig. 5.2 which shows a cross-section of the heart. We describe the chambers as being those of the left or right. This means the left or right of the organ as if it were positioned in your body. It does not mean the left or right side of the page on which the organ is drawn.

5.3 The valves of the heart

Normally there is no connection between the left and right chambers of the heart. A hole in the wall separating these chambers will probably need surgery if the patient is to live a normal, active life. This sort of heart surgery is possible because two things are available – the heart–lung machine (Fig. 5.3) and a sufficient supply of blood through the transfusion service (page 72).

The auricles and ventricles on each side of the heart are connected. However the connection between them can be opened or closed by the action of *valves* (Fig. 5.2).

The *bicuspid valve* lies between the left auricle and left ventricle. The *tricuspid valve* lies between the right auricle and right ven-

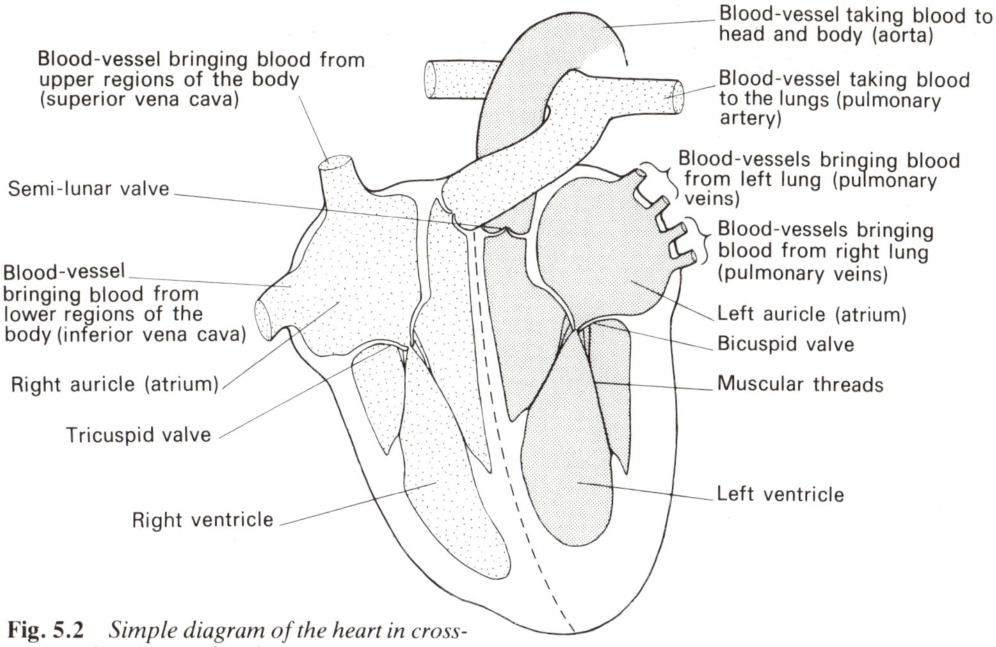

Fig. 5.2 *Simple diagram of the heart in cross-section showing its chambers*

Fig. 5.3 *The heart–lung machine*

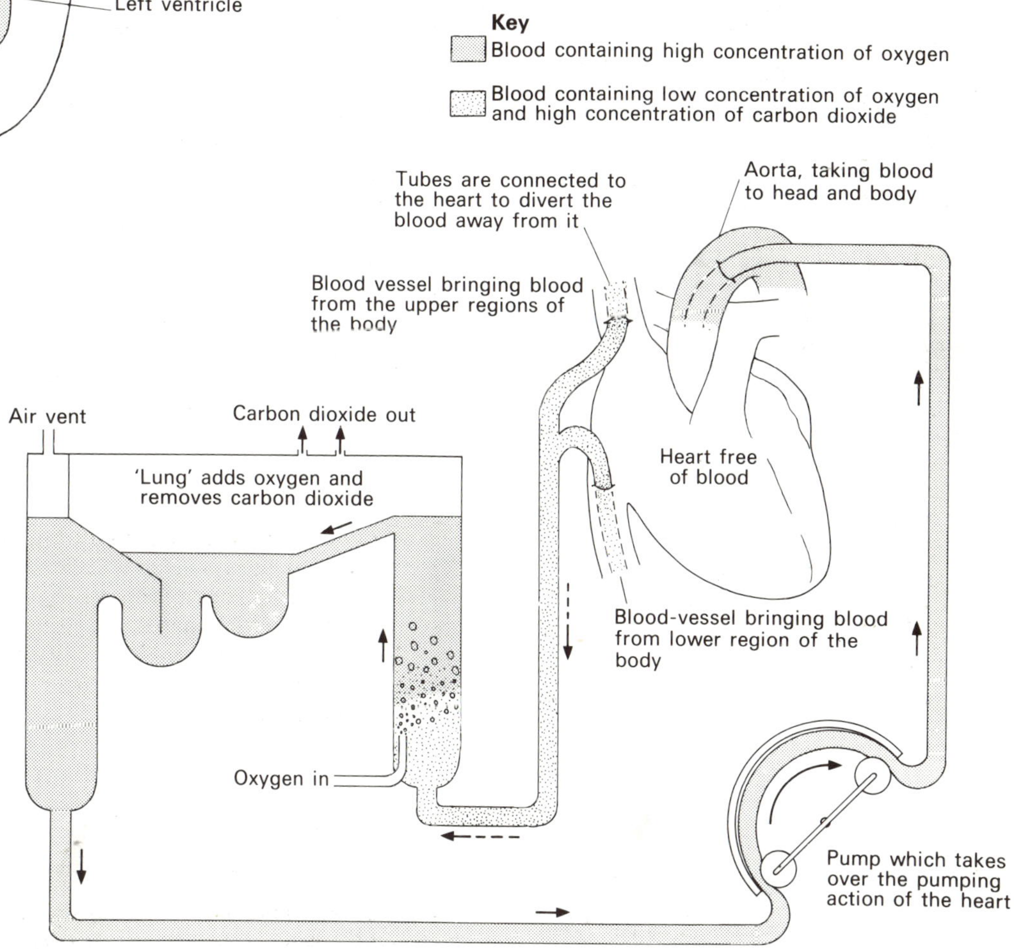

tricle. These, valves, like any other valve in a pump, allow liquid to flow in only one direction. The blood in the heart flows from the auricles to the ventricles. This means that blood must enter the heart at the auricles.

In the case of the right auricle the blood flows in through two main inlet blood-vessels. Smaller inlet blood-vessels supply the left auricle. These blood-vessels will be described in greater detail in Section 5.7.

The muscular walls of the auricles contract so that the blood is squeezed out through the valves into the ventricles. Because of their shape, the valves allow the blood to flow only in this direction. They are triangular flaps of tissue with muscular threads attached to their edges, the other ends of these threads being attached to the walls of the heart. You can imagine these valves as small umbrellas. Now try to imagine what would happen if you stood under a heavy fall of water with an umbrella without its catch on (Fig. 5.4(a)). As we can see in Fig. 5.4(b), the umbrella would fold up. This is exactly what happens to the bicuspid and tricuspid valves when the auricles contract, forcing blood into the ventricles (Fig. 5.4(c)).

The auricles then relax and the ventricles start to contract. No blood must re-enter the auricles, and once again it is the bicuspid and tricuspid valves which prevent this happening. They do this by filling out with blood, closing the passageway between ventricle and auricle. You can understand this action if you again imagine the valves as tiny umbrellas which open out in a gust of wind (Fig. 5.5). Thus as the ventricles contract, blood is forced out of the heart into the two main blood-vessels that carry the blood away to the rest of the body. These blood-vessels are discussed in greater detail in Section 5.7. However, if you look again at Fig. 5.2 you can see that there are other valves at the entrance of these blood-vessels: these are called *semi-lunar* valves. The pockets of tissue that make up these valves (Fig. 5.6(a)) are pushed flat against the walls of the blood-vessel as the blood flows past (Fig. 5.6(b)). However, if the blood tries to flow back it fills the pockets of skin, opening them out and closing the blood-vessel (Fig. 5.6(c)).

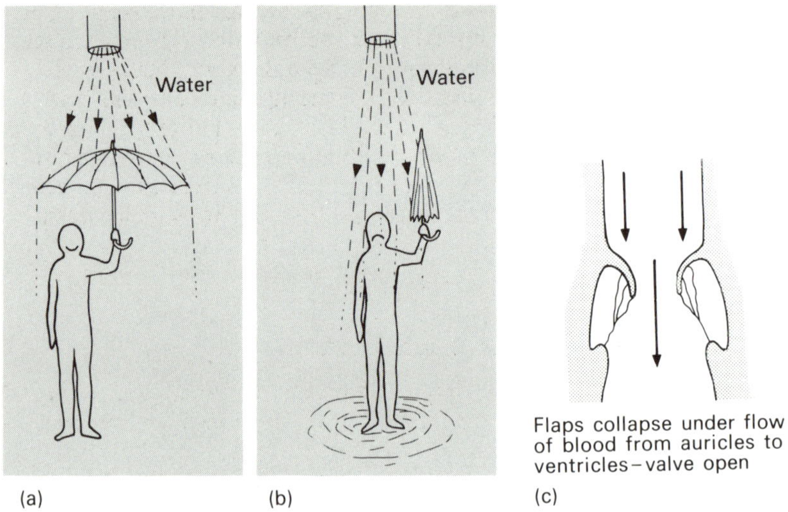

Fig. 5.4 *The action of the valves in the heart – valves open*

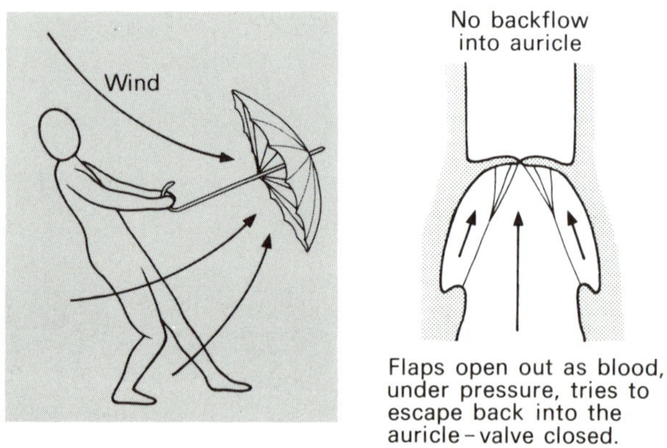

Fig. 5.5 *The action of the valves in the heart – valves closed*

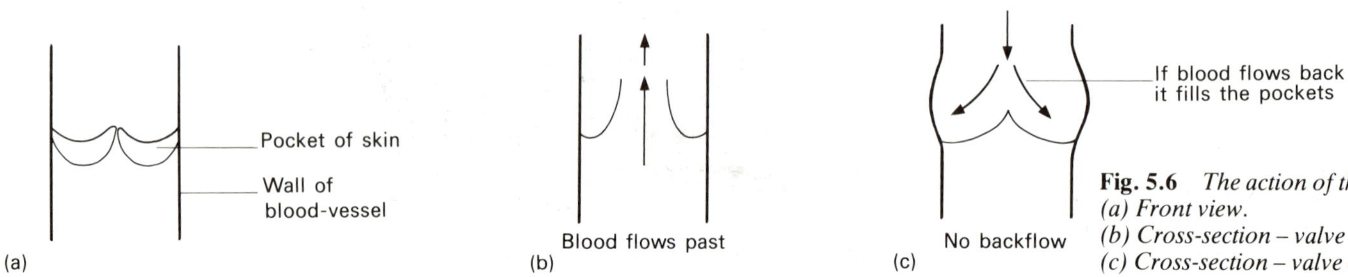

Fig. 5.6 *The action of the semi-lunar valves.*
(a) Front view.
(b) Cross-section – valve open.
(c) Cross-section – valve closed

5.4 **The heartbeat**

Diastole describes the stage of the heartbeat when the auricles and ventricles are relaxed. When they are contracted this stage is called *systole*. The contraction of the auricles is followed a split second later by contraction of the ventricles. This can be heard as a 'lub-dup' sound. 'Lub' is the sound made by the bicuspid and the tricuspid valves closing when the ventricles contract. 'Dup' is the sound made by the semi-lunar valves closing when the ventricles relax. The time between one contraction of the auricles and the next is 0.8 seconds.

If you count your heartbeat as suggested on page 60 you will probably find that it is between 65 and 80 beats per minute. An average beat would be about 72 beats per minute. The heart has a natural rhythm, which is automatic (page 232). Within the tissue of the right auricle there are specialized cells that trigger the wave of muscular contractions that can be seen spreading over the heart. These cells set the pace of the heartbeat so they are called the *pace-maker* of the heart. When these cells fail to work properly an artificial pace-maker may be inserted into the chest. This artificial pace-maker produces a small electric current which makes the muscles of the heart contract. The heart can respond to changes in the body's internal or external environment. Whether the heartbeat becomes faster or slower will depend upon the demands being made on the body (see Section 5.5). Any response made by the heart is brought about by the body's nervous system (page 233).

Question 2 Give *two* examples of experiences which made your heart beat faster.

The beat of the heart also speeds up when the body has a fever during an illness. To try to explain why this happens, imagine a person who has to carry a large number of parcels from A to B. He can do this in two ways: by moving quickly between A and B, carrying one parcel at a time; or by moving more slowly, carrying all the parcels in one go. You can see that the second method will produce more strain. In a similar way, the heart is strained less if it pumps smaller quantities of blood but at a faster rate. However, a rate rising above 90 minutes per minute can also strain the heart; under these circumstances the person must rest so that the heart has to do as little work as possible.

5.5 **The heart and exercise**

In Chapter 3 we learned that muscles need to be exercised if they are to keep their strength and are not to be strained easily. The cardiac muscle of the heart is no exception. If the blood supply to the heart increases then the heart will respond with a stronger contraction. So (within limits) the greater the stretching of the heart, the greater will be its contraction. Exercise produces these conditions. Thus the heartbeat of a cycling champion (Fig. 5.7) when he is at rest may be as low as 32 beats per minute. His heart is beating more slowly but it is contracting more powerfully and pumping a large quantity of blood at each beat – around 5½ litres, compared with about 3½ litres in an average heart.

Such a powerful heart can easily cope with the demand put on it during exercise. For instance, during a cycle race a cyclist's heartbeat can reach 230 beats per minute and this may be kept up for as long as 20 minutes. Regular exercise therefore helps to keep the heart fit and less likely to be strained or to collapse under conditions of stress. Nowadays more people are aware of this so they may take up jogging, walking, carrying out a programme of exercises or playing a sport. Such actions are not only physically rewarding but are enjoyable and bring people together socially.

Fig. 5.7 *Reg Harris, Britain's famous cycling champion*

It has already been said in Chapter 1 that caring for our bodies must be a continuous process. Taking no exercise for many years and then trying to run a mile is asking for trouble! Most exercise programmes will emphasize this point and will suggest a gradual build-up to a daily routine. A school timetable will provide periods of exercise for the pupils. When we leave school and are in charge of our own timetables we should consider continuing with some form of regular exercise if we wish to keep our physical fitness.

On page 66 you will find an exercise designed to test physical fitness. You may like to see if a daily exercise can improve your performance. In this exercise the heartbeat is recorded by counting the person's pulse.

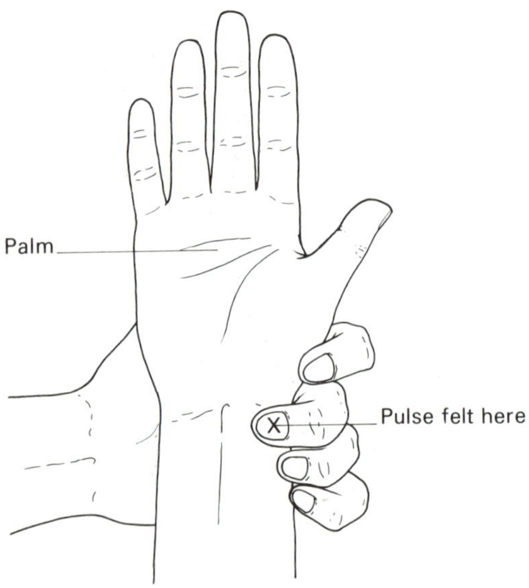

Fig. 5.8 *Taking the pulse*

5.6 Pulse

The blood flows away from the heart in blood-vessels called *arteries*. When the ventricles contract the blood is forced along in a series of surges or *pulses*. The arteries swell a little, then contract to help force the blood along between heartbeats. If an artery is pressed against a bone the pulsing of the blood can be felt. Such pressure points are easily found where arteries are near the surface of the skin, such as on the temple or on the inner side of the wrist at the radial artery (Fig. 5.8). A reading of pulse rate gives a measure of heartbeat.

5.7 The circulatory system of the body

Now that we have seen how the pump of the human circulatory system works it is worth looking at the system itself. The blood can flow in only one direction in a blood-vessel. The direction of this flow allows us to divide blood-vessels into two main groups:

(a) *arteries*, which carry the blood flowing away from the heart
(b) *veins*, which carry the blood back towards the heart.

Fig. 5.9 shows the circulatory system in a very much simplified form. You will notice that it is made up of two circuits: one connecting the heart with the lungs, the other connecting the heart with the rest of the body. In Chapter 4 we have already seen that in the lungs the blood disposes of waste carbon dioxide and takes up oxygen. The left auricle therefore receives blood which is rich in oxygen. This blood is then pumped round the body by the left ventricle to distribute the oxygen to the cells of the body.

Blood containing waste carbon dioxide is returned from the cells to the right auricle. The right ventricle pumps this blood into the lungs, where it gives up the carbon dioxide and takes up oxygen. In this way the blood circulates around the body continuously and at a rapid speed. The average time taken for a red blood cell to travel round the whole system once is 45 seconds.

Fig. 5.10 shows the circulatory system in more detail. In it are given the names of the main blood-vessels and the organs that they serve. Most organs are served by one artery and one vein. The liver is an exception and we will discuss the reason for this in Chapter 12. You will notice that two kinds of shading have been used in this diagram. This is to help distinguish between blood containing a high concentration of oxygen (oxygenated blood) and blood containing a low concentration of oxygen (de-oxygenated blood). Fig. 5.10 is still a simplified diagram representing a very complex system. One thing that makes the system so complex is the continual branching of the arteries and veins to form very fine blood-vessels, the *capillaries*. These capillaries spread out through all the tissues and form the links between

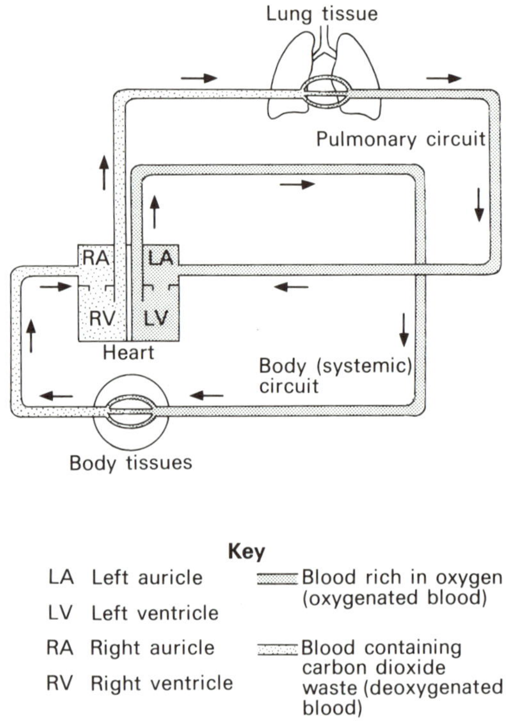

Fig. 5.9 *The circulatory system – simplified diagram*

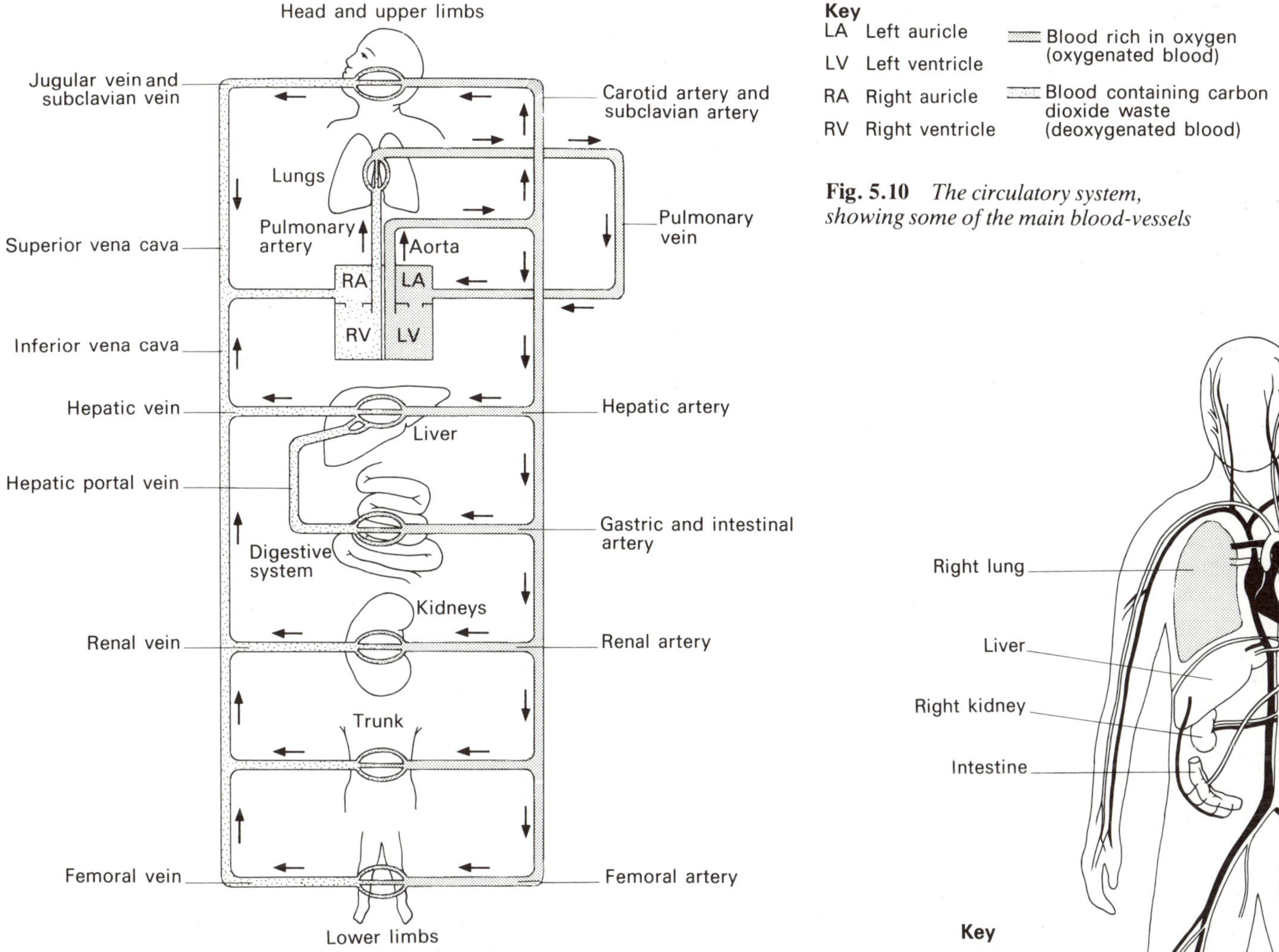

Fig. 5.10 *The circulatory system, showing some of the main blood-vessels*

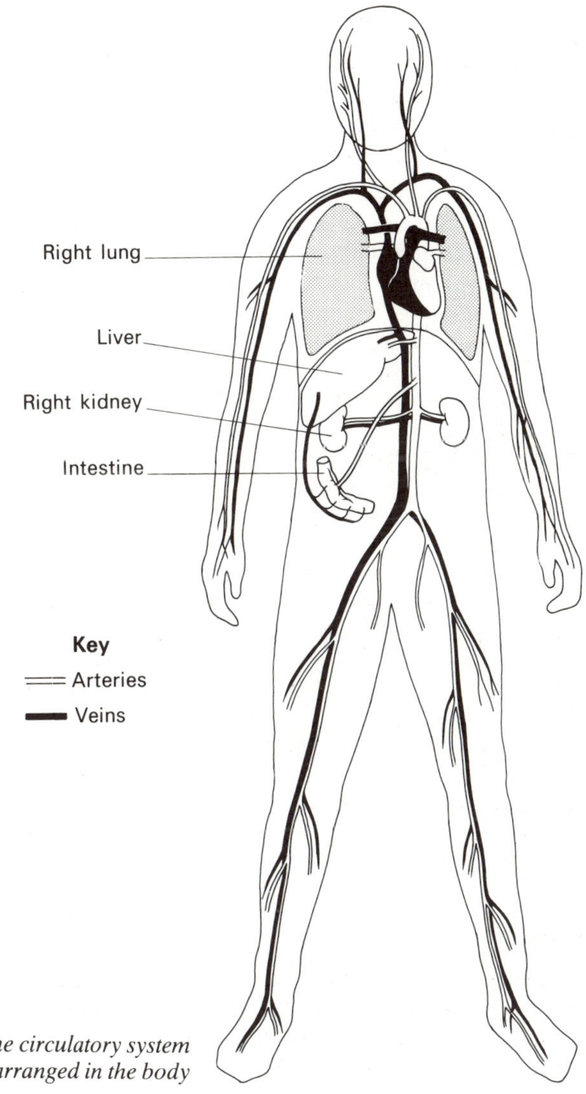

Fig. 5.11 *How the circulatory system is arranged in the body*

arteries and veins, as shown in Fig. 5.10. Fig. 5.11 shows more clearly the branching of the blood-vessels and how this is arranged within the body. Capillaries are so narrow that even a single blood cell may have to squeeze through. They have walls that are one cell thick and because there are so many capillaries the total surface area formed by these walls is very large. This surface is just what is needed for the passage of materials between blood and tissue cells by such processes as diffusion and *osmosis*. Osmosis is like diffusion in that it is concerned with the movement of molecules between regions of different concentrations. In osmosis, however, the cell membrane seems to play an active part by apparently allowing some molecules to pass through it more easily than others. As it appears to select the molecules it is called a *selectively permeable membrane*. Osmosis, as explained in Fig. 5.12, is the movement of water across such a membrane from a weak to a concentrated solution.

5.8 **Lymph**

Lymph is a colourless liquid which seeps through the capillary walls – because of the pressure in the blood – and bathes all the cells of the surrounding tissues as *tissue fluid*. It consists mostly of water and contains many substances that are found dissolved in the blood plasma (page 69). In fact its contents and strength are closely linked with the concentration of these substances in the blood. In their turn the contents of the blood are monitored and regulated by the brain, kidneys and liver. If one of these organs, such as the kidneys (page 173), fails to function correctly the concentration of substances dissolved in the blood will change, causing corresponding changes in the concentration of the tissue fluid. The cells bathed in this fluid could then be damaged as they could lose or gain too much water by osmosis (Fig. 5.12(c)). Patients with kidney failure have this problem and need a kidney machine (page 174) to control the concentration of substances in their blood.

Within the tissues, lymph acts as the 'middle-man' in the movement of food, oxygen and waste materials between blood and tissue cells, as shown in Fig. 5.13. The shopkeeper is the middleman in the movement of goods between manufacture and purchaser. In this way the manufacturer can supply every person without being in direct contact himself. Similarly, the blood can serve the cells without being in direct contact with every cell.

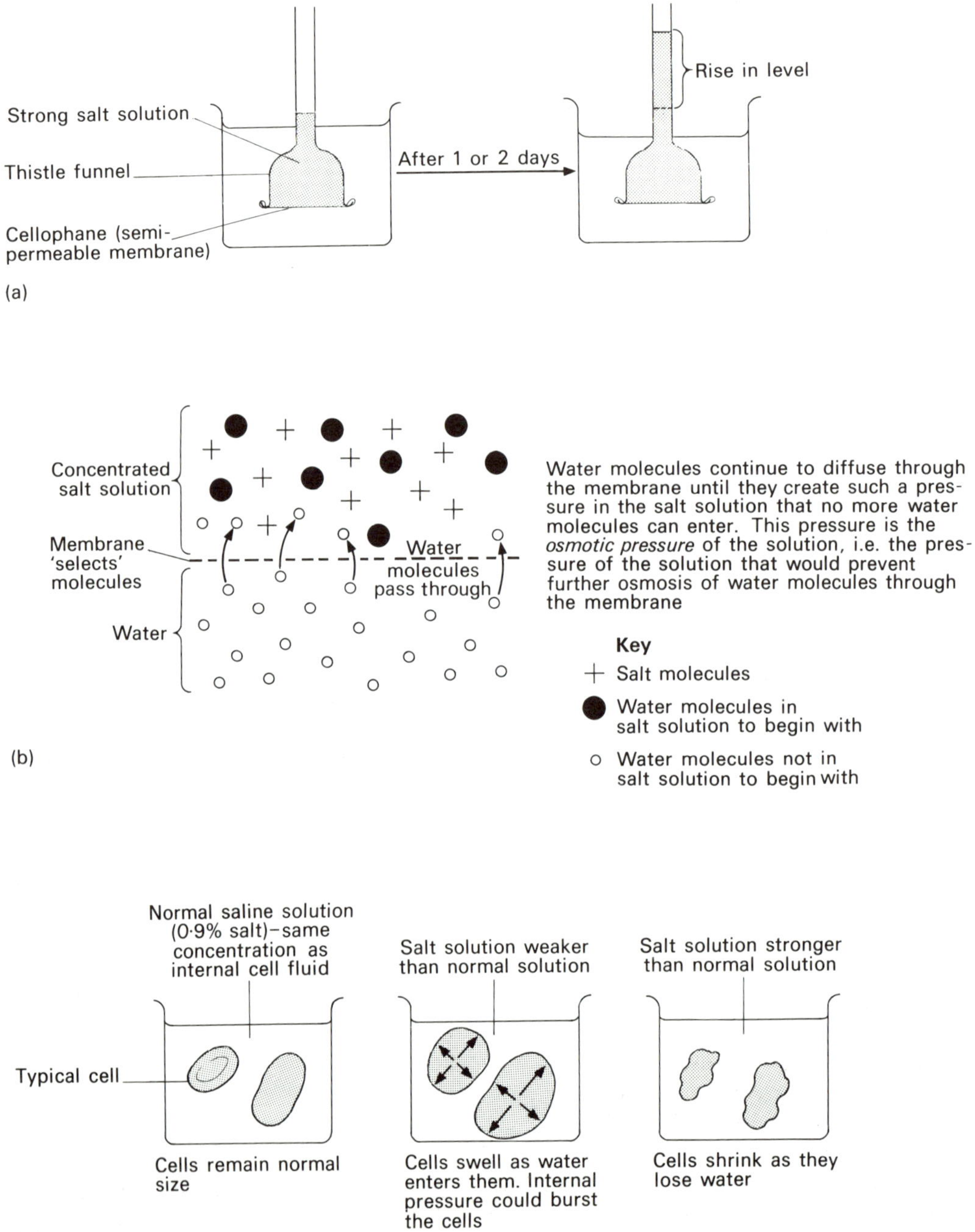

Fig. 5.12 *Osmosis. (a) Experiment which demonstrates osmosis. (b) What has happened. (c) The sort of effect this can have on body cells*

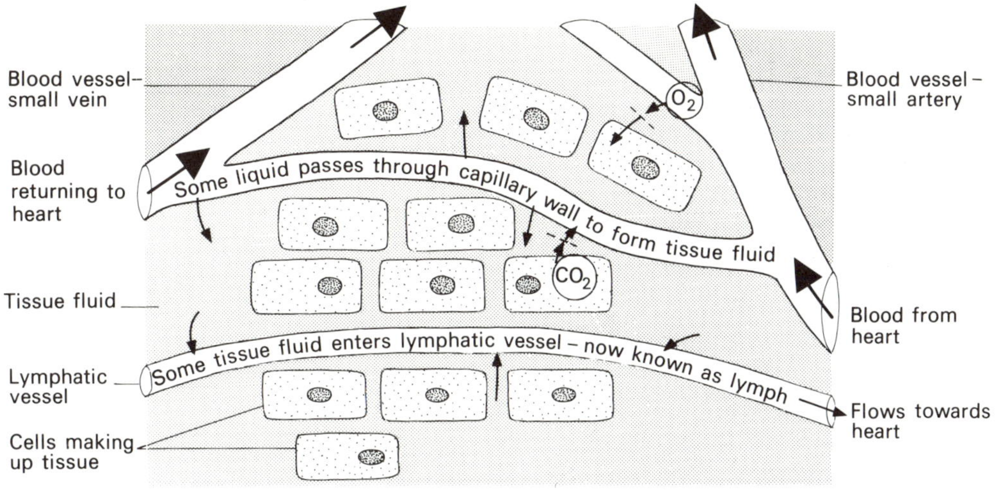

Fig. 5.13 *How tissue fluid acts as a 'middle man' in exchanges between blood and tissue cells*

Some of the lymph returns to the blood capilliaries while the excess drains into lymphatic vessels, which are colourless tubes present in the tissues (Fig. 5.13). These *lymphatic vessels* are the smallest 'twigs' of the lymphatic system. All over the body they join together to form larger branches which then meet, making two main 'stems' or *ducts*. The ducts enter the main veins of the right auricle of the heart. Thus the lymph flows from the tissues towards the heart, where it re-enters the blood system through a vein. There is no pump in the system. The lymph is under pressure from the tissues and there are valves (which function like those on page 58) in the vessels to prevent backflow.

The lymphatic system is not drawn here but appears on page 92 in Chapter 7, where another important function of the lymph is described. This function arises because the body uses the lymph as a way of bypassing organs with materials which might do them harm. This is the case with:

(a) fat, which is taken into the lymph from the intestine and bypasses the liver (page 160)
(b) 'invaders' such as germs, which are prevented from spreading within the body by *lymph nodes* or *lymph glands*; the position of these glands is shown in the diagram on page 92.

5.9 The structure of arteries and veins

Fig. 5.14 shows a typical artery and vein in cross-section. These blood-vessels are compared in Table 5.1. Veins do not have the strong elastic walls which arteries need to help push the blood along. The blood in veins does not flow with such force; as a result, the pulse is smoothed out. Because of this there is more danger in the veins of the blood flowing backwards. This is especially true in the lower parts of the body such as the legs. It is here that the earth's gravity is pulling downwards on the blood, encouraging it to flow back along the veins.

Semi-lunar valves are present on the walls of veins to prevent the backward flow of blood. Their shape and action are similar to those of the valves in the arteries leaving the heart (page 58). Exercise, with muscles

TABLE 5.1 **COMPARISON OF ARTERIES AND VEINS**

Arteries	Veins
Take blood away from the heart	Return blood to the heart
Have a pulse	Have no pulse
Have no valves along their length	Have pairs of valves at intervals along their length to prevent backflow of blood
Have thick walls (see Fig. 5.14)	Have thin walls (see Fig. 5.14).

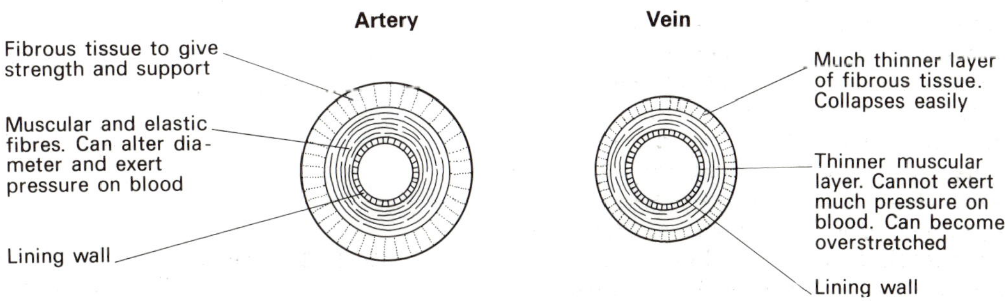

Fig. 5.14 *Structure of the walls of arteries and veins*

squeezing the veins together as they contract, and a slight suction effect from the heart both help the blood along in the right direction. The importance of exercise is seen when we do without it and stand still for a long time. We may faint, which is the body's way of getting itself into a position where blood flows easily back to the heart. If the walls of the veins or the valves weaken blood can collect in pools. This makes the veins swell and a condition known as *varicose veins* develops (Fig. 5.15). These are more often seen in legs, but they can also develop in the rectum (page 158) and are then called *piles (haemorrhoids)*. Piles may develop with constipation if there is straining to pass a motion (page 160).

Question 3 Try to explain the difference that you see in the veins on the back of your hand after

(a) holding your arm and hand straight above your head for several minutes
(b) holding your arm and hand down by your side for several minutes

The pooling of blood in the lower half of the body is a natural effect of life on earth and the heart can cope with the situation. However, on a space station there is zero gravity, so pooling of the blood no longer takes place. Now the heart does not have to work so hard to force the blood around the body; scientists say that the heart has become *deconditioned*. Deconditioning shows how easily the body adapts to life in space. But the disadvantage of this is that if an astronaut stays in space for a very long time his heart may not be able to cope on returning to earth. Such physical changes may force man to accept a limit to the time he spends in space, or he may have to create artificial gravity aboard his space stations.

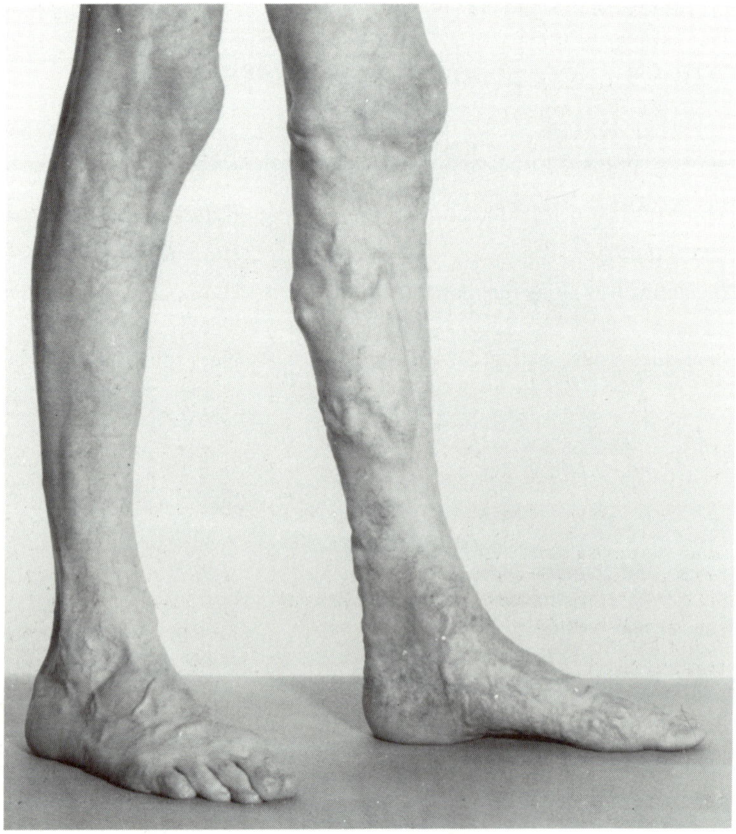

Fig. 5.15 *Varicose veins*

5.10 **Damage to the circuit**

The circulation of blood must be continuous. Any break in the circuit can be dangerous because of the loss of volume and pressure (see page 73). The area in which the injury occurs often affects the seriousness of the situation.

Bleeding from an artery is very serious because the blood is quickly pumped out of the cut vessel; the wound spurts blood. Bleeding from a vein occurs more slowly.

Any wound from which blood is flowing needs immediate attention to stop the bleeding. In an emergency like this, a knowledge of first aid could save a life – perhaps your own.

In general the following actions can all help to stop bleeding:

(a) Apply pressure to the wound.
(b) If the cut is on a limb, lift the limb above the level of the heart.
(c) Make the person lie still (Fig. 5.16).

Bleeding can also take place internally. The simplest form of this is when capillary walls are broken by banging or knocking the skin. Blood escapes into the tissues, forming a red swelling. This changes colour as the blood undergoes various modifications and we see the black and blue shades of a common *bruise*. If a person suffers from haemophilia (page 69) such bruising can be serious and extremely painful. If severe internal bleeding is suspected the person should lie still and an ambulance must be sent for. A shallow (superficial) cut damages the capillaries of the skin. We can usually cope with this bleeding ourselves. Yet it is important to remember that any break in the skin is a break in the body's first natural line of defence against disease. Although the body can fight infection at a break in the skin (page 96) we can help make sure it wins the fight. One way of helping is to clean the wound. Other actions we can take are discussed in Chapter 7.

On average there are 6 litres of blood in an adult male's body. It becomes dangerous when the body loses one-third of this. If such a loss occurs the patient will have to be given another person's blood. This is called a *blood transfusion*. But can an injured person receive blood from any other person? We would be foolish to try it without knowing more about blood itself, and this is dealt with in the following chapter.

Answers to questions in the text

1 You should place your fingers on the centre of your chest then move them slightly to the left (to the left of the sternum).

2 Undergoing exercise, like running, will make the heart beat faster. During exercise, muscles need greater supplies of oxygen and produce more waste carbon dioxide. The heart increases its beat to supply enough blood to meet this demand.

Being frightened or excited will also make the heart beat faster. This is the way the body 'gets ready for action' (page 59).

3 (a) The veins can be seen as thin blue lines under the skin.
(b) The veins swell and bulge, and in some cases the position of valves can be seen (Fig. 5.17).

In position (a) gravity is helping the blood to flow back to the heart, so there is no backflow. In position (b) gravity is pulling the blood downwards and it is having to 'struggle uphill' back to the heart.

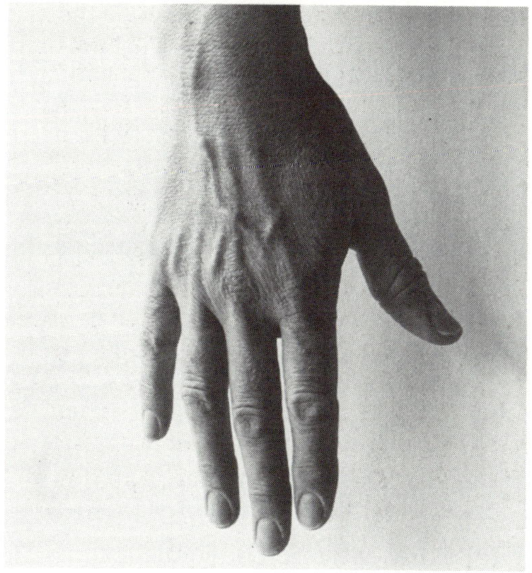

Fig. 5.17 *The semi-lunar valves of the veins in the hand filling with blood*

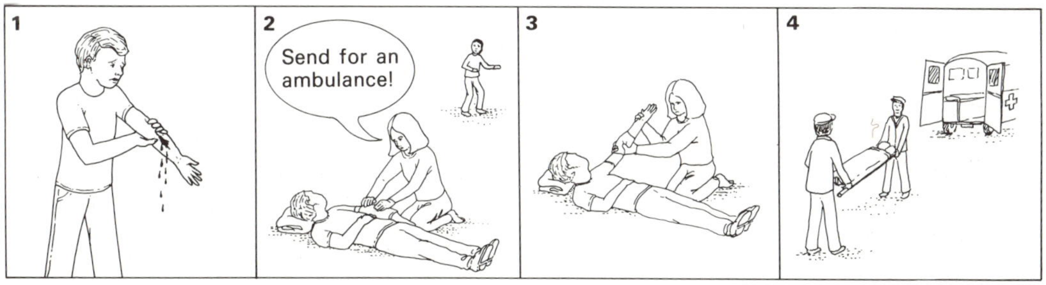

1. Accident! What action should be taken?
2. Get the person to lie down. Press firmly on or around the wound with a pad of clean linen or bare hands or fingers. It should be obvious if expert services are needed. If so, send for them
3. Keep pressure on and if possible lift the wounded area above the level of the heart. Keep person still
4. Get the casualty to hospital and into expert care

Fig. 5.16 *How to deal with severe external bleeding*

THINGS TO DO

1 Obtain from your butcher a complete sheep's heart. Under the guidance of your teacher you can examine both its external and its internal structure.

2 Listen to the heartbeat of another person by placing your ear close to their chest. Try to construct your own stethoscope using a plastic funnel and pieces of rubber tubing.

3 If you have a microscope examine the capillaries in the thin skin at the base of your finger-nails.

4 (a) Record the pulse rate of another person while they are
(i) lying down
(ii) sitting doing nothing
(iii) sitting and slowly drinking a glass of water
(iv) standing

(b) Record another person's pulse rate after they have
(i) walked slowly for 50 m
(ii) run for 50 m
(iii) carried out the 'step-up, step-down' exercise for three minutes (page 38)

(c) After each exercise continue to record the person's pulse rate at one-minute intervals and record how long it takes the pulse rate to return to normal. (Remember that physical characteristics vary considerably: this includes our pulse rate. Therefore a person's pulse rate may vary considerably from the average but still be considered normal.)

5 Collect newspaper articles on anything relating to heart surgery, for example heart transplants.

6 Study first aid and learn how to deal with minor and major cuts.

THINGS TO WORK OUT

In Questions 1 and 2 write down the *one* answer which you think is correct.

1 An adult has an average pulse rate of
A 80 beats per minute
B 60 beats per minute
C 140 beats per minute
D 72 beats per minute
E 40 beats per minute

2 The bicuspid valve prevents blood from flowing into the
A left ventricle
B left auricle
C pulmonary vein
D right auricle
E right ventricle

3 Osmosis is the term used to describe the movement of water . . . . . across a . . . . . permeable . . . . . from a . . . . . solution to a . . . . . solution.

4 By setting your answer out under the two headings 'arteries' and 'veins', state *three* differences between these two types of blood-vessel.

5 What information would you use to persuade a person that regular exercise is an important way of maintaining their health?

6 (a) Describe the function of semi-lunar valves.
(b) What is meant by varicose veins?
(c) How does exercise help to return blood to the heart?
(d) What signs would convince you that a person had cut an artery?

7 (a) Blood which leaves the left ventricle of the heart could pass through the structures listed below before returning to the left ventricle. List the structures in order according to the movement of the blood. Start and finish your list with the left ventricle:
posterior vena cava, hepatic vein, aorta, pulmonary artery, right ventricle, liver, left auricle, hepatic artery, lungs, right auricle, pulmonary vein.
(b) Name the veins which, like the aorta, carry blood with a high concentration of oxygen.
(c) How does exercise affect pulse rate?
(d) In which direction does blood flow in arteries and in veins?

8 (a) Describe how tissue fluid is produced from the blood.
(b) Give *one* function of tissue fluid.
(c) State *two* methods by which tissue fluid re-enters the bloodstream.
(d) How do extreme changes in the composition of the surrounding tissue fluid upset the tissue cells? Name *one* disease which causes these changes to occur.

6 Blood

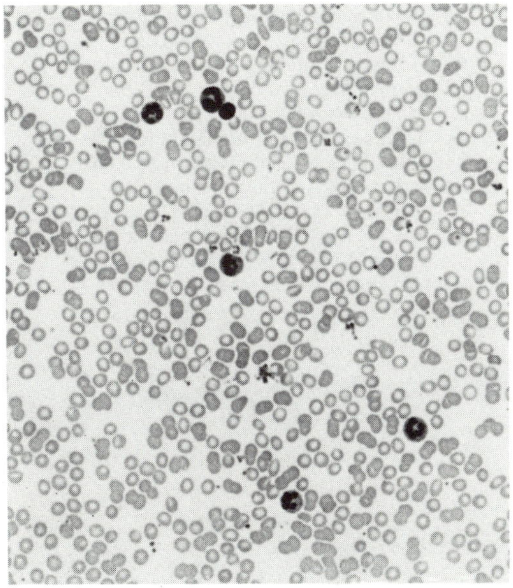

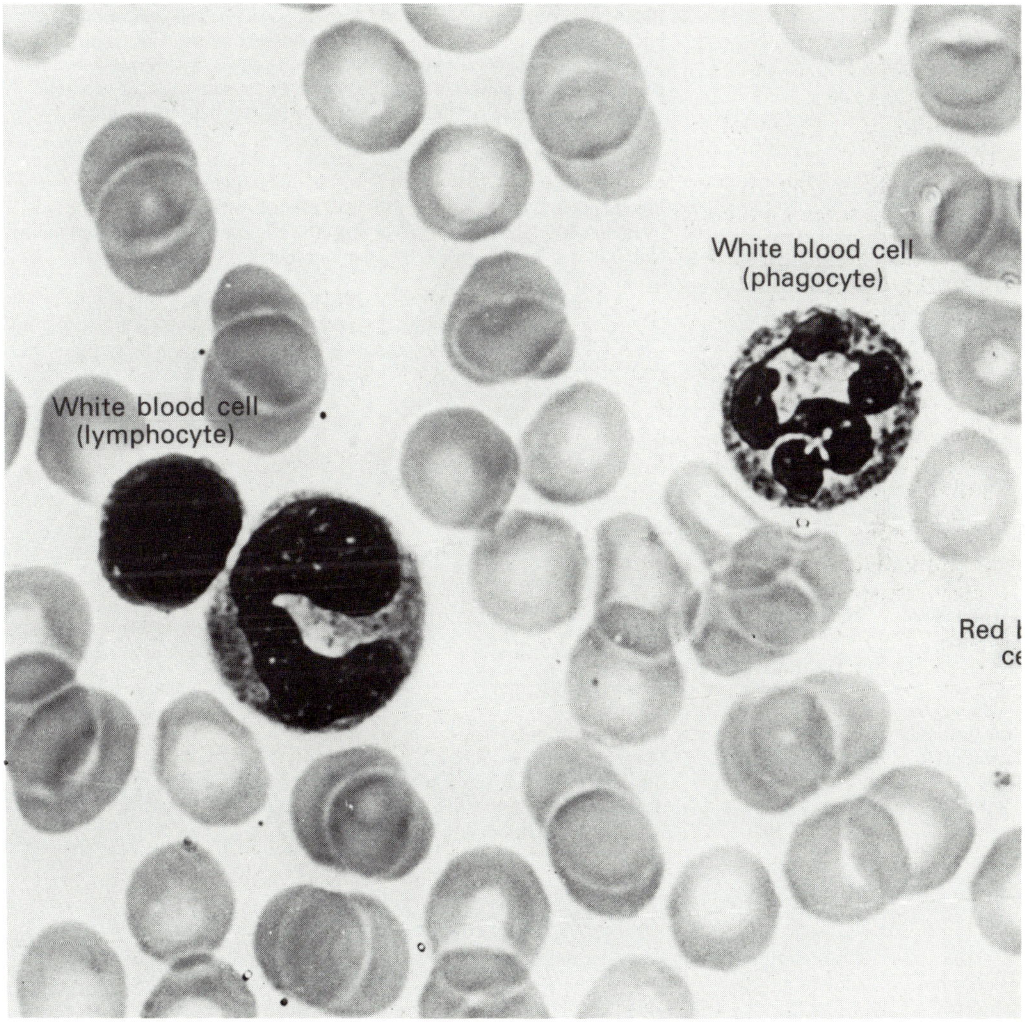

6.1 Blood – its composition

If a sample of blood is placed in a machine called a *centrifuge*, it can be spun round very fast. We find that this spinning separates the blood into approximately two halves: one half is a fluid; the other half is a solid mass of very small particles. The straw-coloured fluid is called *plasma*. The solid particles, which are normally suspended in the plasma, consist of several types of *blood cell* (Fig. 6.1). By far the most numerous are the *red blood cells*. Their presence gives blood its characteristic colour. There are also different types of *white blood cell* and particles called *platelets*. Table 6.1 describes and illustrates these components and their functions (see pages 68/9).

Fig. 6.1 *(a) Human blood viewed under a microscope at low power. (b) Human blood viewed under a microscope at high power*

TABLE 6.1 **COMPOSITION OF THE BLOOD**

Name	Quantity present in 1 mm^3 of blood (adult male)	Description	Where made and destroyed	Function
Red blood cells	5 million	Diameter 0.0075–0.0085 mm Thickness about 0.002 mm Biconcave single cells. No nucleus after release into bloodstream. Single cells are pale in colour; in a mass they are red. Protoplasm contains a substance called *haemoglobin*. Iron is essential for the manufacture of this chemical. It is a pigment within the haemoglobin that gives red blood cells their colour.	Made in red bone marrow of certain bones, e.g. sternum If there is not enough iron, fewer red blood cells are made. This results in a condition known as *anaemia*. Anaemia can also be caused when the red bone marrow fails to make enough cells Destroyed in liver and spleen. Iron is stored in the spleen for further use. The liver releases other remains of the cell into the bile (page 157). These remains colour the bile If too many red blood cells are destroyed, the excess waste colours the skin yellow, a condition known as *jaundice* (page 161) Lifetime 2–3 months	**Transportation of oxygen** If oxygen is present, haemoglobin combines with it to form oxyhaemoglobin. Oxygen does not dissolve easily in blood to form a simple solution. However, by using the action of haemoglobin the blood can carry all the oxygen the body requires. The blood picks up oxygen from the lungs, where it is highly concentrated, and transports it round the body as oxyhaemoglobin. The chemical attachment of oxygen to haemoglobin is a very weak one, so when the blood reaches body cells which have a low level of oxygen, it releases its oxygen to those cells
White blood cells	8000 in total			**Protect the body against disease** (see also Chapter 7, page 78)
(a) Phagocytes (or granular white cells)	Three-quarters of the total	Diameter 0.010 mm Have no particular shape. Move by a flowing action (like amoeba). Nucleus is irregular. Cytoplasm is 'spotty', hence the term *granular*	Made in red bone marrow. More are made when the body is fighting off infection Destroyed in the spleen. We see them at the site of infection as white matter called *pus* Lifetime a few days	They can squeeze through capillary walls to reach and attack any foreign body which has entered the body. They 'flow' around the foreign body and absorb it They use the same method to get rid of dead or damaged body tissue
(b) Lymphocytes	One-quarter of the total	Diameter 0.008 mm Large circular nucleus. Not 'spotty' in appearance	Made in lymph glands and spleen Too many white blood cells with fewer red blood cells produce a blood condition known as *leukaemia* Lifetime a year or more	**Produce chemical substances** which: (a) destroy invading cells (b) make the invaders mass together (c) prevent the invaders multiplying (d) counteract the chemical poisons produced by the invaders

TABLE 6.1 **continued**

Name	Quantity present in 1 mm^3 of blood (adult male)	Description	Where made and destroyed	Function
Platelets	300 000	Diameter 0.002 mm No nucleus. Look like fragments of red blood cells	Made by specialized cells in bone marrow Used up in clotting	**To prevent loss of blood by assisting clotting – aid to survival** Platelets are damaged by an injury. They then release a substance which starts a chain of reactions between substances in the blood. The end result is that one of these substances, known as *fibrinogen*, solidifies as very fine threads. These act like a net, catching blood cells and plugging the wound. The 'plug' gradually dries and hardens to form a scab. Healing goes on underneath the scab. *Haemophilia* is a disease in which the blood will not clot (page 15)
Plasma	55 % by volume of the blood is plasma; the rest is cells	Straw-coloured fluid whose composition is controlled by the body. Any great change in its composition could bring serious consequences. Contains: (a) 91–92 % water (b) 7 % blood proteins (c) 0.9 % salts of elements, mainly sodium and calcium (blood has a salty taste) (d) food material, e.g. blood sugar (page 160) (e) waste products, e.g. urea (page 160) and carbon dioxide		**Acts as a 'heat fluid'** (transporting heat around the body and maintaining its temperature (page 186) **Carries chemical messengers within the body** **Carries food materials to the tissues** (page 158) **Removes waste products from the tissues and takes them to certain organs for disposal** (pages 47 and 173)

6.2 **Blood groups – A, B, O groups**

So far all that we have said about the structure of blood is true for everyone. However, there are important differences and these arise because of substances which may or may not be present in the red blood cells and plasma. The following explains in simple terms what is a very complex subject.

The four main blood types are decided by the chemicals that may be present on the surface of the red blood cells. These chemicals are called *antigens* and the groups are known as A, B, AB and O. This means that the red bloods cells of a person in group A contain the A antigen, those for a person in group B the B antigen, those of a person in group AB both the A and the B antigen and those of a person in group O neither the A nor the B antigen.

The plasma contains certain chemicals which are part of the body's defence system (page 91). Two of these substances are the *antibodies* (page 93) anti-A and anti-B. The function of the anti-A and anti-B antibodies is to react against any red blood cells which contain the wrong antigen. This reaction makes the 'invading' blood cells clump together or *agglutinate* (Fig. 6.2). Table 6.2 shows which antibodies are present within the plasma of each blood group. You can now see why it is so important for a person to receive only blood which matches his own.

In a transfusion the donor's (giver's) blood must be *compatible* with the blood of the *recipient*. If an incompatible type is used the donor's red cells are regarded by the recipient's body as invaders and are agglutinated. It is this clumping of the donor's blood that causes distress to the recipient. (The recipient's red blood cells do not appear to be harmed if the donor's antibody does not match.) Table 6.3 shows how the different blood groups may be used in a transfusion.

Question 1 Use Table 6.3 to explain why a group O person is called a *universal donor* and a group AB person is called a *universal recipient*.

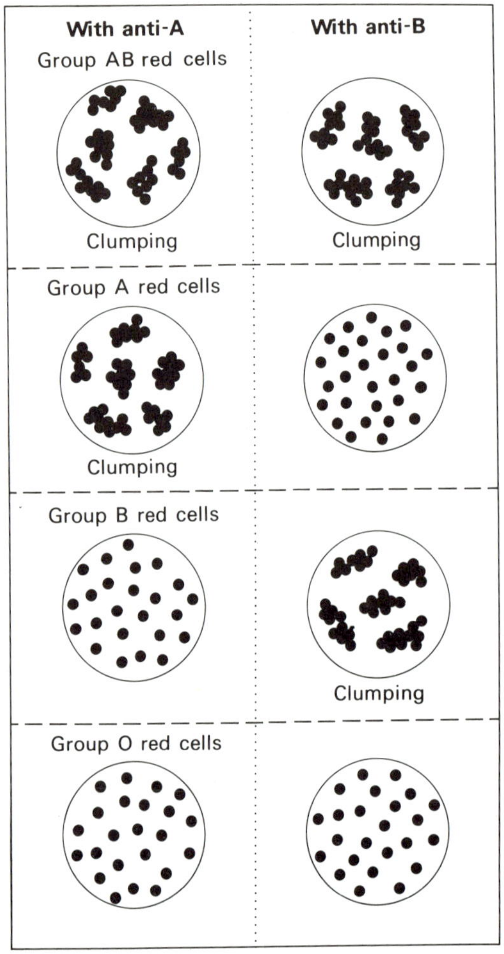

Fig. 6.2 *The action of anti-A and anti-B antibodies on red blood cells*

Antigens are inherited (page 13). Research has shown that the percentage of different blood groups within the population varies from country to country (Table 6.4). For example, in England on average 47 per cent of the population has O group antigens, 8 per cent B group, 42 per cent A group and 3 per cent AB group. This sort of information gives some clues to our racial origins.

6.3 **Blood groups – the Rhesus factor**

Another factor which is inherited and which is also important in blood typing is the *Rhesus (Rh) factor* (so called after the rhesus monkeys whose blood was used in the original research). There are two Rhesus blood groups: *Rhesus positive (Rh+)* and *Rhesus negative (Rh−)*.

As with A, B and O groups it is important in a transfusion that the donor's Rh group matches that of the recipient. If Rh+ blood is given to a Rh− recipient then his blood will probably form an antibody against it. This may not harm the recipient during the first transfusion but the antibody will remain in his blood. Thus if a further transfusion of Rh+ blood is given the antibodies will go into action, destroying the foreign blood cells with serious consequences.

This antibody reaction is very important for an Rh− woman, especially if she intends to have children. If the father is Rh− there will be no problem, but if he is Rh+ there is a

TABLE 6.2 **BLOOD-GROUP CHARACTERISTICS**

Blood group	Antigen in red cells	Antibody in plasma
A	A	Anti-B only
B	B	Anti-A only
AB	A and B	No antibody
O	None	Both anti-A and anti-B

TABLE 6.3 **MATCHING BLOOD FOR TRANSFUSIONS**

Donor's blood group	Recipient's blood group			
	A (anti-B)	**B** (anti-A)	**AB**	**O** (anti-A and anti-B)
A	Yes	No	Yes	No
B	No	Yes	Yes	No
AB	No	No	Yes	No
O	Yes	Yes	Yes	Yes

good chance that the baby will also be Rh+. Usually there are no complications with the first baby (unless the mother has previously had a transfusion of Rh+ blood, and this will be checked by the antenatal clinic). When the first baby is born there is a chance that some of its blood may leak across the placenta into the mother's blood system. If this is Rh+ blood the mother will make antibodies against it. These antibodies will remain in her blood. The antibodies are able to cross the placenta; thus if the woman conceives another baby and it has Rh+ blood, her antibodies will cross the placenta and destroy her baby's blood. The baby may not survive the pregnancy, or it may be born anaemic and *jaundiced*. The yellow colour of jaundice is caused by substances that are released from the destroyed red cells. Such a baby will need immediate attention. It will receive a transfusion to replace its damaged blood. Until the 1960s one in every 200 babies born was in this condition. Nowadays it is not so common, as an Rh− mother can be given an injection of anti-Rh antibody after the birth of an Rh+ baby, which will destroy any Rh+ blood cells that may have leaked into her blood system. This will prevent her blood making the antibodies which could destroy the blood of any future Rh+ baby.

The ABO and Rh groupings are not the only blood groups. So far a dozen different groups have been found and the numerous combinations that this gives makes blood very individual! Luckily these other groups do not react against one another in a way that might cause problems during or after transfusion. However, there are people of rare blood types or people who have had numerous transfusions that may need to be given blood that is very similar to their own. Finding the correct blood for such transfusions is part of the work of the National Blood Transfusion Service and may involve links that are world-wide.

TABLE 6.4 **PERCENTAGE DISTRIBUTION OF BLOOD GROUPS**

Population	Blood group			
	O	**A**	**B**	**AB**
English	47	42	8	3
Welsh	49	38	10	3
Scots	51	35	11	3
Irish	54	32	11	3
Gipsies in Europe	31	27	35	7
Indians	33	24	34	9
Maoris (Polynesians)	40	56	3	1
Blackfoot N. American Indians	23	76	0	1
Most S. American Indian tribes	100	0	0	0

Probably a high percentage of early inhabitants of England were group O
Probably a high percentage of Anglo-Saxon invaders were group A

6.4 The National Blood Transfusion Service

The giving of blood voluntarily has made the National Blood Transfusion Service in Britain possible. Anyone between the ages of eighteen and sixty-five may decide to donate blood.

They can contact their regional transfusion centre or attend a local donor session. At first their blood will be analysed for grouping and they will be told the result (Fig. 6.3). A person who has had a disease such as jaundice or malaria (page 104) cannot donate blood. Anyone whose blood is suitable will be registered as a donor. Donors are asked to attend a local donor session usually no more than twice a year. Each time they will give up to 420 cm^3 of blood through a vein in the arm. Donors are given awards as recognition of their public service when they have given 10, 25, 50 or 100 donations (Fig. 6.4). Giving blood (Fig. 6.5) is neither painful nor harmful, as a normal healthy person soon replaces what has been given. The blood collected at a donor session is taken in a specially refrigerated van to the regional transfusion centre. Here it is placed in a regional blood bank, where it is stored as whole blood at a temperature of 4 °C for up to three weeks. When the blood is needed it is taken to blood banks within the hospitals. From there it can be used at a moment's notice by doctors to meet emergencies.

There are many reasons why a transfusion may be needed, for example:

(a) when a person has lost a lot of blood in an accident
(b) during surgery
(c) because of an illness

Fig. 6.4 *Gold-top award for a special pinta man. This milkman has given 50 pints of blood and has been awarded his gold badge from the Blood Transfusion Service.*

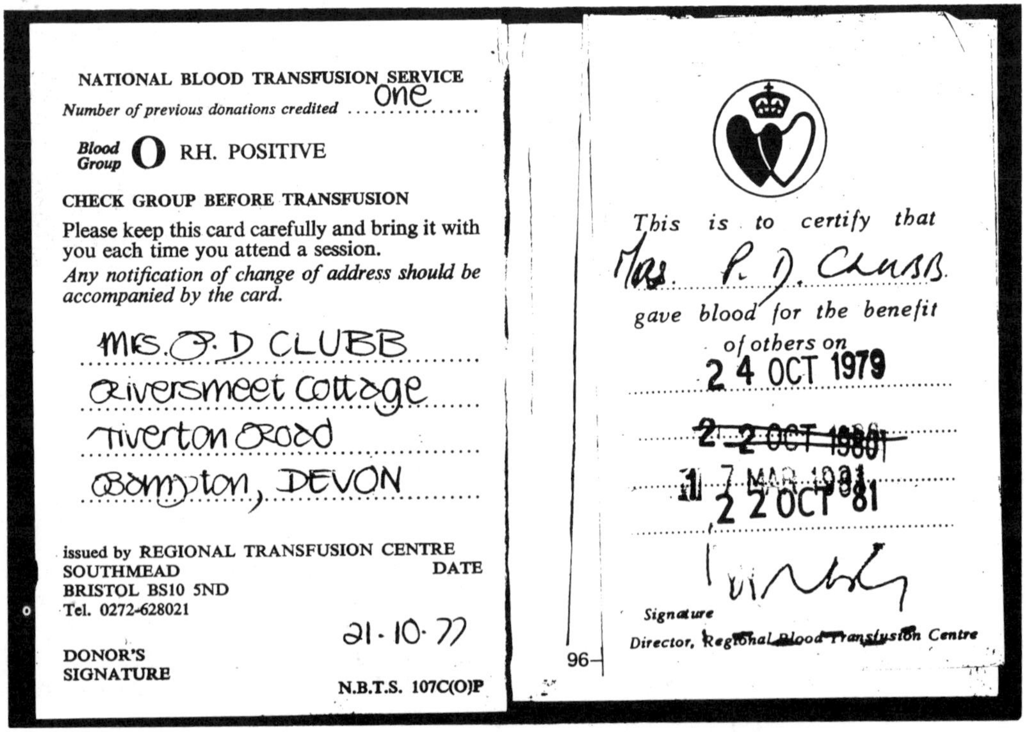

NATIONAL BLOOD TRANSFUSION SERVICE
Number of previous donations credited one
Blood Group O RH. POSITIVE
CHECK GROUP BEFORE TRANSFUSION
Please keep this card carefully and bring it with you each time you attend a session.
Any notification of change of address should be accompanied by the card.
MRS. P. D CLUBB
Riversmeet Cottage
Tiverton Road
Bampton, DEVON
issued by REGIONAL TRANSFUSION CENTRE
SOUTHMEAD
BRISTOL BS10 5ND
Tel. 0272-628021
DATE 21-10-77
DONOR'S SIGNATURE
N.B.T.S. 107C(O)P

This is to certify that Mrs. P. D. Clubb *gave blood for the benefit of others on*
24 OCT 1979
~~22 OCT 1980~~
17 MAR 1981
22 OCT 81
Signature
Director, Regional Blood Transfusion Centre
96

Fig. 6.3 *Personal record card which a person can carry to show his blood group*

If a lot of blood has been lost then a transfusion is needed to restore the volume of blood and to replace the red blood cells. For certain illnesses only part of the blood may be needed, such as the red cells or the platelets. In this case the patient does not need a large volume of whole blood. To meet this sort of demand the blood from donors may be processed to obtain just one substance from its contents. This is the case with plasma, which is separated from whole blood that has not been used within the first three weeks. It is dried after separation and can be stored for several years. When it is needed it is mixed with sterilized water.

Plasma is used in the treatment of burns because it replaces the fluid that has been drawn off in forming blisters. It also helps to prevent shock, and is used in many emergencies.

Question 2 Can any blood group receive a transfusion of plasma?

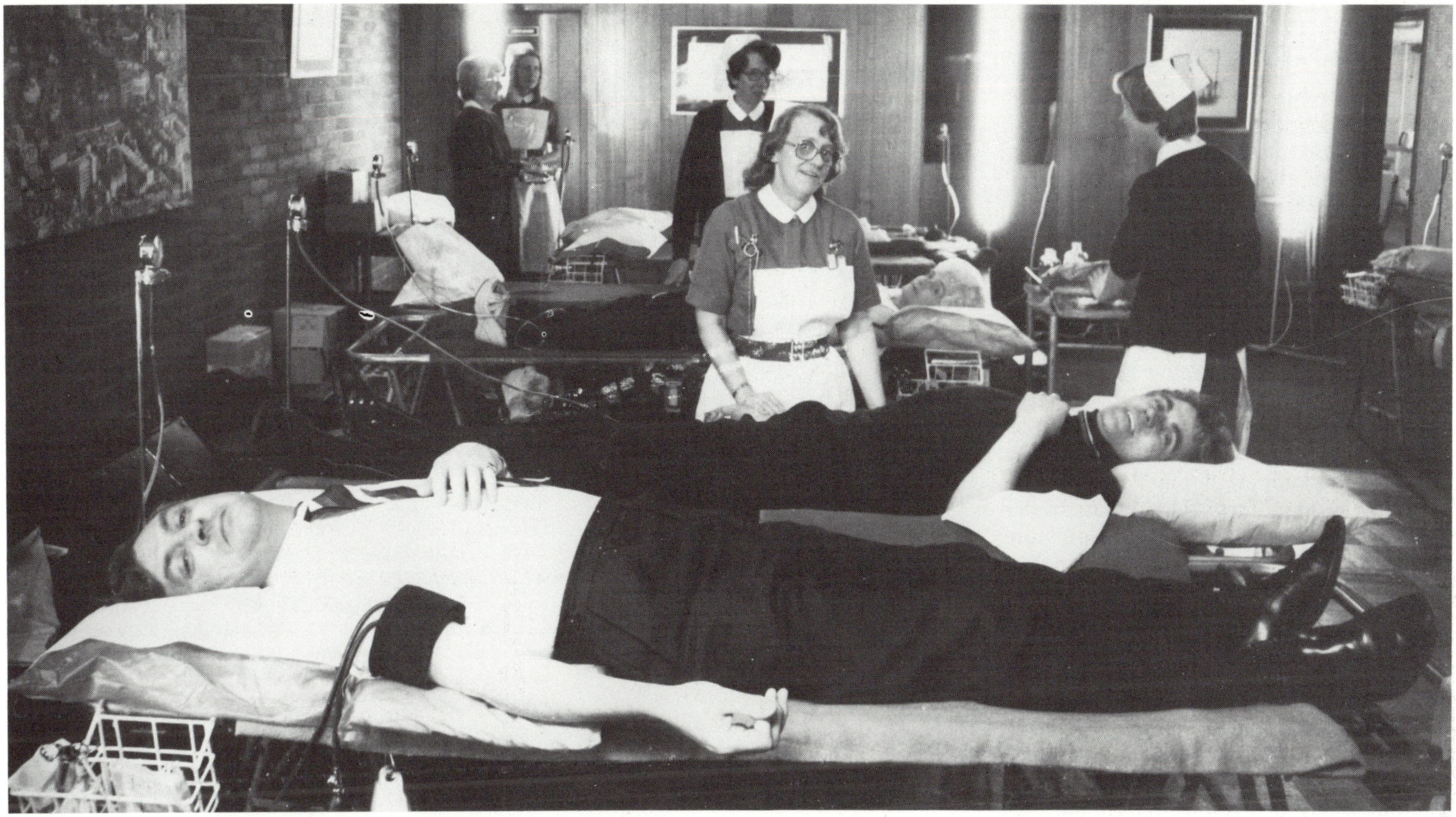

Fig. 6.5 *Donors giving blood at a mobile clinic held in a restaurant*

Fig. 6.6 *A manometer*

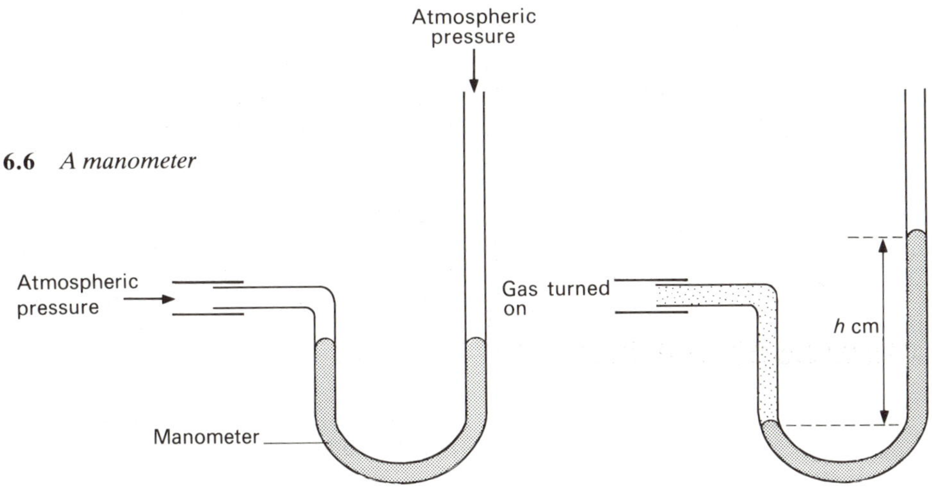

6.5 Blood pressure

Blood must be under pressure if it is to flow along the blood-vessels. Pressure is built up by the pumping action of the heart. In other science lessons you may have measured the pressure of a gas using a manometer (Fig. 6.6). The gas pressure is measured by the length of the mercury column that it can support, *h* cm in Fig. 6.6.

We cannot connect a manometer directly to the blood system! Therefore the manometer is connected to a long 'air-bag', which can be blown up. The manometer records the pressure of the air in this bag. The whole instrument is called a *sphygmomanometer* (Fig. 6.7). The doctor wraps the bag, as an armband, around the upper part of the arm. The bag within the armband is blown up to exert so much pressure on the arm that it stops the flow of blood. At this point the pressure of the air in the armband is greater than the blood pressure. Gradually air is let out of the bag, and the doctor listens with a stethoscope for the first trickle of blood to come through an artery below the armband. At this point the pressure of the blood is taken as being equal to the pressure of the air in the armband, and this pressure can be read off on the manometer in millimetres of mercury. This reading is the pressure of the blood at its maximum level. It is the pressure built up when the ventricles contract and is called *systolic blood pressure*. Systolic pressure in an adult is normally about 120 mm of mercury. Blood continues to flow in the blood-vessels even when the ventricles are not contracting. However, at this time the pressure is much less. The pressure of the blood when the ventricles are not contracted is called *diastolic blood pressure*. Diastolic pressure is about two-thirds or half the value of systolic blood pressure. The two pressures are written like this:

$$\frac{120}{80} \quad \text{or} \quad 120/80$$

Kidney Donor Card

Your kidneys could help someone to live after your death

Keep the card with you at all times in a place where it will be found quickly

YOUR KIDNEYS COULD SAVE LIVES

Although hundreds of successful kidney transplants are carried out each year, many more kidneys are needed. A successful kidney transplant means a new lease of life for somebody who has fallen ill because their kidneys have stopped working. People with kidney transplants can live a full and active life. They can work. They can go on holiday. They can enjoy life and living far more than when survival depends on the regular use of a kidney machine.

I have decided to help—what can I do?

Make your wishes known to your next-of-kin and carry this card when it has been signed by both of you.

Are all kidneys suitable?

Kidneys for transplantation should be removed from the body within half an hour of death. Delay means that they suffer serious damage. Therefore kidneys are only taken

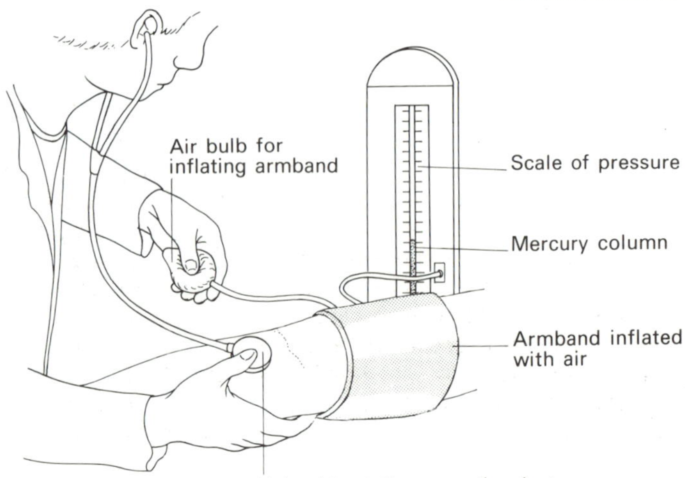

Fig. 6.7 *A doctor uses a sphygmomanometer to measure a person's blood pressure*

6.6 What happens when the circulatory system fails?

The cells of the body depend upon the continuous 'delivery and collection' service of the circulatory system. If this service fails the cells of the body stop their life functions. If they cannot resume these functions when the circulation starts up again then they are considered *dead*. When the circulatory and respiratory systems stop working, the speed at which the cells of the body die varies from tissue to tissue. Generally, the more special-

I (full name) ....

Request that my kidneys be used after my death for transplantation or, if found unsuitable, for research.

Signature of donor .... Date ....

The above-named person has discussed his/her wish with me and I do not object to that wish being complied with.

(Name of next-of-kin) ....

of (address) ....

....

.... Tel ....

Signature of next-of-kin .... Date ....

from people who die in hospital, usually following admission in an emergency.

Success also depends on good "tissue matching". Donated kidneys are compared with the tissue characteristics of all those in the country waiting for a transplant and offered to the best matching recipients. The more donors there are, the wider the choice, the better the chance of a good tissue match – and a successful transplant.

Did you know that your eyes could also be used after your death to restore somebody else's sight? Information can be obtained from the Royal National Institute for the Blind, 224, Great Portland Street, London W.1.

Prepared by the Health Departments & The Central Office of Information 1974
576881/6000M/2.78/LOM *KD1A*

Fig. 6.8 *A kidney donor card*

THE STORY of kidney girl Yvette touched the heart of Britain yesterday.

Offers of help for the thousands of kidney patients flooded in from all over the country.

Between 4000 and 5000 sufferers – 100 or so of them children – die needlessly each year.

The reason: a chronic shortage of people prepared to donate their kidneys after death and an equally serious shortage of the expensive dialysis machines on which victims of kidney disease must depend until a transplant can be carried out.

Eleven-year-old Yvette is one of the lucky ones. The once pale and sickly little girl is now beginning a new life after receiving a kidney from an 18-year-old accident victim.

And in a deeply moving letter from her hospital bed she wrote:

'I only wish more people would give their kidneys so my friends on kidney machines would have a chance to feel better like I do now.'

Fig. 6.9 *How transplants can mean a new lease of life*

ized cells die more rapidly: for example, the cells of the brain will die in a few minutes, yet the cells of fibrous and fatty tissue can survive for a few hours. We can consider a cell dead once it is unable to recover after it has ceased to function.

Some people have accepted the fact that an organ can continue to function for a time after the death of other vital organs such as the brain or heart. This organ, if removed in time, can be used in a *transplant*, where it will replace a defective organ in an otherwise healthy body. By signing a form like the one shown in Fig. 6.8 a person may *donate* an organ, in this case their kidneys, for transplantation. When such a form has not been signed, near relations of the dead person may be asked by doctors if they will give their permission to remove an organ required in a transplant. There is no doubt that individuals have been given new leases of life by such transplants (Fig. 6.9).

6.7 Why the circulatory system can fail

If your central-heating system at home stops working it may be because the electricity supply has been cut off, or the pump has broken or the pipes have become blocked. Similar problems can arise in the body's circulatory system.

The heart can stop working when the supply of blood to its own muscular walls is cut off. The cardiac muscle is supplied with its blood by the *coronary arteries*, which branch out from the main artery (*aorta*) leaving the heart. If these become blocked the muscle stops working and the person experiences a *coronary thrombosis* or heart attack. Speed is essential in restoring the action of the heart within minutes of its stopping.

A blood-vessel may become blocked because of a clot (thrombosis) which forms in the blood. Clots are likely to form when arteries become less elastic and deposits form on their inner walls similar to those found in a domestic water-pipe. This hardening and narrowing of the arteries also means that blood pressure increases. You can see how this happens if you squeeze a garden hose when it has water running through it. The pressure in the water builds up and sometimes forces the hose off the tap. To force the blood through the narrower arteries the heart must work much harder. This increase in the pumping force of the heart raises the blood pressure, and a person is said to have *high blood pressure*. Obviously under these conditions there is a chance of the heart becoming strained and damaged as it tries to cope with this extra demand together with the normal activities of life. Hardening of the arteries can occur with old age, but many medical people are concerned by the fact that heart attacks are becoming more common in middle age. In fact diseases of the respiratory and circulatory systems account for half the deaths that occur in middle age. Research is being carried out to try to find if there are any common factors which increase the risk of an individual suffering a coronary thrombosis. At present a variety of factors appear to have some effect: these are heredity, sex (more men suffer attacks than women), lack of exercise, being overweight, smoking and stress. One major association appears to be the eating of large quantities of *saturated animal fats* (page 118), which contains a substance called *cholesterol*. Cholesterol is found in most body tissues and is used in the building up of deposits on artery walls (Fig. 6.10). You may like to discuss how these factors affect the circulatory system. Everyday literature found in newspapers will help you (Fig. 6.11). Some illnesses can be the direct result of another organism living in the blood. How these organisms are introduced into the blood stream is discussed in the next two chapters. We shall also consider how the blood acts as one of the body's defences against disease.

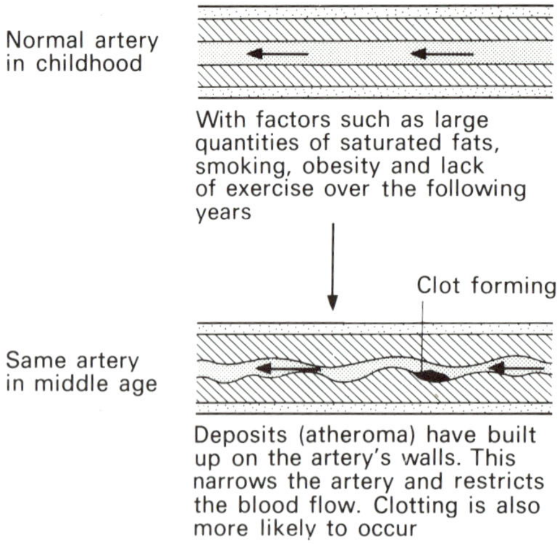

Fig. 6.10 *How cholesterol deposits form on artery walls*

Fig. 6.11 *Press advertisement from the Health Education Council's 'Look After Yourself' campaign*

Answers to questions in the text

1 Group O blood can be given to people with any other blood group. A person with Group AB blood can receive any other group of blood.

2 Yes, because there are no red blood cells present in the donor's plasma to be agglutinated. So plasma may be used while a person's blood groups is being identified.

THINGS TO DO

1 From family and friends collect any information about their blood groups. How do your figures compare with those given on page 71.

2 Find the location of your local blood transfusion centre. Where are blood transfusion clinics held in your area? Collect information on the National Blood Transfusion Service.

3 If you have a spot or scratch on the surface of your skin make a record of the healing process. Look for any area of inflammation, any pus, the release of the pus, the formation of a scab, and when the scab drops off.

4 Find out why it is dangerous to sit in an enclosed space with a car engine ticking over. What regulations cover the safe installation of gas fires in the home?

THINGS TO WORK OUT

In Questions 1–3 write down the *one* answer which you think is correct.

1 The four main groups of blood are
- **A** A, BO, AB, O
- **B** A, B, AB, O
- **C** A, B, Rh+, Rh−
- **D** A, B, C, O
- **E** AO, B, AB, O

2 Blood is able to clot through the action of the
- **A** red corpuscles
- **B** white corpuscles
- **C** platelets
- **D** plasma
- **E** antigens

3 You are the first on the scene when a man has badly cut his hand on a broken window pane. Which of the following would you do first?
- **A** Give the man artificial respiration.
- **B** Go to a telephone and phone for an ambulance.
- **C** Give the man a drink of tea.
- **D** Restart the man's heart.
- **E** Attempt to control the bleeding.

4 (a) What is meant by the condition of anaemia?
(b) Give *two* examples of how a person may become anaemic.
(c) What disorder of the blood is present when a person has leukaemia?
(d) Describe how the blood acts against micro-organisms that enter the body when the skin is damaged.

5 (a) Describe the appearance of a red blood corpuscle (diagrams may be used).
(b) Draw a diagram of *one* type of white blood corpuscle.
(c) What pigment is present in red blood cells?

6 (a) What regions of an adult's body manufacture (i) red blood cells, (ii) white blood cells?
(b) In what region of the body are red blood cells destroyed?
(c) Name *two* substances that are transported within blood plasma.

7 (a) What is meant by the term semi-permeable membrane? Give an example of such a membrane.
(b) Describe what would happen to a red blood cell if it were put into (i) strong salt solution, (ii) pure (distilled) water.
(c) Name the biological process that is at work in (b).
(d) Give a simple explanation of this biological process.

8 Give *two* examples of how each of the following may be caused in the home:
(a) cracked skull through a fall
(b) scald
(c) burn
(d) asphyxia
(e) poisoning
(f) carbon monoxide poisoning

9 (a) Give *two* examples of when a blood transfusion may be necessary.
(b) What is meant by the terms donor and recipient?
(c) What substance other than *whole* blood may be used in a transfusion. Give *one* example of when such a transfusion may be given.
(d) Explain why the red blood cells of a person of blood group A agglutinate when mixed with blood of group B.
(e) Construct a chart that shows the reactions between the various blood groups. Use this chart to explain the terms universal donor and universal recipient.

10 (a) Write about the various functions of the blood.
(b) Describe the circulation of lymph.
(c) Compare the structure of lymph with that of blood.

7 Micro-organisms

7.1 **Early observation of micro-organisms**

Diseases that can be passed on from one person to another are classified as *communicable* or *transmissible* diseases. The fact that communicable diseases such as *plague, typhoid* and *cholera* are caused by *micro-organisms* explains why so little was known about them in early times.

We need to magnify a micro-organism to be able to see it. To magnify we use a lens. Optical lenses were first ground by Galileo (1564–1642). In 1590 Jenssen in the Netherlands used lenses to construct a *microscope*, but it was his countryman Anton van Leeuwenhoek (1632–1723) who was the first to observe micro-organisms through such a device. He observed the 'little creatures', as he called them, in dirty water and in material removed from between his teeth.

The 'little creatures' of Leeuwenhoek became known as *microbes*. Scientists continued to study them throughout the following years but although their microscopes improved their understanding of microbes did not. They knew that microbes appeared in great numbers in anything that was decaying and rotting, like bad meat and stale milk. So they assumed that when something went bad it gave rise to microbes, as people in the sixteenth century believed that frogs developed from mud! Then 200 years after Leeuwenhoek, a Frenchman called Louis Pasteur (1822–95) set out to disprove the idea that a living microbe could suddenly develop from non-living matter (spontaneous generation).

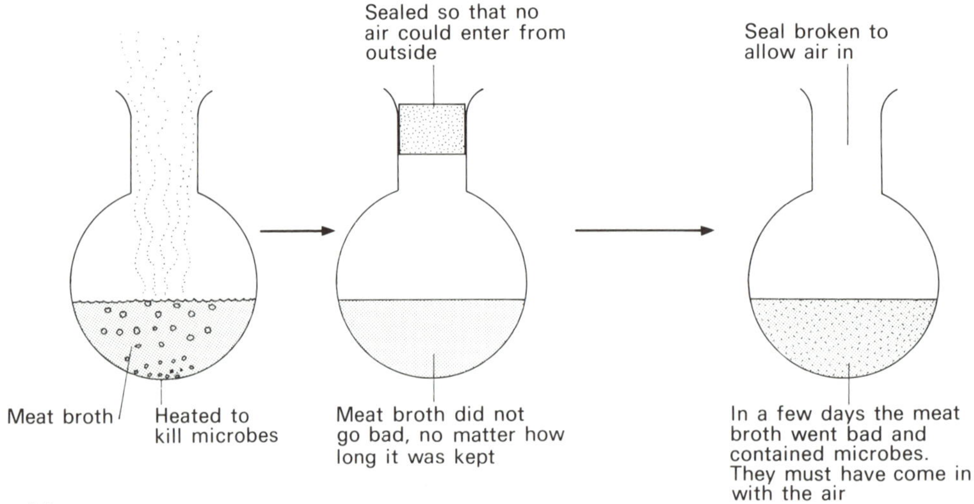

(a)

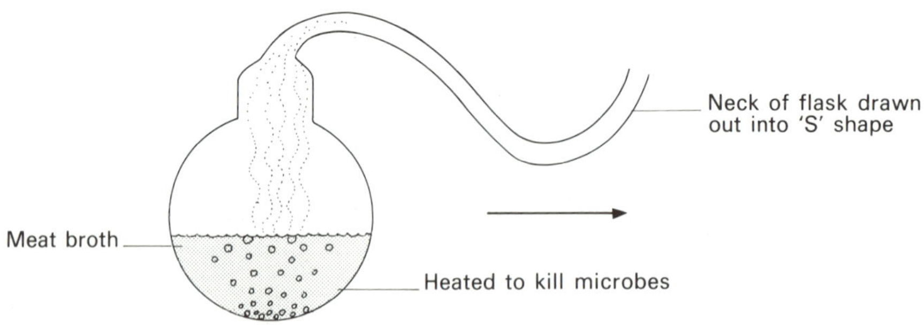

(b)

7.2 **Pasteur's 'germ theory' and 'germs' as we know them today**

Pasteur believed that microbes were present in air and that they caused decay by feeding and multiplying in decaying material. His experiments, using the apparatus illustrated in Fig. 7.1, produced results that confirmed his belief and disproved the idea of spontaneous generation.

Gradually his work was accepted by other scientists. They realized that living organisms had to have living 'parents' and that microbes were the cause of decay. Pasteur then went one step further: if microbes cause decay, might some of them also cause disease? This is known as *Pasteur's germ theory*. Once again his numerous experiments and painstaking investigations into certain diseases of silk-worms, cattle and poultry proved that he was correct.

By 1880 there were very few people who doubted the evidence that when some microbes invade the body they cause disease. Work by other scientists over the following years gradually revealed more about the relationship between disease and micro-organisms. It became clear that not all micro-organisms cause disease. Those that do are classified as *pathogens* (germs). Each pathogenic micro-organism causes a particular disease in its host.

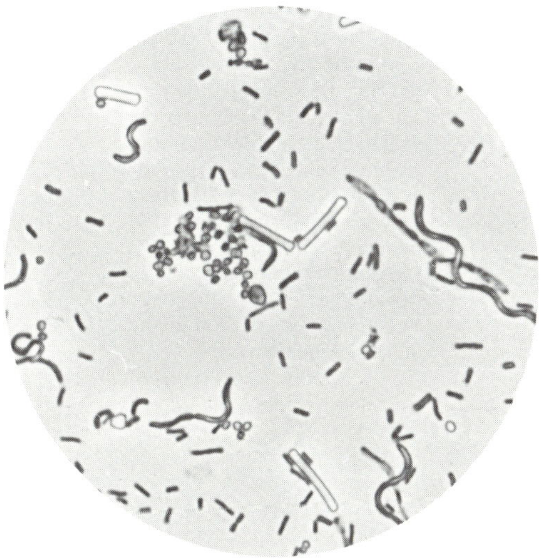

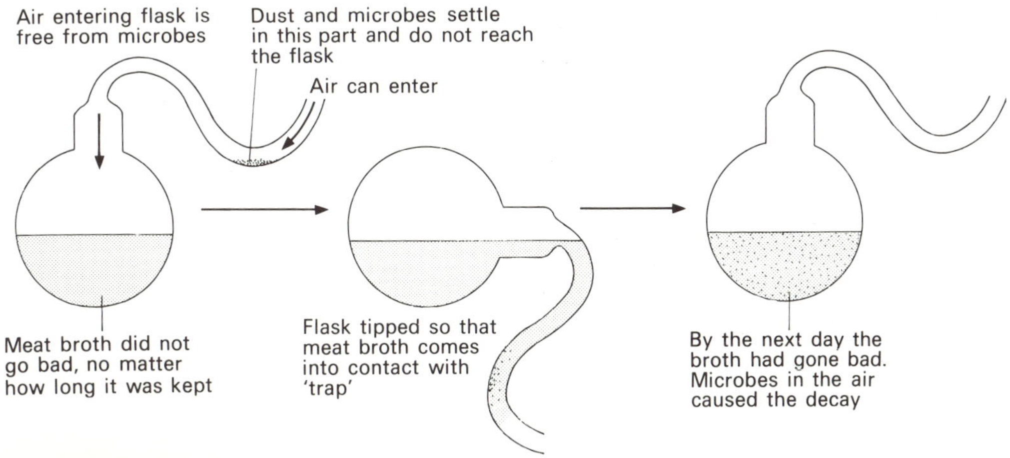

Fig. 7.1 Left *(a) Pasteur's first experiment. (b) Pasteur's second experiment, which showed that microbes are present in the air.* Right *(c) Microbes.*

Nowadays most pathogenic micro-organisms can be identified as belonging to one of four main groups. These are *bacteria*, *viruses*, *fungi* and *protozoa*. The general features of these groups are described in Table 7.1 on page 80. Examples of micro-organisms within each group are also given, as well as the diseases that they cause in man.

7.5 **How diseases are transmitted**

With the knowledge that micro-organisms caused disease came an understanding of how diseases were transmitted. If the transmission of the micro-organism could be prevented, the spreading of the disease could be halted. So began the research and development of techniques aimed at stopping the transmission of disease within the community. For example, cholera and typhoid epidemics were eliminated in this country once raw (untreated) sewage was prevented from entering a communal water-supply (Fig. 7.2).

Fig. 7.2 *Old houses in London Street, Dockland, about 1810. The stream was the only water-supply and means of sewage disposal*

TABLE 7.1 **TYPES OF MICRO-ORGANISM**

Appearance	Characteristics	Reproduction	Diseases that they can cause
BACTERIA 1 Spherical bacteria (cocci) (a) Staphylococcus (clusters)  (b) Streptococcus (chains)  (c) Diplococcus (groups) 	0.0005–0.002 mm in length. A million would fit on a pin-head Single cells. Many can move with the aid of tails called *flagella* Some are pathogens, and live in our body tissues between the cells. These can protect themselves when they are outside the body by forming a *spore*. The spore has a thick outer wall to create sterile conditions we must kill the spores as well as the live bacteria. For growth they need food, warmth and moisture. Some need oxygen (aerobic), others do not (anaerobic) Found in water, air, soil, living and decaying tissue	Asexual reproduction: each bacterium divides into two. If the conditions are right this can happen every 20 minutes with some bacteria. Therefore: (i) a new strain can develop in a short time (this could be resistant to antibiotics) (ii) the strength (virulence) of the bacteria could alter	1 (a) boils, styes, impetigo (b) blood poisoning, scarlet fever, sore throat (c) pneumonia
2 Spiral-shaped bacteria (spirochaete) 			2 Syphilis
3 Rod-shaped bacteria (bacilli) 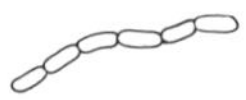 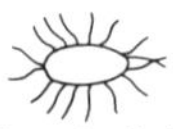 (a) Chains (b) Free individuals			3 (a) anthrax, diphtheria (b) dysentery, gastro-enteritis, tetanus, typhoid, T.B.
4 Comma-shaped bacteria (vibrio) 			4 Cholera
VIRUSES   (a) (b)	Can be seen clearly only under an electron microscope. A billion (million million) would fit on a pinhead Some resemble crystals in structure Each virus invades a specific host and tissue. Lives inside the body's cells. Found only in living tissue	Takes over a cell's metabolism. The cell manufactures more virus material instead of cell material. The host cell is often destroyed	Chicken-pox, mumps, influenza, rabies, yellow fever, measles, smallpox, German measles, poliomyelitis, warts (verruca), herpes blisters

TABLE 7.1 **continued**

Appearance	Characteristics	Reproduction	Diseases that they can cause
FUNGI 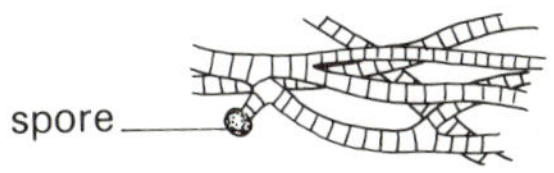	The cells of the fungus form a 'web' of very fine threads. These threads penetrate the food material and absorb nourishment from it. Moisture encourages growth Live on decaying organic matter or in living tissue	Can reproduce asexually and sexually. Sexual reproduction leads to spores, which can be transmitted to another host	Skin infections, e.g. ringworm, athlete's foot Mouth infections, e.g. thrush Vaginal infections
PROTOZOA 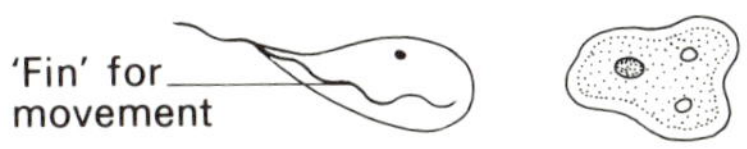 (a) (b)	Single cell like amoeba. Absorb nourishment from their surroundings through the cell membrane. Some are free living, others live within another organism such as man. Not all protozoa that live within man are pathogenic	Can reproduce sexually and asexually within their life cycle. Asexual reproduction gives rapid multiplication	Malaria, sleeping sickness (a) amoebic dysentery (b), vaginal disorders

The main methods by which pathogenic micro-organisms are spread are listed in Table 7.2 on pages 82/3. Some of the diseases transmitted by these means are given and these may be linked with Table 7.1. Table 7.2 also gives some of the measures that can be taken to prevent transmission. These are discussed in greater detail in the following sections and in other chapters.

Question 1 What action should you take to prevent the spreading of disease in the following instances?

(a) before handling food
(b) after going to the lavatory
(c) on seeing a cracked cup in your cupboard
(d) when coughing
(e) on seeing evidence of mice or rats in your home

TABLE 7.2 THE SPREADING OF DISEASES CAUSED BY MICRO-ORGANISMS

Sources

- A person showing symptoms
- A person who is infectious
  (i) before he has the symptoms
  (ii) after the symptoms have gone
- A person who is a 'carrier' and shows no symptoms
- If the disease has another host other than man this can act as a 'reservoir' of the disease (e.g. a fox can be the 'reservoir' of rabies)
- The soil (e.g. tetanus and botulism)

Means of entry

- Through damage:
  (i) bites
  (ii) scratches and grazes
  (iii) deep cuts
- Through body openings:
  (i) eyes
  (ii) nose
  (iii) mouth
  (iv) skin (through pores)
  (v) reproductive and excretory passages

Ways of transmitting diseases

(a) Through contaminated food

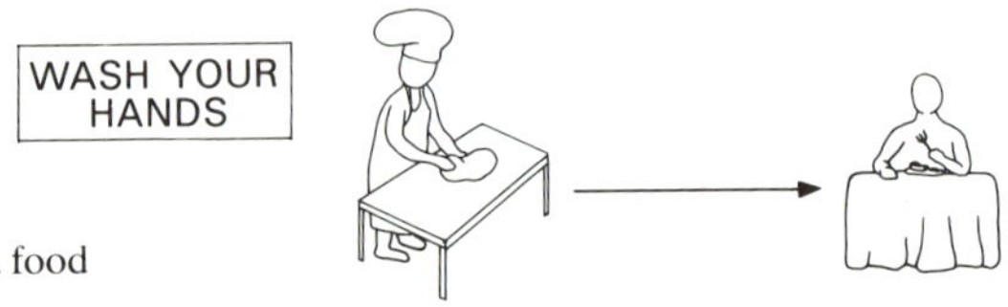

(b) Through contaminated water

(c) Through the air
(i) droplet infection
(ii) dust

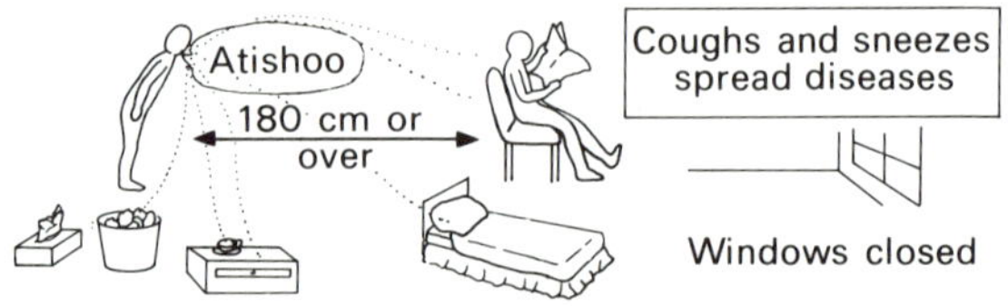

(d) Through direct contact (e.g. skin contact)

(e) Through indirect contact (e.g. personal belongings)

(f) Through a vector
(i) direct contact

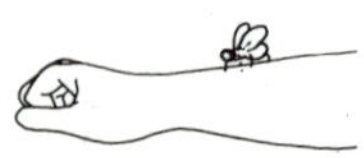

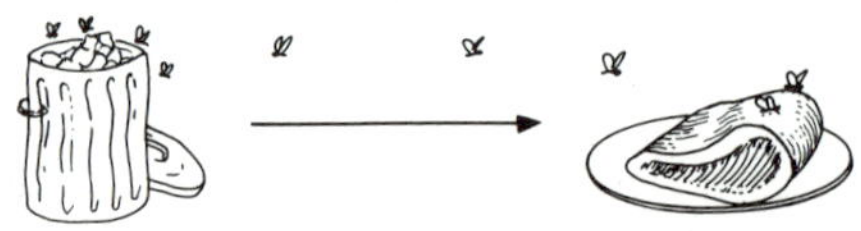

(ii) indirect contact
(e.g. mice droppings on food)

. . . EXAMPLES OF THE DISEASES AND METHODS OF PREVENTION

Examples		Prevention
Typhoid Cholera Food poisoning Botulism Gastro-enteritis (epidemic)	Diarrhoea T.B. (bovine) Poliomyelitis Amoebic dysentery	Food inspection (page 00) Inspection of shops, restaurants Washing soil off crops Cleanliness of working surfaces and utensils Personal hygiene Use of disinfectants – sterilization (Section 7.4)
Typhoid Cholera Food poisoning Gastro-enteritis (epidemic)	Diarrhoea Poliomyelitis Amoebic dysentery	Purification of water supply (page 135) Testing water-supply Responsibility of everyone not to contaminate water-supplies
Whooping cough Diphtheria T.B. Scarlet fever Sore throat Colds	Influenza Measles Mumps Chicken-pox Smallpox Poliomyelitis	Good ventilation Personal care over disposal of handkerchiefs and sneezing Removal of dust using vacuum cleaner Hygiene in the home Isolation of patient
Impetigo Brucellosis Boils Blood poisoning (septicaemia)	Smallpox Chicken-pox Ringworm Vaginal infections Sexually transmitted diseases (V.D.)	Personal hygiene Avoiding contact – isolation Use of dressings to avoid transmission Sexually transmitted diseases can be prevented only by not having a sexual relationship with an infected person
Warts (verruca) Vaginal infections Ringworm Athlete's foot		Avoiding contact – with infected personal belongings Cleanliness Disinfecting communal areas Not sharing clothing or footwear
(i) Plague Typhus Yellow fever Malaria Sleeping sickness Psittacosis (from pet birds – respiratory infection) Food poisoning	(ii) Food poisoning Infective jaundice Cholera Epidemic diarrhoea	Control of the vector by insecticides and pesticides Quarantine of animals Good housing and cleanliness in the home to prevent pests Care and hygiene when handling pets

7.4 **Preventing disease by killing the micro-organism**

Robert Koch (1843–1910) worked out a method of growing a culture of bacteria, and we use a similar method today (Fig. 7.3(a)). A culture can be used to find ways to stop the growth and reproduction of the micro-organism (Fig. 7.3(b)). In Table 7.3 some of these ways have been listed.

The word *sterile* is used to describe any situation where all life forms have been killed or removed. The first attempt at keeping anything that came into contact with a wound sterile was carried out by a surgeon in Glasgow called Joseph Lister (1827–1912). He had read reports on Pasteur's work and saw the connection between pathogenic micro-organisms and the 'going bad' of wounds, i.e. the wound going *septic*.

TABLE 7.3 **WAYS OF CONTROLLING OR KILLING MICRO-ORGANISMS**

Method		**Example of use**
Temperature	High:	
	(i) moist heat (boiling, steaming)	Cooking food, sterilizing
	(ii) dry heat	Flaming needles, burning infected material – clothes, tissues, dressings, etc.
	Low	Freezing food
No moisture		Drying food for storage
Rays	Sun's rays (ultraviolet) Also dangerous rays (X-rays, gamma rays, etc.)	Sterilization of air
Chemicals	(a) Some kill *all* forms, e.g.	
	acids	Antiseptic use
	alkalis	Antiseptic use – detergents
	alcohols	On skin before injection or operation
	aldehydes	Formaldehyde – disinfectant, fumigation
	halogen family	Chlorine compounds – disinfectants; Tincture of iodine – wounds. Iodoform antibacterial used in hospitals ('hospital smell')
	(b) Others kill *one* form (they are said to be *specific*)	Medicines used internally, such as antibiotics (see Section 7.6)

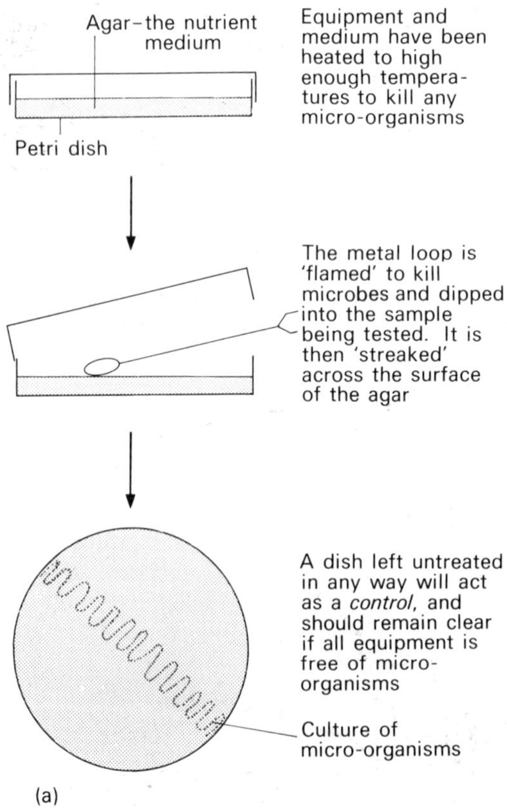

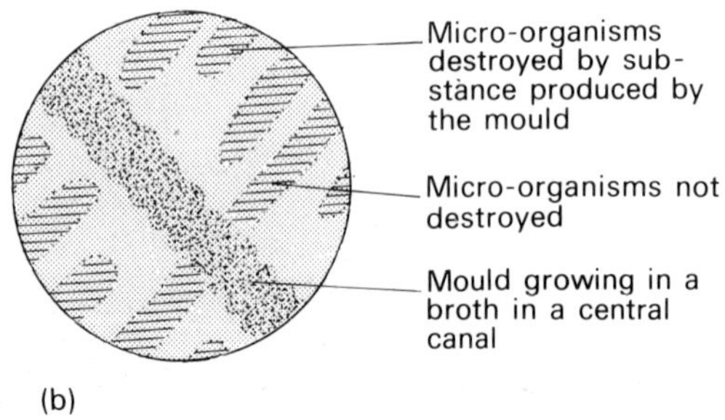

Fig. 7.3 *(a) Culturing microbes. You need supervision for these experiments because the cultures could be dangerous to your health. Care must be taken that the culture containers are not opened after infection, and that they are safely disposed of according to official recommendations. (b) It is in such an experiment that Alexander Fleming first noticed the action of the mould* Penicillium *(page 86) on his bacteria cultures.*

He experimented by using a solution of carbolic acid on a wound. The wound healed without going septic. By using carbolic acid on wounds, as a spray in hospital rooms (Fig. 7.4) and as a wash for instruments and hands Lister found that few patients died from septic wounds.

Nowadays everything around the patient during surgery is made sterile, even the air. This is known as *aseptic surgery*. Figs 7.4 and 7.5 both show an operation in progress. The advances in aseptic surgery that have taken place since Lister first attempted it are very clear.

Question 2 Compare Figs 7.4 and 7.5 and explain what measures are now taken to stop harmful microbes being transferred to the patients from:

(a) the surgeon and his assistants
(b) the instruments
(c) the surroundings

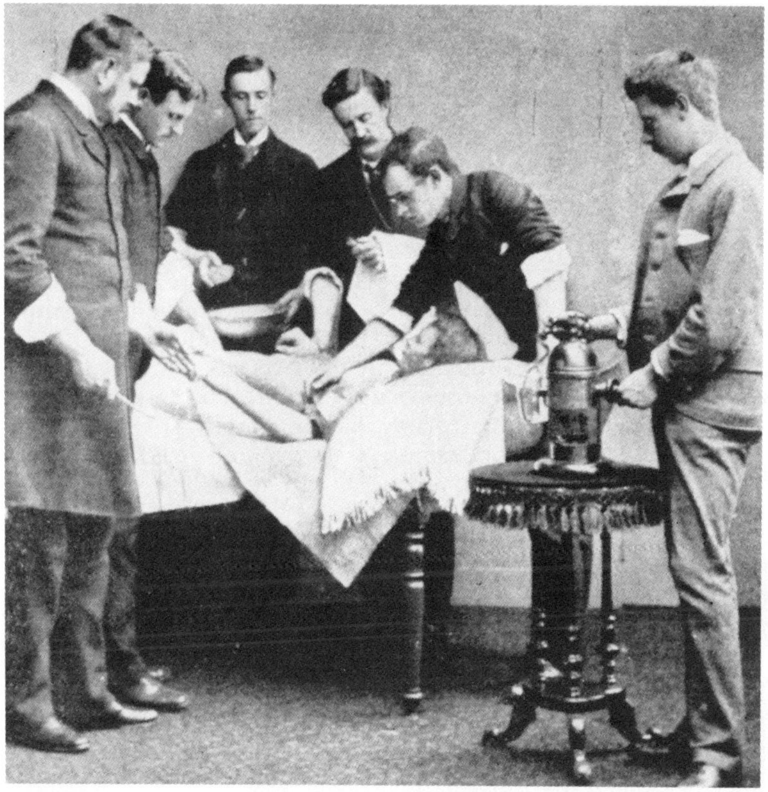

Fig. 7.4 *An operation at Aberdeen in Lister's day. An assistant attends to the carbolic spray*

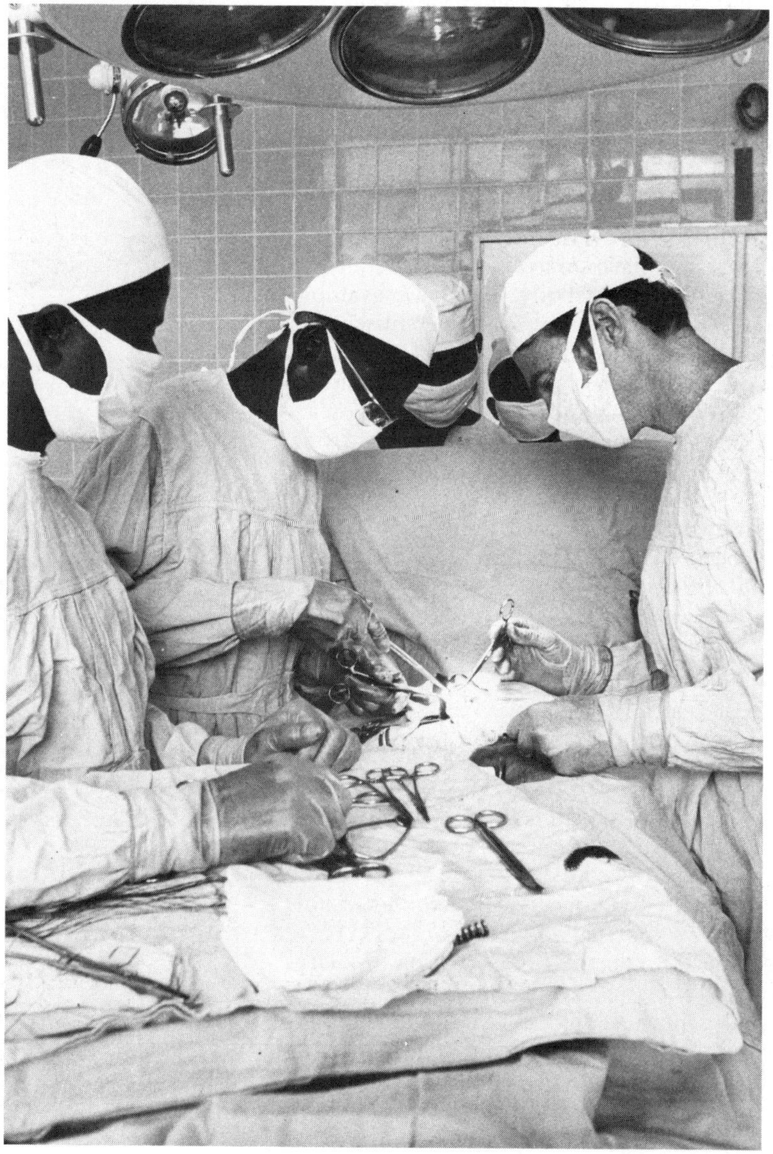

Fig. 7.5 *Modern operation in progress*

7.5 **Killing micro-organisms by the use of chemicals**

The chemicals that are used to attack harmful micro-organisms outside the body are either *disinfectants* or *antiseptics* (see Table 7.3).

A disinfectant is a chemical which kills harmful microbes. The strongest and most effective disinfectants are fairly corrosive chemicals which contain chlorine, like bleach (calcium hypochlorite). They may burn the skin and the housewife should handle them with caution, storing them safely, away from young children. They are essential in the fight against harmful micro-organisms in the home.

An antiseptic is usually weaker than a disinfectant. It prevents the growth and multiplication of harmful microbes. Generally it can be used on the body surface, for example some soaps and detergents have antiseptic properties (Fig. 7.6). Antiseptics like that being used in Fig. 7.7 are found in the first-aid kit of most homes. When smeared on a wound the antiseptic should prevent it from going septic.

Question 3 How should you treat a wound (a) before applying and (b) after applying antiseptic?

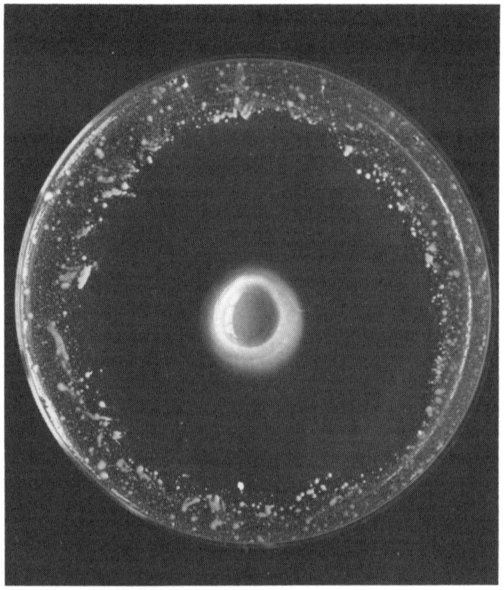

Fig. 7.6 Above *(a) The central reservoir contains a particular soap with antiseptic properties. The clear zone shows the effect it is having on the bacteria.*
Below *(b) The soap is particularly used in hospitals, surgeries, food factories, kitchens, canteens and shops.*

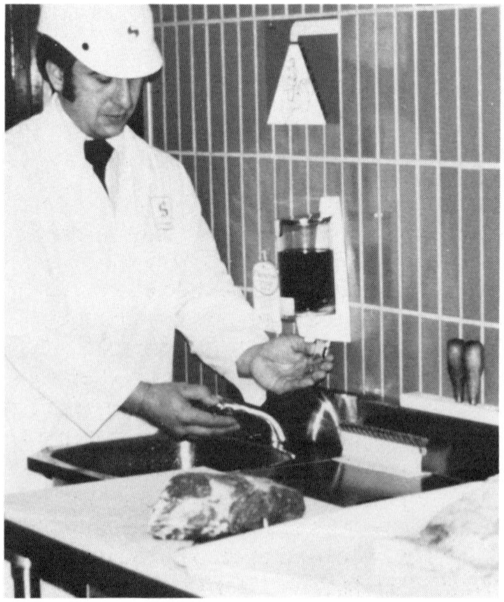

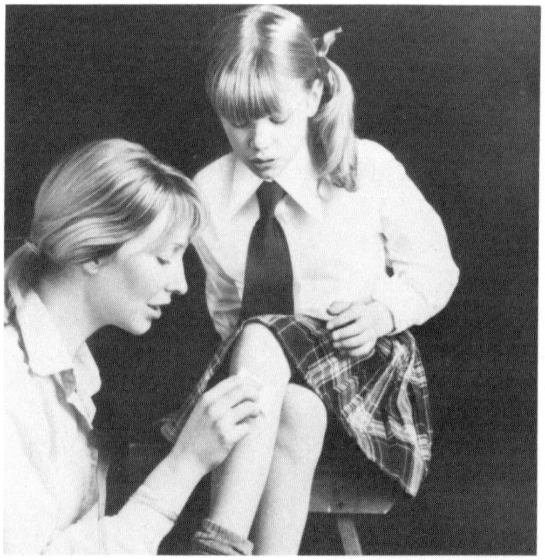

Fig. 7.7 *Antiseptics help us to fight infection of wounds*

7.6 **The use of chemicals within the body**

Only a short time ago, in 1930, *chemotherapy* was thought to be impossible. Chemotherapy is the treatment of disease using chemicals which kill the pathogen but not the host. In 1935 chemicals called *sulphonamides* were developed. The success of these chemicals in the treatment of diseases such as pneumonia, and the demand for medical treatment of soldiers during the Second World War encouraged research into chemotherapy.

In 1929 Alexander Fleming had published a paper on the results of his work, describing 'the antibacterial action of a culture of *Penicillium*' (Fig. 7.3(b)). *Penicillium* is a mould and Fleming found that it produced a substance which could kill bacteria. This substance was called *penicillin*. Sir Howard Florey decided to investigate penicillin in 1939. By 1941 satisfactory methods of growing and extracting penicillin on a large scale had been found and the results of the first medical trials proved its great value in the fight against disease. Penicillin is classed as an *antibiotic*. This is a substance which is produced by one organism and acts against another organism, stopping its growth or killing it.

Penicillin can be used against many different strains of bacteria, such as streptococci, staphylococci and the pneumonia bacillus. It is a *broad-spectrum* antibiotic. *Tetracyclines* also fall into this group and are used to treat typhoid fever, salmonella poisoning, gastroenteritis and urinary-tract infections. Other antibiotics are more specific and may attack only one or a few strains of bacteria; they are *narrow-spectrum* antibiotics. Streptomycin, for example, is used to treat T.B.

We have already stated that chemotherapy aims to kill the pathogen without harming the host. Viruses live within body cells. To attack the virus the drug would need to destroy the body tissues. So at present chemotherapy is of no use against viruses. A medicine may be prescribed in cases of virus infection for another reason, for example to lower the patient's temperature or to relieve pain. However, a breakthrough in our fight against virus infection may not be far away.

Recent work on a substance called *interferon* suggests that it may be as important to medical advance as penicillin. Interferon was discovered in the 1950s by Dr Alexander Isaacs, an Englishman, but he died in 1957 and no further research on interferon was done until the late 1960s.

Interferon is a natural substance made by the nuclei of human body cells when a virus invades them. The interferon is released to warn other cells of the invasion of the virus. These cells can then produce an antiviral protein to protect themselves.

The supply of interferon is limited at present because it cannot be produced on a large, commercial scale. Further research into its manufacture is needed before its full impact on medicine can be felt.

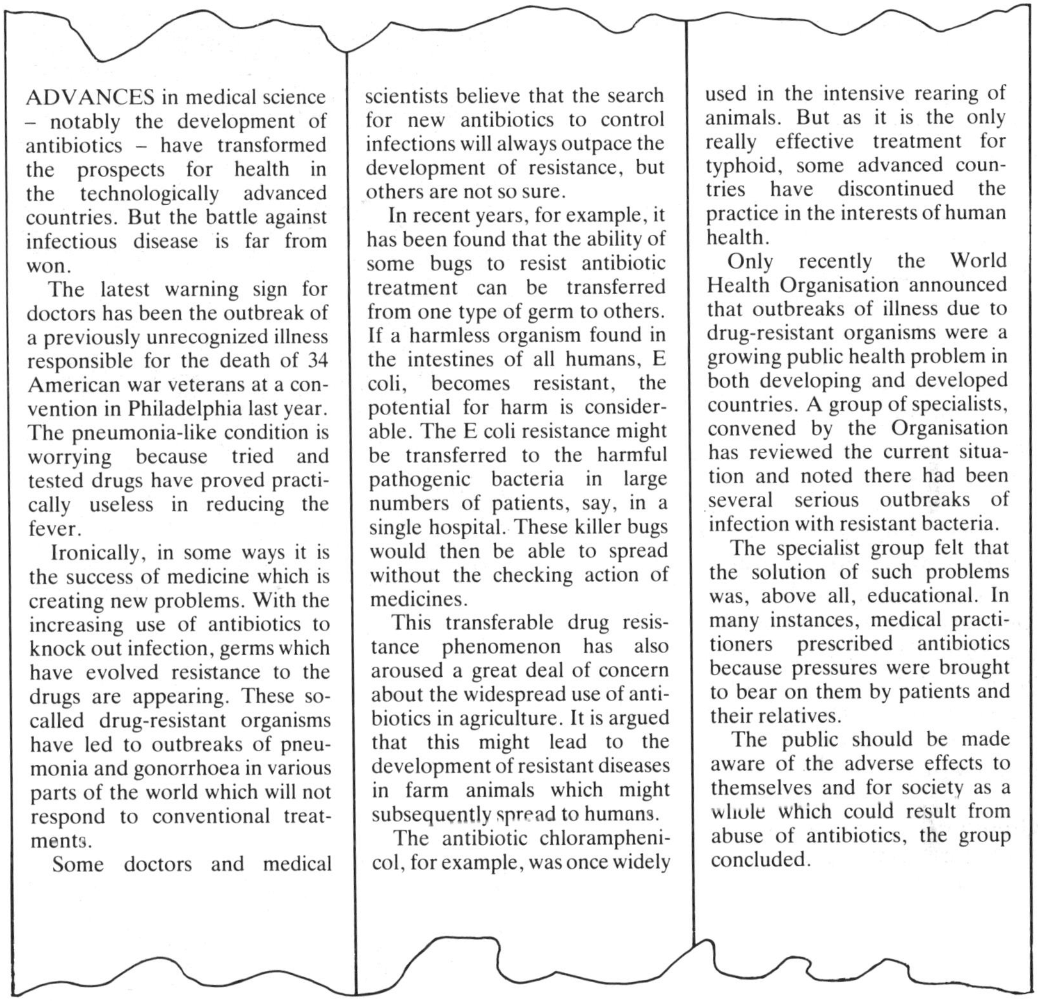

ADVANCES in medical science – notably the development of antibiotics – have transformed the prospects for health in the technologically advanced countries. But the battle against infectious disease is far from won.

The latest warning sign for doctors has been the outbreak of a previously unrecognized illness responsible for the death of 34 American war veterans at a convention in Philadelphia last year. The pneumonia-like condition is worrying because tried and tested drugs have proved practically useless in reducing the fever.

Ironically, in some ways it is the success of medicine which is creating new problems. With the increasing use of antibiotics to knock out infection, germs which have evolved resistance to the drugs are appearing. These so-called drug-resistant organisms have led to outbreaks of pneumonia and gonorrhoea in various parts of the world which will not respond to conventional treatments.

Some doctors and medical scientists believe that the search for new antibiotics to control infections will always outpace the development of resistance, but others are not so sure.

In recent years, for example, it has been found that the ability of some bugs to resist antibiotic treatment can be transferred from one type of germ to others. If a harmless organism found in the intestines of all humans, E coli, becomes resistant, the potential for harm is considerable. The E coli resistance might be transferred to the harmful pathogenic bacteria in large numbers of patients, say, in a single hospital. These killer bugs would then be able to spread without the checking action of medicines.

This transferable drug resistance phenomenon has also aroused a great deal of concern about the widespread use of antibiotics in agriculture. It is argued that this might lead to the development of resistant diseases in farm animals which might subsequently spread to humans.

The antibiotic chloramphenicol, for example, was once widely used in the intensive rearing of animals. But as it is the only really effective treatment for typhoid, some advanced countries have discontinued the practice in the interests of human health.

Only recently the World Health Organisation announced that outbreaks of illness due to drug-resistant organisms were a growing public health problem in both developing and developed countries. A group of specialists, convened by the Organisation has reviewed the current situation and noted there had been several serious outbreaks of infection with resistant bacteria.

The specialist group felt that the solution of such problems was, above all, educational. In many instances, medical practitioners prescribed antibiotics because pressures were brought to bear on them by patients and their relatives.

The public should be made aware of the adverse effects to themselves and for society as a whole which could result from abuse of antibiotics, the group concluded.

Fig. 7.8 *Are the bugs fighting back?*

7.7 How should we use antibiotics and medicines?

Today an enormous range of medicines and antibiotics is manufactured. In Britain most are available only through a prescription issued by a doctor. It is the doctor with all his specialized knowledge who will decide whether or not his patient will benefit from their use. If a prescription is given it is the responsibility of the patient to see that he uses the medicine as instructed. For example, if a patient is instructed to take a dose of a medicine four times a day until it is all gone, it is not much use if he takes it three times a day and throws half the bottle away. Action of this sort with antibiotics such as penicillin is likely to encourage the development of a *resistant* strain of bacteria. Resistant strains are those bacteria that are unaffected by the antibiotic. While other bacteria are destroyed by the antibiotic the resistant ones grow and multiply. They are also able to transfer their resistance to other bacteria. If these other bacteria are pathogens then that antibiotic will no longer be of use against them. Thus the patient's symptoms continue and a different antibiotic will have to be used to destroy the resistant strain. The dangers of the over-use of antibiotics and the development of resistant pathogens is outlined in the article in Fig. 7.8. A medicine prescribed for one individual is for him alone! Different people react in different ways to particular medicines. A doctor's advice before taking any medicine is essential; medicines can react with one another, with certain foods and particularly with alcohol. It is therefore essential to follow instructions (Fig. 7.9).

TREATMENT CARD

Carry this card with you at all times. Show it to any doctor who may treat you other than the doctor who prescribed this medicine, and to your dentist if you require dental treatment.

INSTRUCTIONS TO PATIENTS

Please read carefully

While taking this medicine and for 10 days after your treatment finishes you must observe the following simple instructions:–

1 Do not eat CHEESE, PICKLED HERRING OR BROAD BEAN PODS.

2 Do not eat or drink BOVRIL, OXO, MARMITE or ANY SIMILAR MEAT OR YEAST EXTRACT.

3 Do not take any other MEDICINES (including tablets, capsules, nose drops, inhalations or suppositories) whether purchased by you or previously prescribed by your doctor, without first consulting him.

NB *Cough and cold cures, pain relievers and tonics are medicines.*

4 Drink ALCOHOL only in moderation and avoid CHIANTI WINE completely.

Report any severe symptoms to your doctor and follow any other advice given by him.

Prepared by The Pharmaceutical Society and the British Medical Association on behalf of the Health Departments of the United Kingdom.

11530/1963 R16S 558905 250m 11/77 AG 3640/4

Fig. 7.9 *Treatment card (above)*

Too large a dose or constant use of any medicine or antibiotic can be very harmful to the body. This includes the brands of medicine that can be bought without a prescription, for example, pain-killers like aspirin, cough medicine and indigestion powders or pills. Any unwanted medicines should be taken back to the pharmacist rather than left in a cupboard or thrown carelessly away. Very young children especially are at risk if medicines are kept in open cupboards, for so many are attractive-looking liquids or sweet-like pills.

Question 4 Can you suggest what else a parent might do to make sure that a child does not eat such medicines or chemicals if they are found outside the home?

* Always read the directions carefully and give or take the exact recommended dose. This is particularly important if you're giving medicines to children.
* If your doctor has prescribed medicine for you, complete the course – even if you do feel better. If in doubt, ask your doctor or pharmacist.
* Tell your doctor or pharmacist of any side-effects you may have noticed.
* Take your own medicine! It was specially prescribed to deal with *your* illness. Don't offer it to a friend — and don't 'borrow' his.
* If you're advised not to drink, drive or operate machinery while taking a medicine — don't. It's dangerous.
* If you're pregnant, ask your doctor or pharmacist about any medicines you take.

* Don't keep unused prescribed medicines or over-the-counter medicines with an expiry date that has passed. Return them to the pharmacist or throw them down the lavatory — not into the dustbin or on the fire.
* Keep all medicines, including aspirin, away from children. Store them in one lockable medicine cabinet and take the extra precaution of keeping them in childproof containers.
* Never let a child get hold of medicines. To children, brightly coloured tablets are just like sweets.

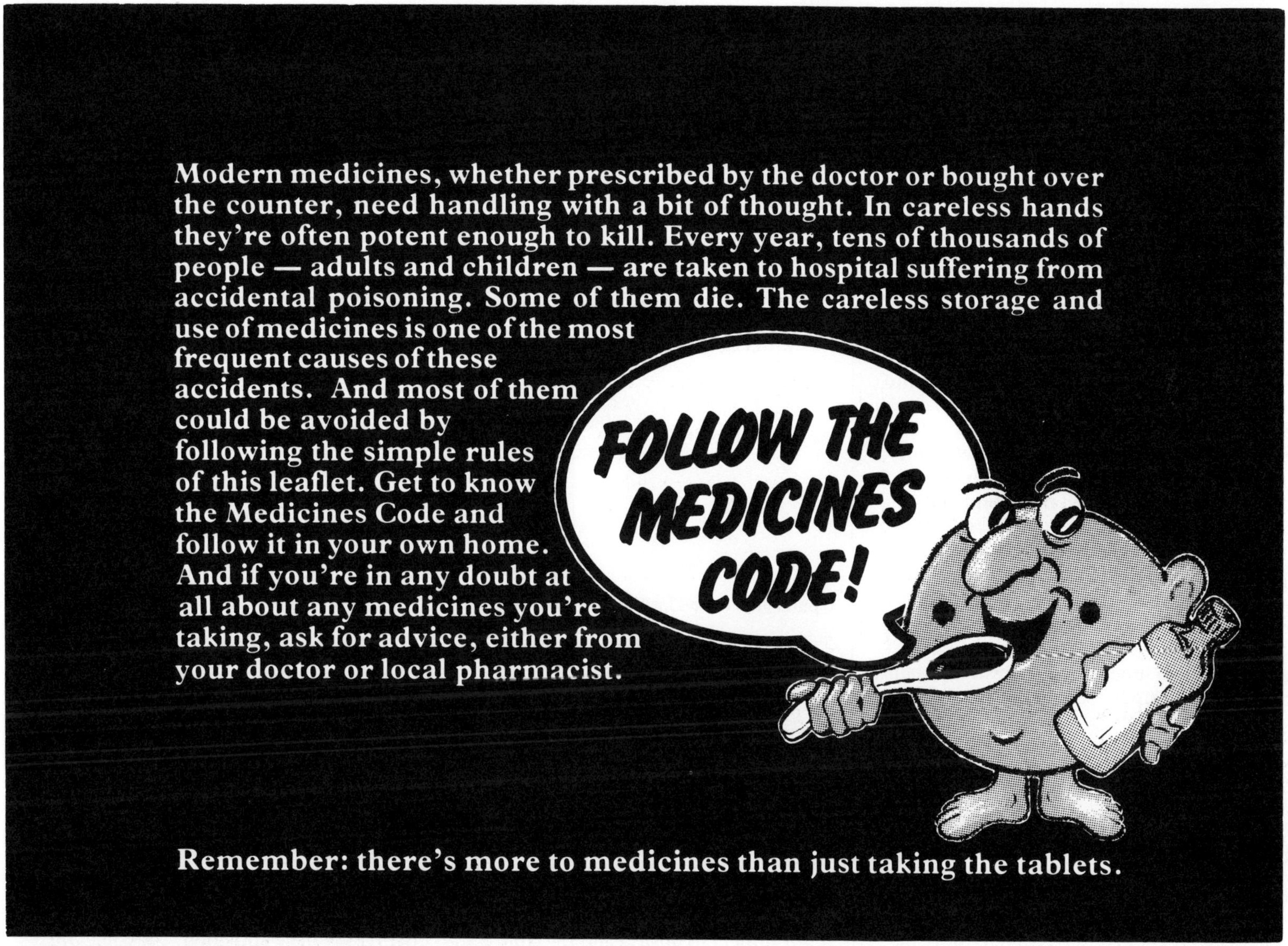

Fig. 7.10 *The Medicine Code (from a Health Education Council leaflet)*

All medicines affect our bodies in some way. We have seen that they can be dangerous if they are not used properly. It is up to every individual to understand and observe a medicine code (Fig. 7.10).

7.8 **The drug habit and drug dependence**

In Chapter 4 we stated that people who find it difficult to give up smoking do so because it becomes a habit to smoke. People can develop a similar habit of taking medicines or drugs. This habit can develop through taking medicines (a) when they are not really needed, (b) to cope with a persistent complaint that the person should really see a doctor about, or (c) to cope with other problems.

Substances that affect how we feel are termed *drugs*. They affect us in this way because they enter the bloodstream and reach the brain (page 226). Tobacco, alcohol and caffeine (found in tea and coffee) are drugs in this sense. *Analgesics* affect the level at which the brain and nervous system register pain. Aspirin is probably the best-known analgesic. When over-used it can irritate and cause bleeding of the stomach wall (page 158). *Sedatives* are drugs that calm the activity of the brain so that the person feels sleepy or drowsy. Sleeping-pills fall into this group and so does alcohol (Table 7.4). *Stimulants* act in the opposite way to sedatives, speeding up the activity of the brain, hence the common terms pep pills and anti-depressants (caffeine is a mild stimulant). Strong sedatives and stimulants can be obtained only on prescription. *Hallucinogens*, which are rarely used medically, affect the senses so severely that the person can no longer make accurate judgements. Drugs are useful to us when they are carefully used under medical supervision. However, when drug use is unsupervised there can be very serious consequences.

Drug misuse usually refers to the taking of a drug illegally without medical supervision just for the effect that it has on a person's mood. If a drug is used constantly for such a purpose the body develops a *tolerance* to it, i.e. larger and larger doses of the drug are needed to produce the same effect. These larger doses produce a physical change in the body. The body's cells get used to the drug and cannot work without it; so the body develops a physical craving for the drug. If a dose cannot be obtained the person experiences *withdrawal symptoms* and feels physically ill. Similarly, he may also rely emotionally upon the drug. This total dependence or *addiction* to drugs causes so many changes in a person's personality and behaviour that the whole pattern of his life is changed. Table 7.5 names some of the drugs

TABLE 7.4 **HOW ALCOHOL AFFECTS PHYSICAL AND MENTAL BEHAVIOUR**

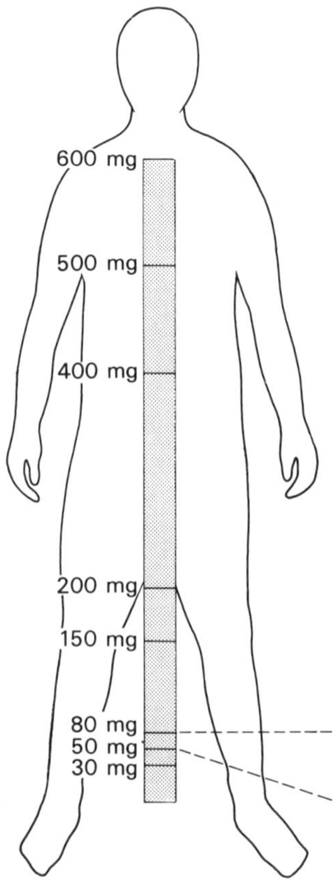

Number of drinks	**Blood-alcohol level (mg per 100 ml of blood)**	**Effects**
	600 mg	Death certain
1 bottle of spirits	500 mg	Death possible
¾ bottle of spirits	400 mg	Oblivion, sleepiness, coma
		Stagger, double vision, loss of memory
6 pints of beer or 13 whiskies	200 mg	Loss of self control. Exuberance, quarrelsomeness, slurred speech. 25 times more likely to have an accident
5 pints of beer or 10 whiskies	150 mg	
2½ pints of beer or 5 whiskies	80 mg	Four times more likely to have an accident than at 30 mg. LEGAL LIMIT
1½ pints of beer or 3 whiskies	50 mg	Cheerful, feeling of warmth. Impairment of judgement, loss of inhibition. Likelihood of having an accident starts to increase
1 pint of beer	30 mg	

that are subject to misuse and the consequences of this. As the misuse of drugs is illegal, large sums of money are needed to buy drugs illegally and addicts are often driven to crime to get it. Death of an addict can result from such causes as an overdose, infection from dirty injections, an accident while he is under the influence of drugs or from the general decline in his health.

There are both public and voluntary organizations that seek to help people to overcome drug addiction. To 'come off drugs' is no easy task and requires a great deal of determination and courage from the addict.

With the development of so many drugs in recent years it has become vital for the community as a whole to make laws to see that they are manufactured correctly and are safe for use. In the same way as there has to be legislation to cover the sale and handling of dangerous chemicals such as dynamite, there

TABLE 7.5 **DRUGS WHICH ARE SUBJECT TO MISUSE**

Type of drug	Examples	Form	Consequences of continual misuse
Narcotics	Opium (from opium poppy) Morphine, Heroin, Codeine – derivatives of opium	Pills or capsules	Tolerance develops rapidly. General health deteriorates. Personal welfare neglected
Stimulants	Cocaine Amphetamines (pep pills, 'speed' – a great many types)	Pills or capsules	Tolerance develops. There is a psychological need for the drug, a feeling of being unable to cope with life without it. After-effects are severe depression and 'bad dreams'
Sedatives	Barbiturates ('barbs' – more than 30 types) Mandrax ('mandies')	Pills or capsules	Tolerance develops. Produces a drunken effect, and if a large dose is taken, or when taken with alcohol, unconsciousness may occur. During withdrawal there is severe physical discomfort and psychological disturbances
Hallucinogens	L.S.D. ('acid')	Liquid, powder, pills or capsules	Can have serious psychological effects. While on the drug, the addict can be at risk through attempting 'the impossible'. Can produce disturbing and horrible experiences
	Cannabis resin (from Indian hemp plant – hash) Cannabis leaves (marijuana – 'pot' or 'grass')	Smoked with tobacco as a 'reefer' or 'joint'	Similar effects to L.S.D. The feeling of well-being and awareness that this drug produces may lead a person to try stronger hallucinogens. As with L.S.D. these may produce serious physical and psychological disturbances

has to be control over the sale and possession of drugs. The Misuse of Drugs Act, which came into force in 1973, makes it an offence to possess, without authority, narcotics, cocaine, cannabis, amphetamines, L.S.D. and other hallucinogens. Barbiturates may be dispensed by a pharmacist only on recepit of a prescription from a doctor. Such controls seek only to protect the community and the individual.

7.9 The body's natural defences against invasion by micro-organisms

The skin

The first defence of a castle against invaders is its outer wall (Fig. 7.11). The body has a similar defence in the form of its outer wall, the *skin*. The fact that bacteria are found in large numbers on the skin is clearly shown in Colour plate 9. We can help our skin to do its job by washing regularly. This helps to remove the bacteria and the substances produced by their activities. It is these substances which give rise to unpleasant body odours, or B.O.! Micro-organisms can infect the skin, as we have already seen in fungal infections like athlete's foot. Care and personal hygiene can also prevent some of these infections. The washing of skin and clothes is more successful if soap or a detergent is used. These have antiseptic properties, and also are able to break the

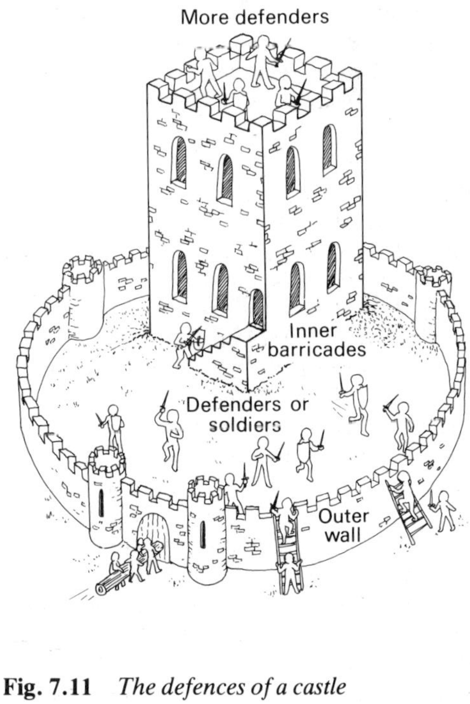

Fig. 7.11 *The defences of a castle*

attachment that the oils and fats have for the skin or textiles. The grease and dirt is therefore lifted off. (A similar action by detergents can be seen when they are used to break up an oil-slick at sea.) Micro-organisms can enter the body where there are natural openings, for instance the nose and the mouth (see Table 7.2) and the openings where material is removed from the body, that is, the openings into the reproductive passages (page 247) and the excretory passages (page 168). If a person is in good health, the moist linings of these passages seem to be able to prevent certain micro-organisms from forming an established colony (page 94).

The blood and lymph

The skin is our castle wall, but what happens if it becomes broken? In a battle attempts would be made to block up the hole in the wall, and soldiers would engage the invaders that had entered (Fig. 7.11). The body reacts in a similar manner with its second line of defence. The blood clots, sealing off the skin, and white blood cells enter the tissues surrounding the wound to attack any micro-organisms that have entered (see Table 6.1). If the defenders succeed all invaders are wiped out and the healing of the skin continues beneath the scab by the natural action of the cells. When the skin has healed the scab falls off.

If some micro-organisms are not dealt with at the point of entry they can spread further into the body. However, there is a third line of defence to get through: groups of 'barricades' called *lymph nodes* (the 'inner barricades' Fig. 7.11). In Fig. 7.12 you can see the position of these lymph nodes within the body.

Lymph nodes contain a system of fibres and white blood cells. As the lymph passes through them, the micro-organisms are trapped by the fibres and attacked by the white blood cells (lymphocytes). By this action the lymph nodes prevent infection spreading all over the body. Sometimes during such action they become swollen and painful – this usually indicates that the person will have to seek a doctor's advice, as an antibiotic may be needed to help fight the infection. Most of us have at some time suffered from tonsilitis. Tonsils are also lymphatic tissue: they 'guard' the passageways into the respiratory and digestive systems. When infected by micro-organisms they become inflamed.

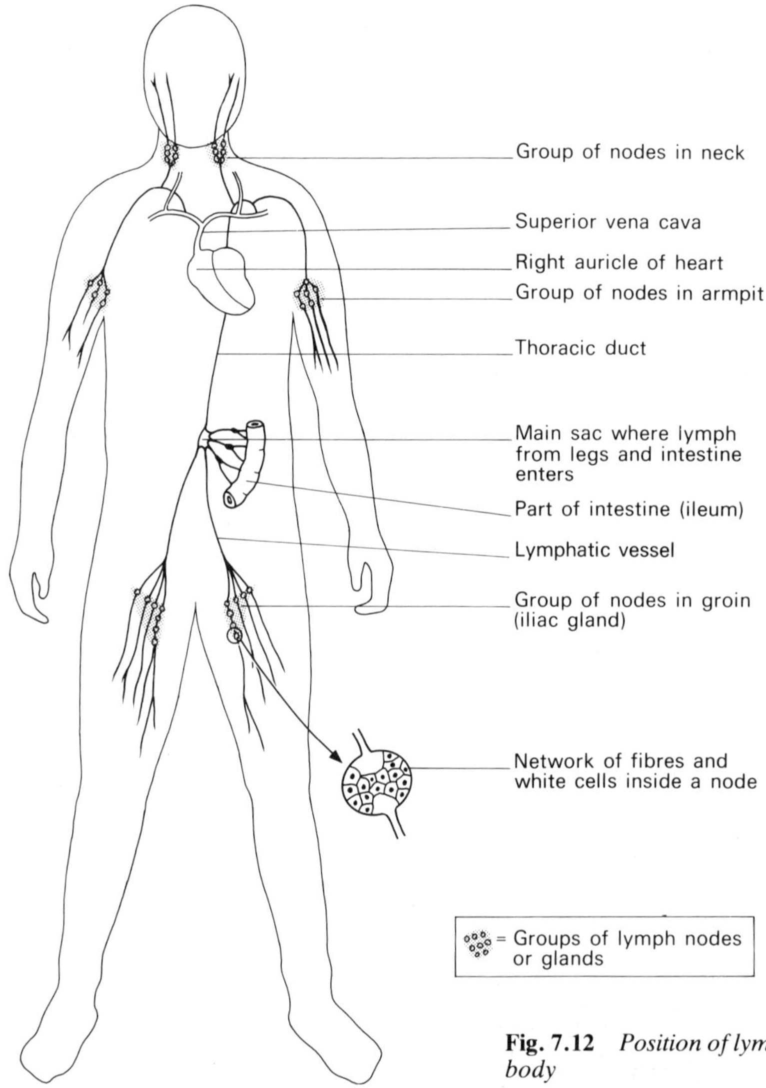

Fig. 7.12 *Position of lymph nodes within the body*

Question 5 If a person shows visible signs of a disease, such as a rash or swollen glands, what do you think has happened?

Antibodies
The signs of a disease are the reactions of the body to the foreign matter of the bacteria themselves and to the poisonous substances that they produce. The poisonous substances which enter the bloodstream are called *toxins*.

The body can make *antitoxins* in the blood to render these toxins harmless. Antitoxins are just one of the many chemicals which are made in the blood to assist the white blood cells in their attack upon foreign matter. These chemicals are known collectively as *antibodies*, even though they may have different methods of attack. Some attack toxins, other the bacteria themselves. They can destroy bacteria or make it easier for the white cells to engulf them.

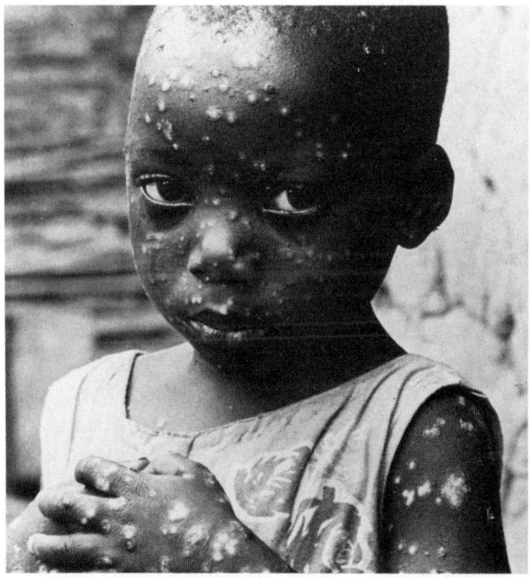

Fig. 7.13 *A victim of smallpox*

Each antibody works against only one sort of organism or substance, and it remains in the blood only for a short time. But once a person's body has produced a certain antibody it is able to make it again much more quickly. The body may be compared with a country that has learned how to manufacture a weapon during wartime. When peace returns production is stopped, but if needed the blueprint and assembly lines are all ready to go into action once again.

The readiness of the body to resist disease in this way is called *immunity*. Thus if we have an attack of chicken-pox in childhood it is very unlikely that we shall ever have it again. Even a very mild attack of the disease, with no visible signs, can produce immunity against further attacks. This fact was behind the idea, used in eastern countries and introduced into England in 1720 by Lady Montagu, for protecting people against smallpox (Fig. 7.13). Liquid from one of the spots of a person suffering from a mild attack of smallpox was scratched into an uninfected person's arm. This method of *inoculation* was not always satisfactory for often severe attacks developed, with many deaths. It was not until 1798, through the work of Dr Edward Jenner, that a safe method of making a person artificially immune against smallpox was developed.

7.10 **Preventing disease by active immunity**

Jenner successfully gave a person immunity against smallpox by scratching into their arm matter taken from a spot of a person suffering from cowpox (Fig. 7.14). The mild strain of cowpox microbes caused the body to manufacture antibodies. In this way immunity was developed even against the stronger smallpox virus.

Vaccination, taken from the Latin word *vacca* (cow), is the term used nowadays to describe this method of giving artificial immunity. When a person is vaccinated a *vaccine* is introduced into the bloodstream. The vaccine can be made up of any one of the following:

(a) dead micro-organisms (e.g. the vaccine against polio)
(b) attenuated strains – these weakened strains can be produced by culturing the micro-organisms in a different host, such as a horse (e.g. the B.C.G. vaccine against T.B.)
(c) harmless toxin of the micro-organism (e.g. the vaccine against diphtheria)

Such vaccines cause the body to manufacture its own antibodies, and this *active immunity* can protect a person for a long time, although several vaccinations may be

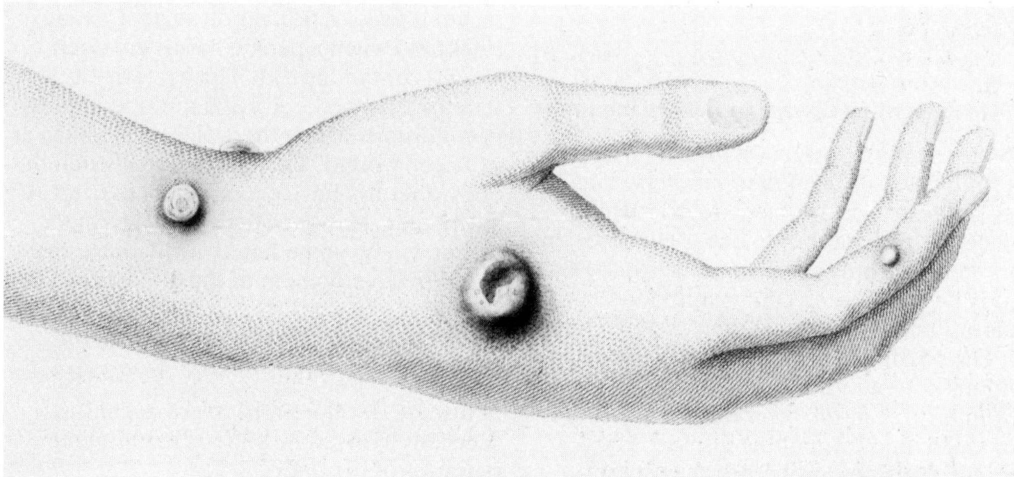

Fig. 7.14 *Jenner's drawing of the hand of Sarah Nelmes, a milkmaid. The pox or spot from which he took the cowpox vaccine is clearly visible. Cowpox is caused by a virus that is similar to the smallpox virus. It is transmitted to people from cows.*

needed to produce and maintain the necessary level of immunity. This can be seen in the vaccination or immunization programme that is available for young children in Britain (Table 7.6). It has been very effective against diseases such as diphtheria, which used to cause many deaths. However, outbreaks of these diseases would still be possible if in a generation the number of vaccinated people within the community fell below effective levels.

Our fight against disease often requires international co-operation, which is possible through an organization such as the World Health Organization (W.H.O.). The W.H.O. was first established in Geneva, Switzerland, in April 1946. In July 1948 it became a United Nations agency. Its aim is to promote the highest possible standards of health for all peoples of the world. Its services include advisory, technical and financial support to countries that are in the process of developing community health services, and social services such as water-supplies and sanitation. The W.H.O. also monitors the outbreak of diseases throughout the world and advises on the action that should be taken to prevent transmission from one country to another. It assists in the measures taken to eliminate diseases such as malaria, and in research into disease and drugs. In 1979 the W.H.O. declared that the world was completely free of smallpox.

TABLE 7.6 **THE VACCINATION PROGRAMME THAT IS AVAILABLE FOR CHILDREN IN BRITAIN**

0–1 year	1–2 years	At entry to primary school	During the secondary-school period
Vaccine given by injection against diphtheria, tetanus, whooping cough Vaccine taken by mouth (oral vaccine) against poliomyelitis Three doses of the vaccines are given: 1st dose at 3 months 2nd dose at 5 months 3rd dose at 11 months	Vaccine given by injection against measles (at about 14 months)	'Booster' injections given against diphtheria, tetanus, poliomyelitis	B.C.G. vaccination against T.B. (at about 12 years) For girls who have never had German measles a vaccine is available (between 11 and 13 years) Vaccination against tetanus and polio-myelitis (at 15–19 years or on leaving school)

7.11 Preventing disease by passive immunity

Babies acquire protection for a time against certain diseases from their mothers. This is possible because antibodies can cross the placenta from mother to baby. However this protection is only short-lived. The baby's body has only received the antibodies, and has not been involved in their manufacture.

The passing of antibodies from one organism to another produces *passive immunity*. A *serum* is produced by giving an animal a mild form of the disease. The animal develops antibodies against the disease and these are present in its blood. Samples of the animal's blood are taken and treated to make the serum. If a person has been in contact with a disease then a serum may be given to prevent it developing. Such a serum is used as protection against tetanus (lockjaw) when a person has a very deep wound containing dirt. The bacteria of tetanus exist in the soil and multiply rapidly in conditions where there is little air, such as in a deep wound. They act so quickly that the body does not have enough time to prepare antibodies. However the antibodies in the serum will be immediately available to prevent the development of the disease.

7.12 The development of a disease in the body

A disease develops if its micro-organisms succeeds in invading and overcoming the defences of the body.

However, you may have already noticed that the natural resistance of your body to disease can vary. When your body is in a low state of general health you are more likely to go down with a cold or some other complaint. Other variables are the number of micro-organisms that are trying to invade the body and their strength (*virulence*, see Table 7.1). The micro-organisms are more likely to succeed if they are large in number and if you are exposed to them for a long time.

Question 6 If there is an epidemic of influenza why do you think it is best to avoid crowded and poorly ventilated places?

After the person has been in contact with the source of the infection it is some time before signs of the disease begin to appear. This is the *period of incubation* and varies with each disease (see Table 7.7).

TABLE 7.7 **DETAILS OF SOME INFECTIOUS DISEASES**

Disease and incubation period	Signs	Preventive measures if thought necessary
Chicken-pox 11–19 days	Rash of raised pimples on trunk which spreads to limbs. Pimples form blisters filled with fluid. Scabs form and drop off after a week. Can scar if scratched and infected	
Diphtheria Within a week	Sore throat, fever. Greyish-white membrane starts to grow over the tonsils. This can cause serious breathing difficulties	Immunization by three injections (page 20) and boosters in childhood. Serum can be given in early stages
German measles 12–21 days	May be a fever. Pink spots cover body, limbs and face. Lymph glands in neck become swollen	A disease best caught in childhood, particularly for girls. Can harm a foetus in the first three months of pregnancy (page 24). Vaccination available to girls at thirteen years if they have not had the disease
Influenza 2 days	Fever, tiredness and depression. Three types: (a) respiratory with all appearances of a heavy cold, (b) nervous with backache, pains in the limbs, (c) gastric with vomiting and diarrhoea	A *pandemic* disease, which means that it spreads rapidly from one country to another. Vaccination is available for those at risk through the nature of their jobs, or when it might complicate another illness
Measles 9–16 days	At the start appears as a heavy cold, also causes discharge from the eyes. Early rash of white spots on an area of redness inside the cheeks. Then spots appear all over the body. Fever. Person is very infectious before the spots appear. Bright light can be uncomfortable to the eyes	Vaccination is available
Mumps 2–3 weeks	Sore throat, hard swelling develops under the ear. This spreads down and over the jawline. Very painful. Patient needs fluid food	Probably best caught in childhood, although a person can have it more than once. In adults it can cause complications. In men the testicles may become inflamed
Poliomyelitis 5–21 days	Fever, possibly vomiting. Tiredness and aching limbs. Paralysis then occurs to a varying degree	Vaccination is available

Table 7.7 continued on page 96

TABLE 7.7 **continued**

Disease and incubation period	Signs	Preventive measures if thought necessary
Rabies 20–90 days after bite	Irritability. Difficulty in swallowing, then breathing. Fear of water. Death in two to ten days. Extremely painful and distressing disease	Vaccination after bite if animal is suspected of being rabid. Quarantine of animals. Use of muzzles in countries where disease is present in wildlife
Scarlet fever 2–4 days	Sore throat, fever, vomiting. Rash of small red spots on a reddened skin, but never on the face. Tongue may be furred, then becomes spotted – 'strawberry tongue'. Skin then scales	This disease is no longer virulent. It can occur as just a mild rash with a throat infection
Tetanus 7 days after suffering a dirty wound	Muscles of face stiffen (lockjaw) and convulsions occur. Breathing becomes difficult. Very serious and painful disease. Death is common	Vaccination in early childhood. Booster doses are necessary, particularly if a person's job makes the disease more likely, e.g. farming, building trade. After a serious deep wound, serum may be given
Typhoid 3–20 days	Pains, tiredness and a cough. Tiny pink spots on abdomen. Diarrhoea	Vaccination if travelling to an area where the disease is present. Purification of communal water
Whooping cough 7–14 days	Heavy cold. Then coughing develops. Spasms of coughing with a child gasping for breath in between and producing characteristic 'whooping' sound. Vomiting also at these times	Very exhausting disease, so considered dangerous to children under two years old. Vaccination available but a doctor's advice must be taken if there is any family history of mental illness or fits

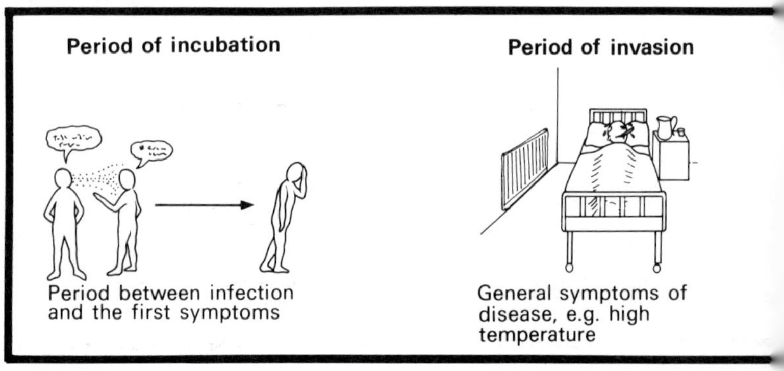

Fig. 7.15
The course of a disease

The other stages of a disease are illustrated in Fig. 7.15. During the *periods of invasion and advance* a doctor will use the patient's symptoms and all visible signs to help him diagnose the disease. It may be necessary for the patient to undergo a more specialized examination such as an X-ray or blood test to reveal the signs that the doctor is looking for.

An increase in the patient's temperature (*fever*) is usually an indication that the body's white cells are at work against invading micro-organisms. This increase in temperature stimulates other systems of the body into working at a faster rate (page 187). The pulse rate and the breathing rate increase so that the supplies and energy needed by the body's defence systems are readily available. The increases in temperature and in pulse rate (page 60) can be recorded on a chart. These recordings can give an accurate indication of the progress of the disease and can be compared with the patient's other symptoms (Fig. 7.15). Rest is essential if the body temperature is very high (page 188) and to help reduce the fever drugs may be prescribed. The person usually has little appetite but a lot of fluid must be drunk to replace that lost in sweating.

Gradually the signs and symptoms disappear, as the body 'wins the fight' and the disease *declines*. However, a period of time will be needed for the body to repair the damage done and return to good health. During this time of *convalescence* the body may be easily infected by other organisms.

Signs of a disease can also be regarded as a warning to others to stay away. This warning is often too late because a person can be releasing microbes to infect others several days before there are any signs of illness. Thus it is difficult to state exactly how long a person is infective. He will remain infective as long as he is releasing pathogens.

7.13 Isolation and quarantine regulations

Keeping a person suffering from a disease away from others, i.e. *isolating* or *quarantining* them, has been used throughout history as a way of preventing the disease from spreading through the community. We can read in the New Testament, St Luke 17:12, that Jesus was met by 'ten men that were lepers, which stood afar off'. In January 1374, at Reggio in Italy, Visconte Bernabo declared the first plague regulation, part of which states, 'Everyone sick of the plague is to be brought out of the town to the fields, there to die or recover. Those who have nursed plague patients are to remain secluded for ten days.'

The town of Marseilles in 1720 built its own plague hospital and enforced quarantine regulations on passengers that arrived on ships from plague-ridden areas. For twenty-one days the passengers remained isolated either on board the ship or in the hospital.

Nowadays communities protect themselves on both a local and a world-wide level by similar regulations.

In this country there is a list of *notifiable diseases*, for example diphtheria and whooping cough. A doctor must report such cases to the District Community Physician. In cases of notifiable diseases such as smallpox and polio, the District Community Physician and other authorities can decide what action should be taken to prevent further transmission. This can involve ensuring isolation of all cases, tracing all contacts and vaccinating people most at risk. This task may be harder if the source of the infection is a *carrier*, i.e. a person who releases the pathogens but shows no signs of the disease.

A person can be prosecuted for ignoring isolation regulations. The periods of isolation from school for children with notifiable diseases are laid down in the 1944 Education Act.

Some notifiable diseases, for example smallpox and yellow fever, have to be reported to the World Health Organization. Such international co-operation is essential nowadays, as so many people travel from one country to another by aeroplane with increasing frequency and speed. Thus contacts can be traced across several countries and quarantine regulations quickly imposed. If a country is declared an infected zone other countries can demand that visitors coming from that country are vaccinated.

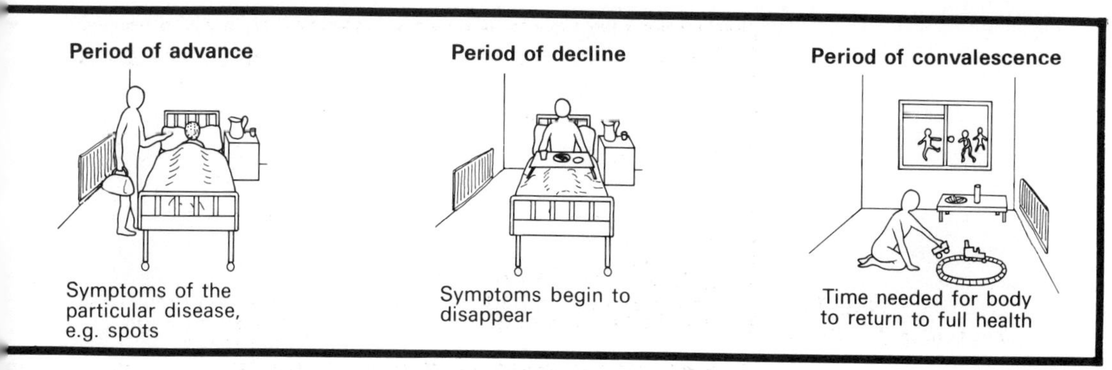

Question 7 A town abroad suddenly has a large number of typhoid cases. This is given as a news item and requests are made for the following two groups of people to visit their doctors: people who have recently visited the town, and people who are going to that area. Explain why this action is necessary.

Many countries require travellers coming from an area that has an *endemic* disease (i.e. a disease that is always present in a proportion of the population) to possess a current certificate of vaccination against that disease. Without such a certificate (Fig. 7.16) entry to the country will be refused. Travellers can find out what certificates of vaccination are needed from the information supplied with a passport, from travel agents or from the embassies of countries to be visited.

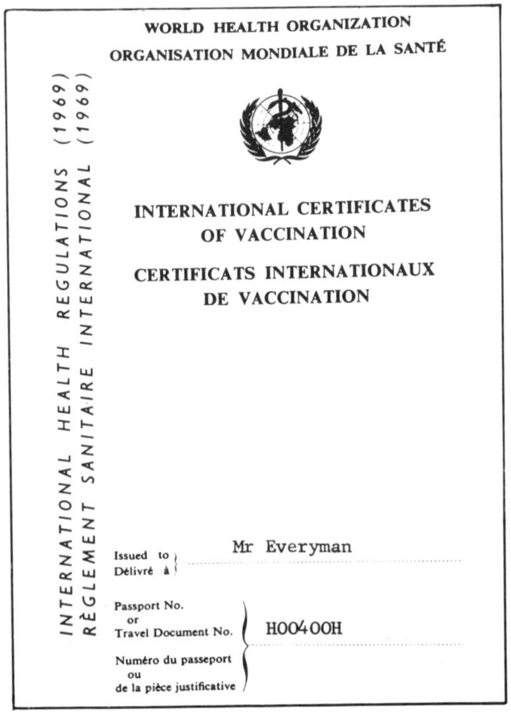

WORLD HEALTH ORGANIZATION
ORGANISATION MONDIALE DE LA SANTÉ

INTERNATIONAL HEALTH REGULATIONS (1969)
RÈGLEMENT SANITAIRE INTERNATIONAL (1969)

INTERNATIONAL CERTIFICATES OF VACCINATION
CERTIFICATS INTERNATIONAUX DE VACCINATION

Issued to / Délivré à: Mr Everyman

Passport No. or Travel Document No. / Numéro du passeport ou de la pièce justificative: H00400H

Fig. 7.16 *A vaccination certificate*

Fig. 7.17 (Left) *An anti-rabies poster*

Many countries control the entry of possible vectors of a disease and impose quarantine regulations. Strict quarantine regulations have so far kept Britain free of rabies (Fig. 7.17). People who break these regulations are prosecuted. To some people prosecution procedures may seem harsh and unnecessary; on the other hand others may demand stricter controls and heavier penalties. This type of conflict is always present in a community that wishes its individuals to have personal freedom yet seeks to protect them from the inconsiderate action of others.

7.14 Other organisms that cause disease and non-infectious diseases

So far we have looked at pathogens which are microscopic in size. However, there are other, larger organisms which are parasitic in man. These are species of worms: roundworms, flatworms and tapeworms. Generally a person becomes infected by food and water contaminated with these organisms: they are discussed in detail in Chapters 8 and 10, which deal with the preparation of uncontaminated food and water.

Not all diseases can be transmitted from one person to another. We saw this in the previous chapters when we indicated how the body systems can fail to function correctly. A machine can break down because of a mechanical fault which may have been there from the start or may have developed through years of use or even because of incorrect use. The body is very similar and we have already discussed how we can use our bodies correctly to avoid disorders through misuse. Table 7.8 lists types of non-infectious disease with some examples.

TABLE 7.8 **NON-INFECTIOUS DISEASES**

Type	Description and examples
Metabolic disorders	Defects that prevent normal working of a body system. The defect might be inherited Examples: haemophilia, diabetes, spina bifida
Nutritional or deficiency diseases	These arise when the body is starved of some essential substance or volume of food Example: shortage of vitamin C produces scurvy (page 119)
Degenerative diseases	These arise as organs and tissues 'wear' and work less efficiently. Such diseases can be advanced by neglect Examples: defects in seeing and hearing, rheumatism, heart attacks
Allergies	Reaction of the body to a substance which it regards as harmful, though the substance is usually harmless Examples: hay fever is the body's allergic response to pollen (page 47)
Cancers	These occur when cell division and cell activities become abnormal (page 10) Examples: cervical cancer, cancer of the stomach and lower bowel, leukaemia
Functional diseases	A disease may arise for which no physical origin can be found. The origin could be psychological rather than physical
Mental illness	This term is used to cover a wide range of disorders from slight depression to more severe mental disorders arising from brain damage, genetic and psychological factors

CERTIFIED COPY OF AN ENTRY OF BIRTH

GIVEN AT THE GENERAL REGISTER OFFICE, LONDON.

Application Number..

Form A 502 (S.358835) Dd. 437122 40M 8/76 Hw. 040-96

REGISTRATION DISTRICT

_____. BIRTH in the Sub-district of__________________in the__________________

Columns:—	1	2	3	4	5	6	7	8	9	10*
No.	When and where born	Name, if any	Sex	Name, and surname of father	Name, surname, and maiden surname of mother	Occupation of father	Signature, description and residence of informant	When registered	Signature of registrar	Name entered after registration
	CANCELLED			SPECIMEN		CANCELLED			CANCELLED	

CERTIFIED to be a true copy of an entry in the certified copy of a Register of Births in the District above mentioned.

*See note overleaf.

Given at the GENERAL REGISTER OFFICE, LONDON, under the Seal of the said Office, the day of 19 .

BC 000000

This certificate is issued in pursuance of the Births and Deaths Registration Act 1953.
Section 34 provides that any certified copy of an entry purporting to be sealed or stamped with the seal of the General Register Office shall be received as evidence of the birth or death to which it relates without any further or other proof of the entry, and no certified copy purporting to have been given in the said Office shall be of any force or effect unless it is sealed or stamped as aforesaid.
CAUTION:—Any person who (1) falsifies any of the particulars on this certificate, or (2) uses a falsified certificate, as true, knowing it to be false, is liable to prosecution.

CERTIFIED COPY OF AN ENTRY OF DEATH

Given at the GENERAL REGISTER OFFICE, LONDON.

Application Number..

REGISTRATION DISTRICT

_____. DEATH in the Sub-district of__________________in the__________________

Columns :—	1	2	3	4	5	6	7	8	9
No.	When and where died	Name and surname	Sex	Age	Occupation	Cause of death	Signature, description, and residence of informant	When registered	Signature of registrar
		SPECIMEN			CANCELLED				

CERTIFIED to be a true copy of an entry in the certified copy of a Register of Deaths in the District above mentioned.

Given at the GENERAL REGISTER OFFICE, LONDON, under the Seal of the said Office, the day of 19 .

DA 000000

This certificate is issued in pursuance of the Births and Deaths Registration Act 1953.
Section 34 provides that any certified copy of an entry purporting to be sealed or stamped with the seal of the General Register Office shall be received as evidence of the birth or death to which it relates without any further or other proof of the entry, and no certified copy purporting to have been given in the said Office shall be of any force or effect unless it is sealed or stamped as aforesaid.
CAUTION:—Any person who (1) falsifies any of the particulars on this certificate, or (2) uses a falsified certificate as true, knowing it to be false, is liable to prosecution.

Fig. 7.18 *Specimen birth and death certificates*

TABLE 7.9 DEATH STATISTICS

Disease	Deaths per million of the population				
	1900	1922	1940	1960	1980
Cancer	829	1229	1764	2158	2651
Diphtheria	290	107	62	0	0
Diabetes	86	119	134	78	97
Influenza	504	563	286	24	10
Measles	394	149	21	1	1
Rickets	40	11	1	0	0
T.B.	1902	119	134	78	12
Whooping cough	356	167	17	1	0

7.15 **Has the pattern of disease changed over the years?**

It is possible for us to look back at the pattern of disease because of the records that have been accumulated over approximately 150 years. The information for these records is drawn from a number of sources: two examples, a birth and a death certificate, are shown in Fig. 7.18. Death certificates record the cause of death and are used to prepare figures or *statistics* on the nature of deaths. Table 7.9 gives such figures for a period of years. If you study these figures closely it is possible to see the changes which have taken place.

Question 8 How have the numbers of deaths caused by infectious diseases changed over the period shown in Table 7.9? Can you give reasons for this change?

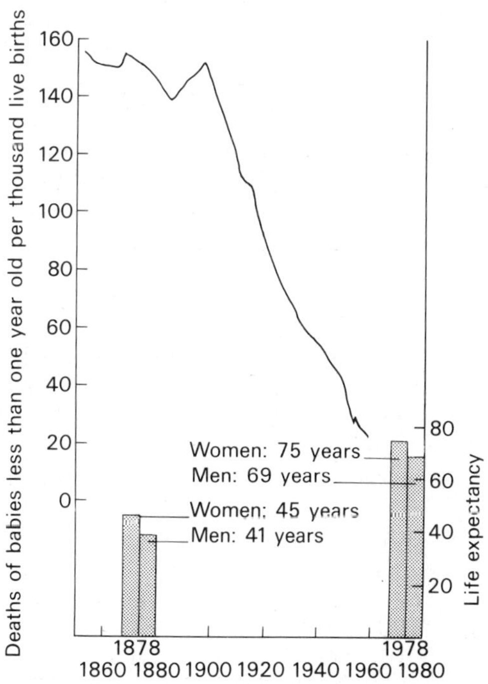

Fig. 7.19 *The drop in infant mortality and the rise in life expectancy*

Table 7.9 indicates which fatal diseases are on the increase, for instance cancer; diseases of the circulatory system are also on the increase. These diseases are more concerned with the malfunctioning of organs and tissues. Other disorders, for example arthritis, rheumatism and mental disorders, are not fatal but are experienced by many people. May one of the reasons for this change from infectious to other types of disease be that our expected lifespan is now much longer (Fig. 7.19)?

Some of the fatal diseases seen nowadays are often linked with the stresses of the modern way of life. However, what is also becoming clear is that with our present understanding of disease we are now more in control of what happens to us. As individuals we can make decisions which directly affect our physical and mental health. This involves us in thinking about our attitudes and our responsibilities towards our own well-being and that of others.

Answers to questions in the text

1 (a) Wash your hands.
(b) Wash your hands.
(c) Throw it away: cracks harbour germs
(d) Hold your hand or handkerchief against your mouth.
(e) Set traps, or ask your local Area Health Authority to send round their pest controller, who will put down poison. There is no charge for his services.

2 (a) Surgeons and assistants wash down before entering the operating theatre.
(b) Gowns, footwear, head coverings, gloves and masks are sterilised.
(c) Masks prevent droplet infection
(d) Gloves prevent infection by direct contact
(e) All instruments are sterilised
(f) All dressings are sterilised
(g) Operating theatres are constructed so that surface areas can be easily cleaned and disinfected.
(h) In many theatres the air may be filtered and sterilised.

3 (a) The wound should be freed of all dirt by washing.
(b) A clean dressing should be applied and changed regularly.

4 Parents can explain that substances other than food can harm (poison) the body. If a child does not recognize a substance as a food he has met before he should ask an adult if it is safe to eat.

5 The invading micro-organisms have managed to overcome the natural defences of the body, and are present in large enough numbers to produce signs of the disease.

6 Crowds of people mean more sources of infection. If the area is poorly ventilated there will be a high concentration of micro-organisms. When people are close together transmission by droplet infection is easy.

7 (a) A person who has visited the town may develop the disease, so a doctor will look for any signs and warn the person of the symptoms. Isolation may be necessary to prevent the disease being spread; certainly that person must not be involved in food preparation for other people.
(b) People will be advised not to travel to the town. If vaccination is possible, the decision as to whether it is used will depend upon the time needed for a course of injections.

8 The number of deaths due to infectious diseases such as diptheria have decreased dramatically. This decrease has been largely due to the success in the fight against disease by the methods already described in this chapter. Better housing, sanitation and education in health have also played a vital role.

THINGS TO DO

1 Find out about the life and work of Louis Pasteur.

2 Carry out a survey of all the diseases which you, your family and your friends have suffered from. Record the numbers in the form of a bar graph.

3 Find out about the foundation and the work of the World Health Organization. Make a scrapbook on disease out of newspaper articles.

4 Visit your local chemist and, with permission, carry out a survey of the 'treatments' which can be purchased and which do not require a doctor's prescription.

5 Use a first-aid book to find out how you should clean and dress a wound.

6 Visit your local churchyard and make a record of the deaths. Record the date of death and the age. By grouping certain ages together (for example the under fives) you can illustrate your findings in the form of a graph. Can you relate your figures to your knowledge of which diseases were prevalent during these years?

7 Carry out a survey among your friends and family of the vaccinations they have received. How many of them have a tetanus booster regularly?

8 When you next suffer from a cold or flu record your temperature regularly in the morning and at night. Display your results in the form of a graph. Does the graph of temperature indicate the various stages of the disease?

9 Carry out a survey to find out how many families observe the medicine code. You can construct your questionnaire from the code given on page 80. For example, one question might be 'Where are medicines kept in your home?'

THINGS TO WORK OUT

In Questions 1–5 write down the *one* answer which you think is correct.

1 Conditions that are free of germs are
- **A** fumigated
- **B** disinfected
- **C** dehydrated
- **D** antiseptic
- **E** aseptic

2 In order to see a virus clearly a person would need
- **A** an electron microscope
- **B** a microscope with a high-power objective lens
- **C** a binocular microscope
- **D** a magnifying glass
- **E** spectacles

3 The first person to develop a successful method of giving a person artificial immunity was
- **A** Edward Jenner
- **B** Alexander Fleming
- **C** Louis Pasteur
- **D** Robert Koch
- **E** Joseph Lister

14 The time between contact with an infection and the onset of symptoms is known as the period of
- **A** isolation
- **B** incubation
- **C** invasion
- **D** advance
- **E** convalescence

5 The lymphatic system includes the
- **A** kidneys
- **B** tonsils
- **C** lungs
- **D** pituitary gland
- **E** sweat glands

6 (a) Explain the terms (i) dependence, (ii) withdrawal symptoms.
(b) Name a drug that is obtained from the poppy. What type of drug is this and what is its effect on the body?

7 (a) What is an antibiotic?
(b) Give *two* possible types of vaccine and in each case *one* disease that it prevents.
(c) Distinguish between passive immunity and active immunity.

8 (a) What is meant by aseptic surgery?
(b) Describe *three* procedures within an operating-theatre that ensure aseptic conditions around the patient.
(c) Briefly describe the work of one of the pioneers of aseptic surgery.

9 Write each word in the second column next to the correct word in the first column.

(a) antibiotic	Dettol
(b) antiseptic	bleach
(c) disinfectant	aspirin
(d) analgesic	caffeine
(e) stimulant	penicillin
(f) sedative	L.S.D.
(g) hallucinogen	alcohol

10 Explain what is meant by
(a) quarantine
(b) a carrier
(c) contacts
(d) a resistant strain of bacteria
(e) a notifiable disease
(f) a broad-spectrum drug

11 (a) Why cannot virus colonies be cultured on agar jelly in the same way as bacteria?
(b) Give *two* diseases of man that are caused by (i) bacteria, (ii) viruses.
(c) Choose *two* virus diseases and for each one describe (i) symptoms, (ii) *one* method of transmission, (iii) *one* method of preventing transmission.

12 (a) State *two* ways in which a person may misuse drugs.
(b) Describe in detail *two* dangers that can arise from the misuse of drugs.
(c) Explain why you think it is important that there should be laws that control (i) the manufacture of drugs, (ii) the sale of drugs to the public.
(d) Imagine you are a patient. Give *three* ways in which you would try to ensure that medicines are used correctly in your family.

13 (a) Explain why in Britain there has been a decline in cases of (i) diphtheria, (ii) typhoid.
(b) Name *two* diseases that are at present on the increase. Give possible explanations for this.
(c) What regulations exist to control the entry of dogs into Britain? Why are these regulations necessary?
(d) How may cases of smallpox still occur in a country where it has been eradicated?

14 (a) Describe *four* ways in which diseases may be transmitted from one person to another. In each case name *one* disease that may be transmitted by that particular method and a way of preventing transmission.
(b) A colony of bacteria was cultured on agar. Three different antibiotics were selected and a drop from each was placed at different positions on the colony. What results would indicate the effectiveness of the antibiotics? Would you be able to compare their effectiveness? (A diagram may be used.)

8 Man's Parasites

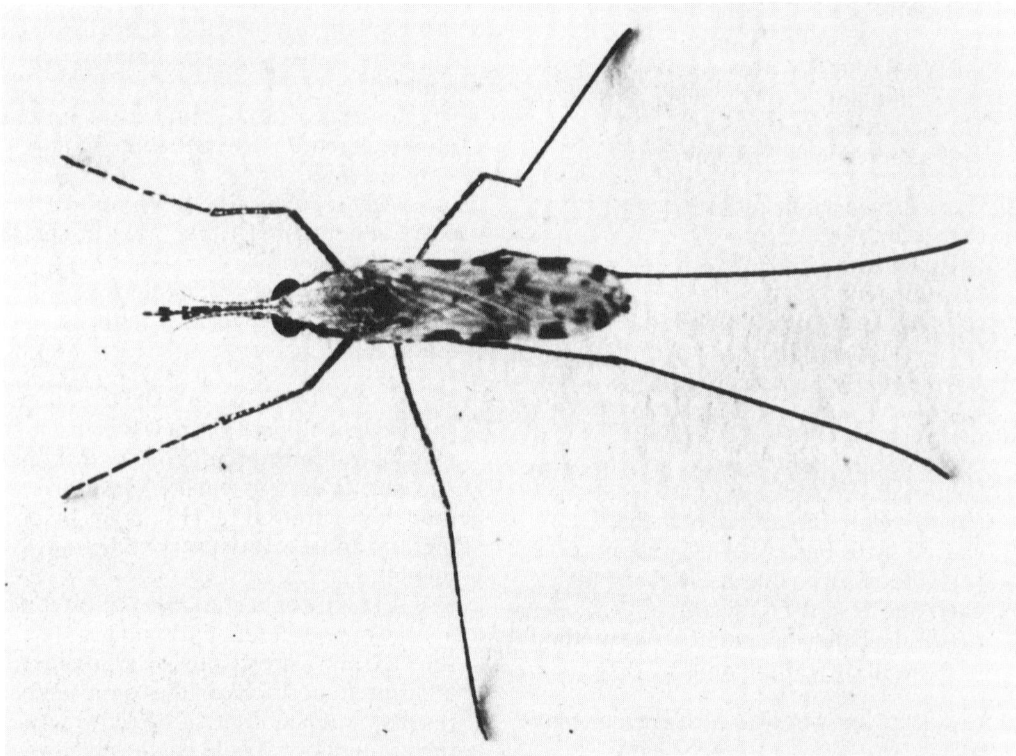

Fig. 8.1 *An adult mosquito (*Anopheles sp.*)*

8.1 Bloodsucking insects – mosquitoes

In the summer months most of us are aware that our bodies are very attractive to small flying insects which we call gnats or midges. We even smear ointments, known as insect repellents, on ourselves in the hope of driving them away. Gnats are members of the mosquito family. An adult mosquito can be seen in Fig. 8.1. Mosquitoes generally feed on plant juices, but the fertilized female must feed on blood before she can produce eggs. In order to pierce flesh the female mosquito has special mouth parts called *stylets*. These can pierce the skin and reach a blood-vessel. Saliva is then produced in the salivary glands of the insect and is passed down the mouth parts into the blood of the victim. A substance present in the saliva prevents the blood from clotting. The blood is then sucked up through the mouth parts and enters the insect's stomach. We call this puncturing of the skin a bite and there is usually a local reaction to the wound which produces redness, swelling and often itching. Scratching to relieve the itching may introduce dirt into the wound and cause a further infection.

In this country it is only the irritation of a mosquito bite that can worry a person. However, in tropical countries such bites can have serious consequences, because some members of the mosquito family carry micro-organisms which cause diseases such as malaria and elephantiasis. These micro-organisms are called *parasites*.

8.2 The malarial parasite

A parasite is any organism which lives on or inside another living organism, taking food from it and giving nothing in return. The living organism which houses the parasite is called the *host*. The host will always suffer in some way and in severe cases may even die. A parasite often has a complex life cycle which depends upon two hosts. This means that a different period of the life cycle is spent in each host. If the parasite is to survive it must be able to transfer from one host to the other. If we are one of these hosts we often refer to the second host as the *carrier* or *vector* of the disease. This is the case with the malarial parasite. The carriers of malaria are the species of mosquitoes which belong to the *Anopheles* branch of the mosquito family.

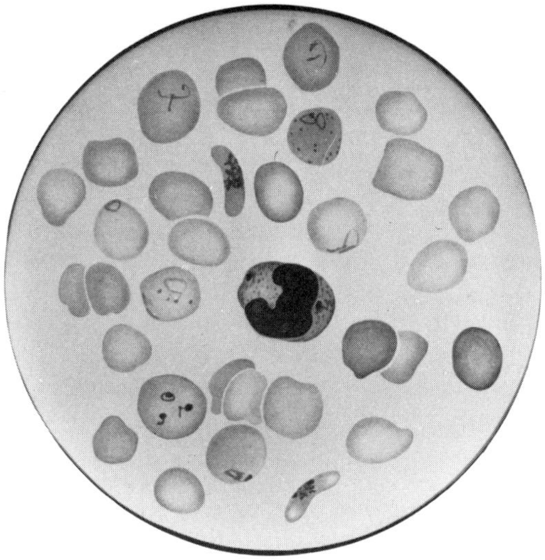

Fig. 8.2 *Malarial parasite in the blood*

The malarial parasite as it appears among human red blood cells is shown in Fig. 8.2. This parasite, which is a protozoon (page 81), invades the red blood cells, growing and dividing inside them. The red cells are destroyed and eventually burst, releasing a greater number of the parasites. These invade more cells and repeat their multiplication process until thousands of parasites are present in the blood. The bursting of the red blood cells seems to produce the attacks of malaria, which consist of shivering, then fever and sweating. These attacks come and go.

The malarial parasite is transferred to the *Anopheles* mosquito when the insect bites and sucks the blood of a person with the disease. In the mosquito the parasite completes the stages in its life cycle. It is then transferred to another person in the saliva of the biting mosquito.

8.3 Where malaria occurs and its control

There are few places in the world that have not experienced outbreaks of malaria. However, in some areas of the world, especially tropical and subtropical regions, it is always present. In such areas we say that malaria is an *endemic* disease. There are a number of factors which encourage it to be endemic, two examples are given here.

First, a mosquito picks up the malarial parasite by sucking the blood of a person who is already suffering from the disease. If a large number of people in the region have malaria, then a large number of mosquitoes will become carriers of the malarial micro-organism and will be able to *transmit* it to other people.

Second, the region's climate and environment may provide the right conditions for the *Anopheles* mosquito to live and breed in large numbers (Fig. 8.3).

Fig. 8.3 *Conditions favourable to the breeding of the* Anopheles *mosquito in Afghanistan (water and a humid climate)*

Fig. 8.4 *Life cycle of the* Anopheles *mosquito*

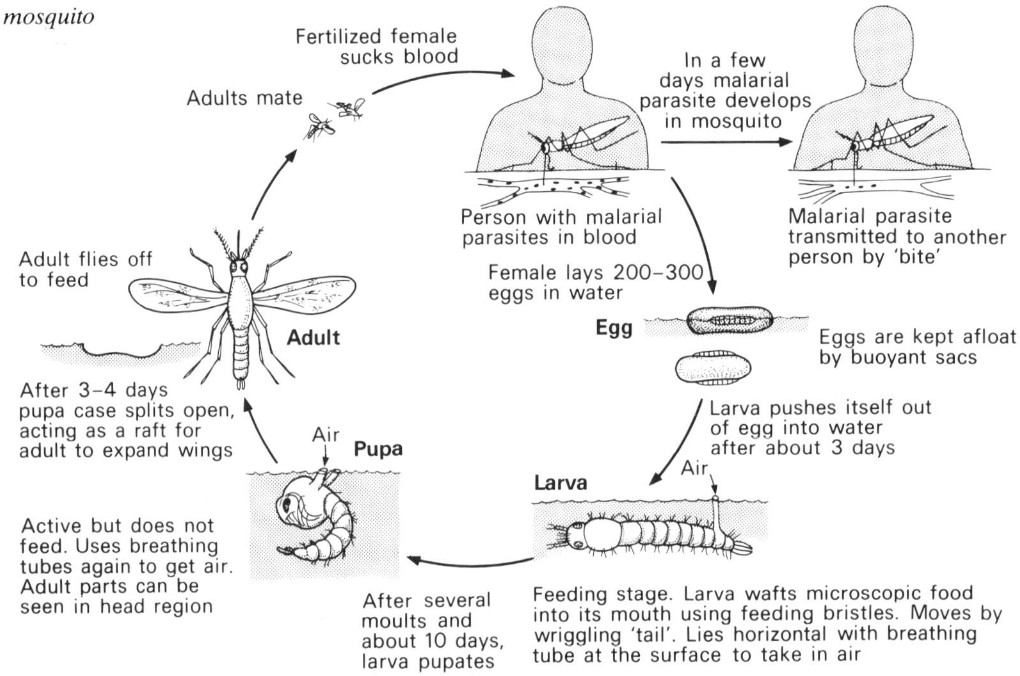

Continual attacks of malaria cause a person to become weak, anaemic and lethargic (unable to work hard or for long periods). In severe cases it may cause death or advance it by lowering the person's general level of health. Since early times it has been a serious disease of man, and methods of controlling it are only a recent development. Between 1880 and 1900 the separate investigations of the scientists Laveran, Ross and Grassi revealed (a) the parasite, (b) the life history of the *Anopheles* mosquito and (c) how the mosquito carries the parasite (Fig. 8.4). This knowledge gave man two main methods of controlling malaria.

The first is by destroying the parasite while it is in the human body. This can be done by the use of antimalarial drugs; the earliest one to be used was quinine. Nowadays travellers to malarial countries can be given a course of a drug which will protect them against infection. A drug which protects us in advance in this way is called a *prophylactic* drug.

The second method is by eradicating the *Anopheles* mosquito. This is not so easy and the problem is so complicated that only an outline of some of the methods used up to now may be given here. All depend upon a knowledge of the life cycle of the mosquito and its habits (Fig. 8.4). Eradication of the mosquito can involve:

(a) spraying buildings, cars, aeroplanes, etc. with a chemical insecticide (Fig. 8.5).
(b) spraying the water in waterlogged areas with chemicals or oil (Fig. 8.3)
(c) stocking water with fish
(d) draining waterlogged areas (Fig. 8.6)

Fig. 8.5 *A spraying team at work in Malaysia*

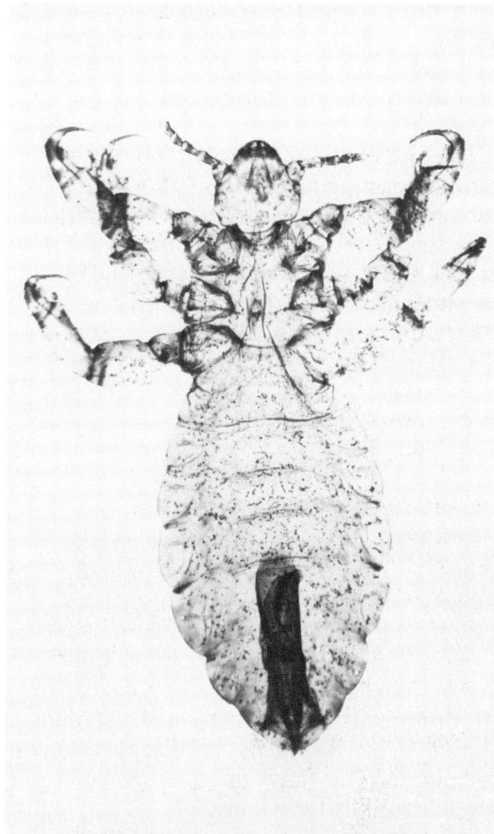

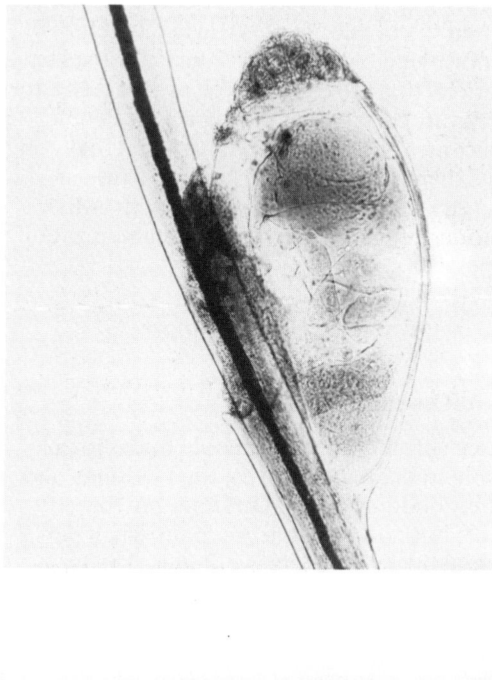

Fig. 8.6 *Draining and irrigation provide rich agricultural land and remove the swamps favoured by the mosquito (Afghanistan)*

Fig. 8.7 (Above) *Head louse.* (Below) *The egg or 'nit' of the head louse which the adult glues to strands of hair*

Question 1 Look again at Fig. 8.4 and state which stage of the mosquito the control methods (a), (b), (c) and (d) above intend to destroy.

8.4 Bloodsucking insects – lice

Lice are external parasites. They live on the outer surface of their hosts and feed on blood by using their mouth parts in the same way as the mosquito. Fig. 8.7 shows clearly how the adult louse found on man has become adapted to its way of life. It is wingless. It has a flattened body so that it can move around close to the skin and not be caught. It has curved claws on the ends of its six legs, which are used for holding on to hairs.

Like other parasites, lice have specific hosts. In fact there are only two types of lice which are parasites of man. These are the human body louse (*Pediculus humanus*) and the crab-louse (*Phthirus pubis*). Human body lice can be subdivided into two specialized forms: one which lives on the body (*Pediculus humanus corporis*) and the other on the head (*Pediculus humanus capitis*). The crab-louse lives on the hair of the lower body around the sex organs.

The life cycles of these lice are very similar and Fig. 8.8 illustrates that of the human body louse. You can see from this diagram that lice live on man throughout their life cycle.

The bite of a louse causes a red, raised, irrritating bump on the skin. This in itself is not serious, but like a mosquito bite there are secondary effects. Irritating bites, particularly at night, cause loss of sleep. The bites may become infected, producing sores and other skin infections. There is a general loss of good health.

8.5 **Diseases transmitted by lice**

Lice can spread serious diseases such as typhus and trench fever. These diseases, like malaria, are caused by micro-organisms (rickettsiae, which have a structure like bacteria but, like viruses, can grow only within living cells). The lice become carriers after biting an infected person. The typhus carried by lice is *epidemic*. A disease is said to be epidemic if it flares up suddenly and spreads rapidly for a limited time. Epidemic typhus is a severe disease causing many deaths. It is most likely to occur when large numbers of people are crowded together in poor living conditions, for instance the conditions that are experienced by refugees after a natural disaster or during a war. People become infected with typhus when they rub the droppings from the lice into a bite or scratch. Treatment is possible with antibiotics (page 86) and through immunization (page 93).

8.6 **The control of lice**

Lice spread wherever people find it impossible to keep themselves, their clothing and their bedding clean. Dirt does not 'breed' lice: both accumulate gradually when personal hygiene becomes difficult. The life cycle (Fig. 8.8) of the louse helps to explain how this happens. Washing our bodies, our clothes and our bedding removes any adult lice that may have been picked up.

Question 2 Why should this washing be done at short and regular intervals? (Clue: look at the life cycle.)

Nowadays better standards of living and greater public awareness about personal hygiene have made the body louse less common.

There are still cases of head lice, and one of the regular health inspections carried out in schools is to look for this parasite. Ordinary combing, brushing and washing of the hair can fail to remove lice eggs (nits) which have become 'cemented' to strands of hair. In this way the lice become established. However, combing the hair with a special fine-toothed steel comb can be more effective in removing lice and eggs. This should be done at least once a week.

If a person does become infested with lice, he should seek advice from a doctor or clinic. Various creams and liquids can be supplied which contain chemicals capable of destroying all stages of the lice. Advice will also be given on how to deal with clothing and bedding that may have become infested.

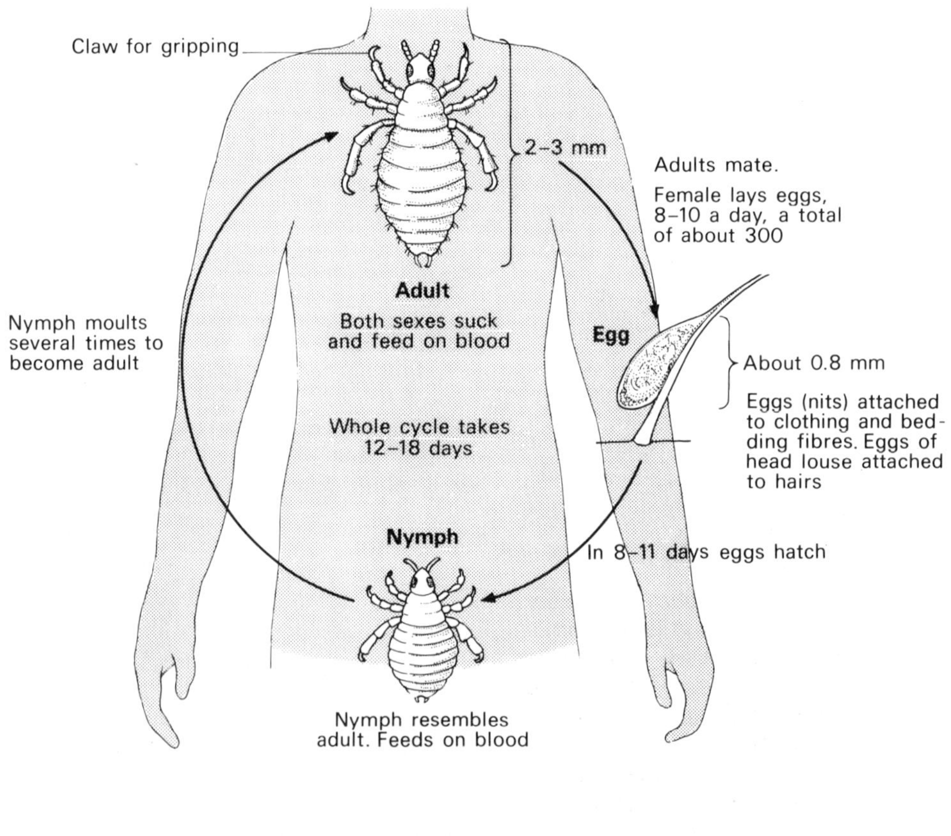

Fig. 8.8 *Life cycle of the body louse*

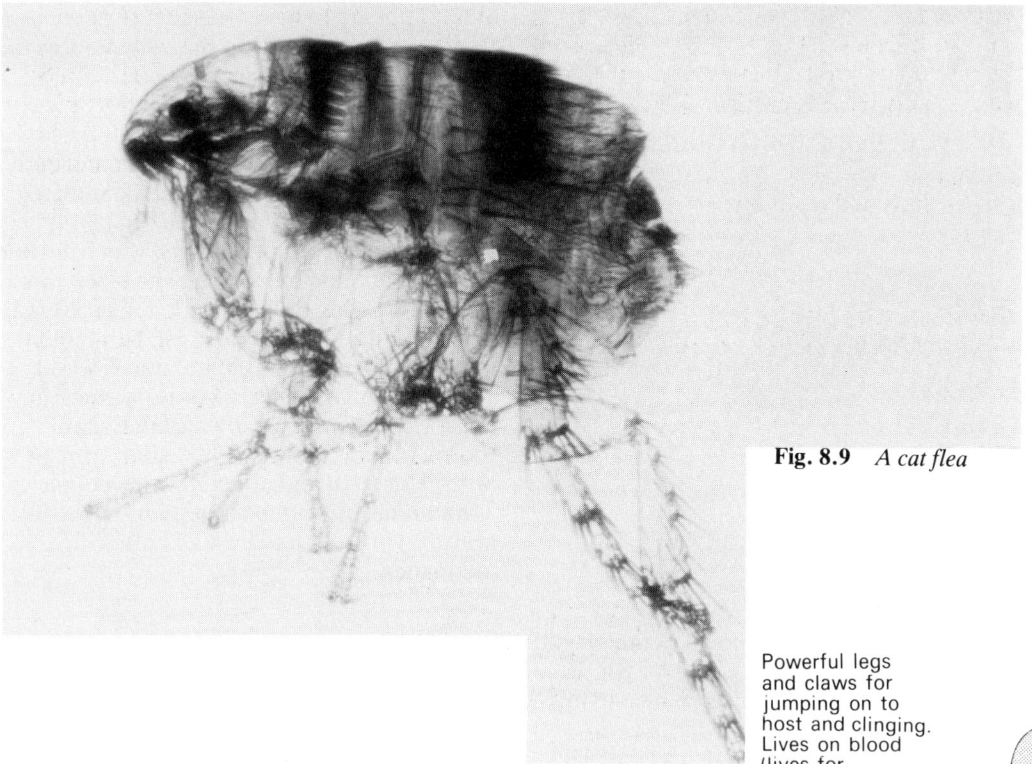

Fig. 8.9 *A cat flea*

Adult fleas feed only on blood. But whereas a louse dies in about two days without blood, a flea can live for 125 days without it. A flea-bite produces a red irritating spot, round which may develop a painful swelling about the size of a 10p piece.

8.8 **The control of fleas**

In the past infestation by human fleas was mainly due to poor social conditions such as overcrowding, poor housing and lower standards of hygiene. Today social conditions have changed but fleas can still be found, as

8.7 **Bloodsucking insects – fleas**

The structure of a typical adult flea can be seen in Fig. 8.9. Like the louse its body is flattened, but in the opposite direction, i,e, from side to side. The legs of the flea are powerfully built for jumping, and there are claws for hanging on once it arrives. Fleas which infest man are capable of jumping a distance of 32 cm.

It is only the adult flea which lives on its host. The rest of its life cycle (Fig. 8.10) is spent in the surroundings of the host, for example in carpets and cracks in furniture. Also the adult flea is not so fussy as the louse about its host. One type of flea can be found on many different hosts. For example, the type which is most commonly found on man, *Pulex irritans*, is also found on dogs, rats, mice, cattle and pigs. Other types of flea common to rats, hens, dogs and cats will also use man as their host.

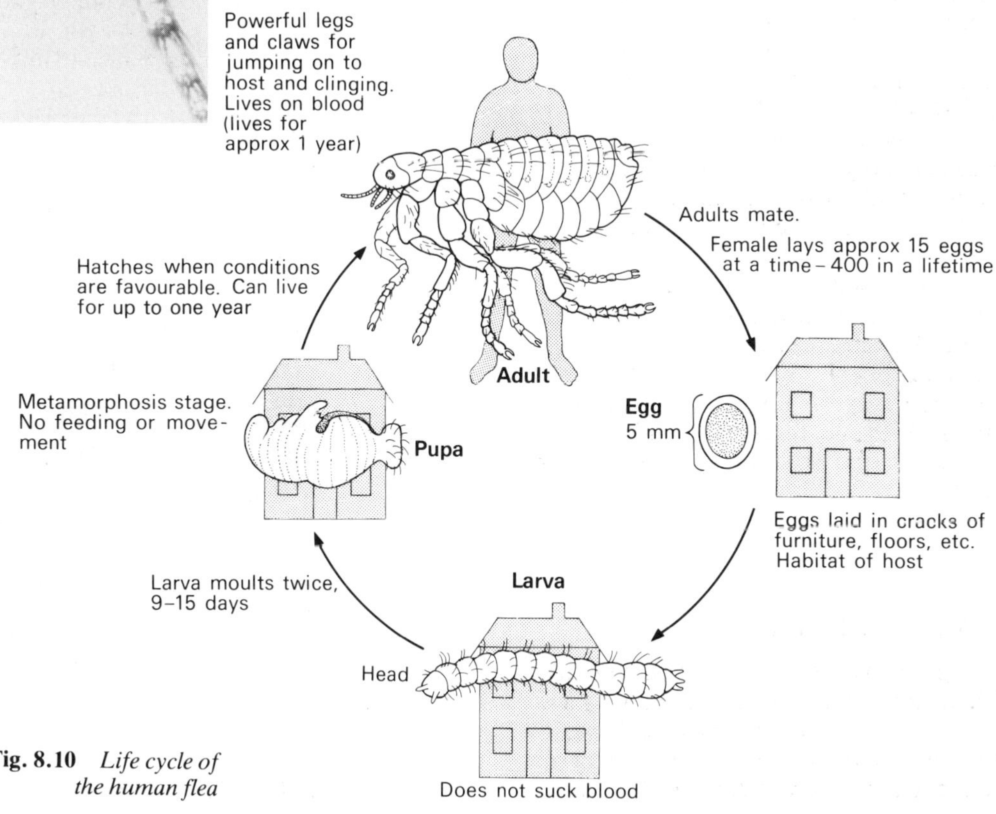

Fig. 8.10 *Life cycle of the human flea*

TABLE 8.1 FLEA INFESTATION: SURVEY OF PROPERTIES 1974

Houses and flats	910
Manufacturing and storage:	
Food	49
Chemical Industry and transport	447
Accommodation: Boarding houses, holiday camps, hotels, hostels	116
Public meeting places: Cinemas, theatres, clubs, cafes, etc.	158
Schools, colleges and convents	31
Hospitals, convalescent homes	119
Shops	105
Offices and laboratories	91
Animal locations: Zoos, farms, poultry houses, kennels, etc.	29
Miscellaneous	56
Total of reported infestations	2111

shown in the figures from a recent survey (Table 8.1). Other figures indicate the culprits. Eighty per cent of flea infestations are of cat fleas, 20 per cent are of bird fleas, and only 2 per cent are of human fleas (rodent and dog fleas are rare). Therefore people suffering from flea-bites often find that the culprit has come from a household pet. Fleas are regarded as a minor pest of man; if they are to remain so we need to be aware of the source of the infestation and how to control it. For example, regular 'dusting' of the pet with anti-flea powder (Fig. 8.11) may be needed.

8.9 Diseases transmitted by fleas

Like other bloodsucking insects fleas can carry micro-organisms that cause disease and transmit infection from one host to another. Endemic (murine) typhus is carried by fleas from rats to man. It occurs when people live in similar conditions to those in the trenches during the First World War.

Fig. 8.11 *Part of an advertisement for a method of controlling cat fleas*

In your history lessons you may have learned about a disease known as the plague and sometimes called the Black Death. It occurred on its largest scale in England in 1348–9. Bubonic plague is a disease of certain species of rodents, such as the black rat (*Rattus rattus*). The plague micro-organism, which is a bacillus, is carried by fleas from the black rat to man. Plague is an endemic disease in certain areas of the world.

Fig. 8.12 *A black rat, the carrier of bubonic plague*

What appears to have helped it to become a world-wide epidemic on the scale seen in the Black Death was the sudden migration of plague-infested rodents, particularly the black rat (Fig. 8.12).

By the late seventeenth century and early eighteenth century in England the infected blacks rats began to be driven out by the larger and stronger brown rats, which did not become infected by plague. The black rats were also losing their struggle for survival as brick and stone began to be used for building instead of their favourite habitat of wood, clay and thatch. Thus bubonic plague is an exceptionally rare occurrence in Britain today. It appeared for a very short time in Suffolk in 1910 and affected a few people. Treatment for plague and typhus is possible nowadays through the use of antibiotic medicines.

8.10 Other insects that are body parasites

A disease that is seen in this country is *scabies*. This appears as a red rash in and around the armpits, around the waist, between the thighs and on the ankles. The rash is extremely itchy and when it is scratched the spots may release a watery fluid. Scabies is caused by a species of mite (*Sarcoptes scabiei*) which is shown in Fig 8.13. The female mite burrows under the skin to lay her eggs. The burrows are often visible under the skin. After five days the eggs hatch, releasing larvae. These extend the burrows and feed on the blood of their host for about six days. Then the larvae change into nymphs, which in two days become adults, so completing the life cycle. Scabies can be transmitted by direct contact with an infected person or by indirect contact through clothing or bedding. Any suspected case of scabies should be seen by a doctor, who will prescribe effective treatment, usually for the whole family. Mange in animals is caused by a similar mite, which can transfer to people under certain conditions. Therefore if your pet has a skin infection you should consult the vet! If a person develops a

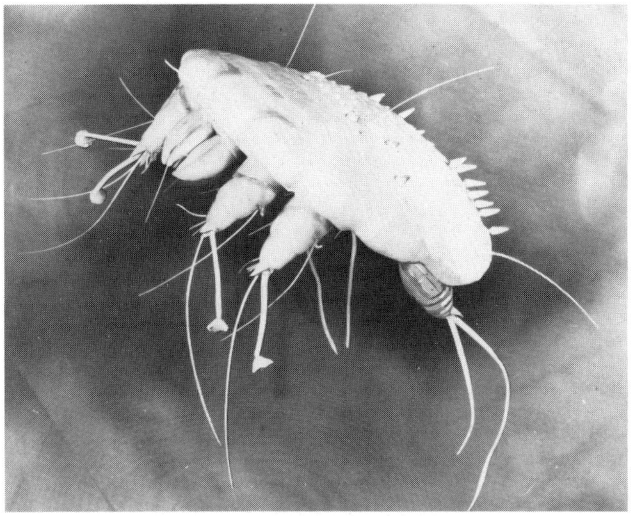

Fig. 8.13 *The mite which causes the skin rash scabies*

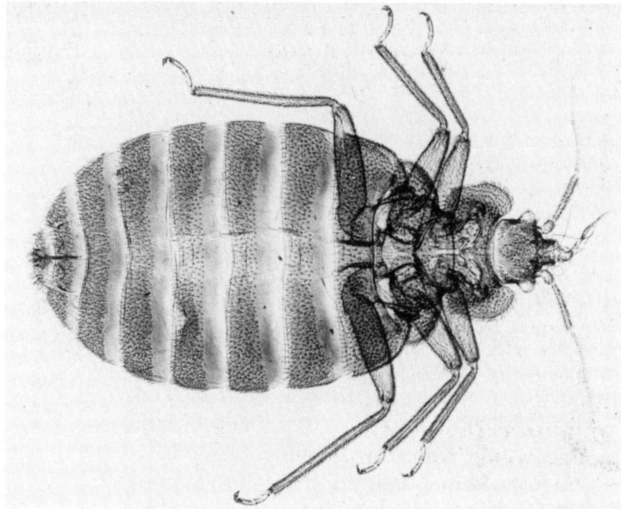

Fig. 8.14 *Bedbug*

skin infection, the doctor may very well ask if any pets are kept by the family. Skin infections like scabies can be infected by bacteria when scratched. Mites also can act as vectors for diseases such as typhus.

Many species of mite are not parasitic. Examples of these invade our crops and cereal foods such as flour. However, all species of *ticks* are parasites and feed upon blood sucked from their host. Some of you may have seen the female hard-bodied tick that feeds on sheep. This tick swells up and then drops off the sheep, gorged with its blood. Other species of tick in various parts of the world feed on people. Their bites can become infected and they also act as vectors for many serious diseases, for example relapsing fever.

Bedbugs (*Cimex lectularius*; see Fig. 8.14) come out from the cracks and crevices in rooms where they have spent the day to feed on a person's blood at night. Man is the bedbug's main host, although it will feed on other mammals and on birds. The female bedbug 'sticks' her eggs to the surface on which they are laid. The eggs hatch, releasing nymphs, which feed on blood like the adults. They grow and moult and eventually reach adult proportions. Bedbugs, like fleas, can live for up to a year without blood. Therefore they can be transferred alive to different places in such things as furniture or personal belongings. Their bites cause much irritation, yet they do not appear to be vectors of particular diseases. A recent survey in Britain collected information on reports of pest infestation over a period of six years. Of the bedbugs infestations that were reported, 60 per cent were in homes, 25 per cent in holiday accommodation and the rest were spread evenly among hospitals, public meeting-places, shops and factories.

8.11 Health education

It is also important to remember that although a society can aim to protect the health of its members it is unlikely to succeed fully if the members themselves are ignorant of the facts. It was stated in 1909 in the Minority Report of the Poor Law Commission that: 'Disease spreads not only by infection and contagion. The habits and practices of people are responsible in even greater measure for the continuance of disease. This cannot be combated by the popular panacea of a bottle of medicine.'

Educating people so that they know and can make personal judgements about health is another essential part of a health service programme. One example of this is the present work of the Health Education Council.

Answers to questions in the text

1 (a) The swampy breeding grounds will be removed.
(b) The larvae will be destroyed. Oil on the surface of the water will prevent the larvae breathing by clogging their breathing tubes.
(c) Fish feed on the eggs and larvae.
(d) This method will kill adults. Different insecticides will have to be developed as the mosquitoes develop resistance to those already in use.

2 This will give no time for the adult lice to begin their breeding and laying of eggs, so they will not become established on the body.

THINGS TO DO

1 Obtain eight sugar cubes. Place four of them on the base of a shallow tray. Put the remaining four on top to form a second row. Pour water coloured with ink into the shallow tray so that it comes half-way up the first row of cubes. Observe how the water rises up through the cubes by capillarity. Repeat the experiment with another eight cubes, but this time place a strip of polythene between the first and second row. You should now observe how the damp-proof course in the wall of a house prevents moisture spreading up from the ground.

2 Draw a chart which compares a modern house with a house of a previous century. Indicate the changes that have taken place and how improvements in living conditions have helped to raise the general standards of health of the community. Your local planning department may be able to provide you with the rules and regulations that cover building. Try to obtain a detailed planning application for a house and study how the builder has to abide by the regulations.

3 Find out about the work of Florence Nightingale. Construct a frieze, using drawings, to show how our hospitals have reached their present standards.

THINGS TO WORK OUT

In Questions 1–4 write down the *one* answer which you think is correct

1 The organism responsible for causing scabies in man is the
   A mosquito
   B flea
   C bedbug
   D mite
   E head louse

2 A life form which lives on or inside another living organism, taking food from it and giving nothing in return, is a
   A vector
   B parasite
   C host
   D micro-organism
   E mutant

3 The cells in our body which are attacked by the malarial parasite are the
   A epithelial cells of the skin
   B cells in lung tissue
   C white blood cells
   D red blood cells
   E cells in liver tissue

4 To prevent epidemics of typhoid it is essential
   A to provide better quality housing
   B to immunize the population
   C to provide a clean water-supply
   D for people to practise good personal hygiene
   E to pasteurize milk

5 (a) Explain (with an example) what is meant by the terms (i) epidemic, (ii) endemic.
(b) What term is used to describe world-wide outbreaks of influenza? In what way is this type of outbreak different to those in (a)?
(c) Describe *one* action that is taken internationally to control the spread of endemic disease.
(d) If an epidemic of typhoid were to break out in a town, describe *four* actions that might be taken at local level to prevent the disease spreading either within the town or to the rest of the country.

6 (a) Name *one* external parasite of man.
(b) Describe:
(i) how it has adapted to its way of life
(ii) how it obtains its food
(iii) its life cycle
(iv) *one* disease its transmits
(v) *one* method of control

7 A building inspector visits a building site where several houses are being built. Each is at a different stage of its construction. Give *five* building conditions which the builder must abide by and which will be checked by the inspector. State why each condition is necessary.

8 (a) Describe a location that would suit the *Anopheles* mosquito.
(b) Describe, with diagrams, the life history of the *Anopheles* mosquito.
(c) How does the *Anopheles* mosquito transmit malaria?
(d) Give *three* methods by which man can prevent malaria.

9 By taking *five* different aspects of life today and comparing them with those of 150 years ago, show how social conditions and provisions can influence the health of the community.

9 Fuel Food

9.1 **The fuel of the body – food**

When a motorist draws up at a petrol station he is confronted with a number of pumps, each delivering a different grade of petrol. If the motorist wants the best performance from his car he will follow the manufacturer's recommendations and supply the engine with the grade of petrol best suited to its design. Similarly, the performance that the human body gives largely depends upon the fuel it receives. However, unlike a car, the body's size and general state of repair are also dependent upon its fuel. The main uses of fuel, that is *food*, within the human body are for energy, growth and repair. These uses are illustrated in Fig. 9.1.

Fig 9.2 shows a sight familiar to anyone who shops for food. The vast array of attractively wrapped foods presents us with a bewildering choice. This is in sharp contrast to the choice given to the people in Fig. 9.3. Yet the people in Fig. 9.3 are receiving food which will adequately meet the demands of their bodies. The quantities and types of food that they need have been worked out exactly so that their intake of food, i.e. their *diet*, meets their special needs. To record your daily diet, you would need to list all the foods and liquids you take in at mealtimes and in between! We need to find out more about food itself in order to see whether your diet is meeting your body's demands for energy, growth and repair.

Fig. 9.2 *The fruit and vegetables that are available in a supermarket*

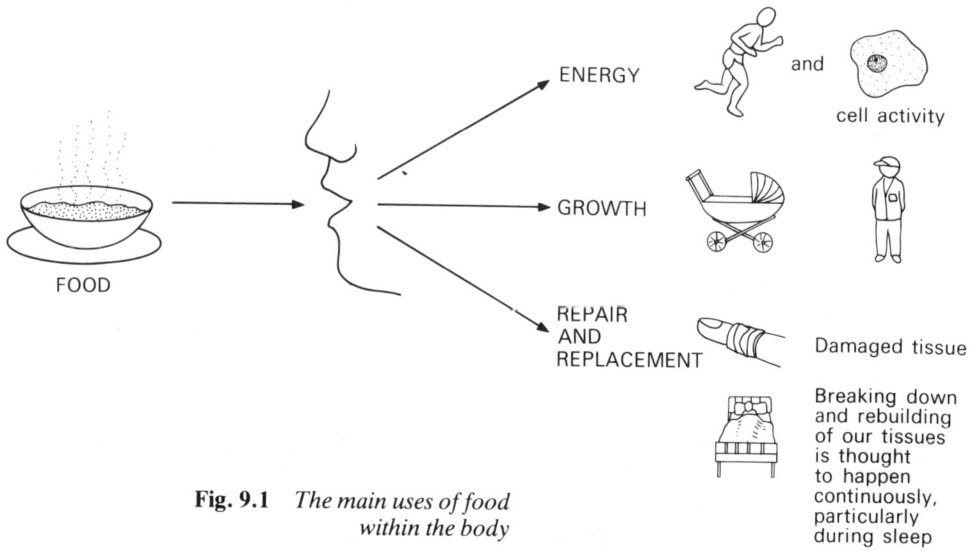

Fig. 9.1 *The main uses of food within the body*

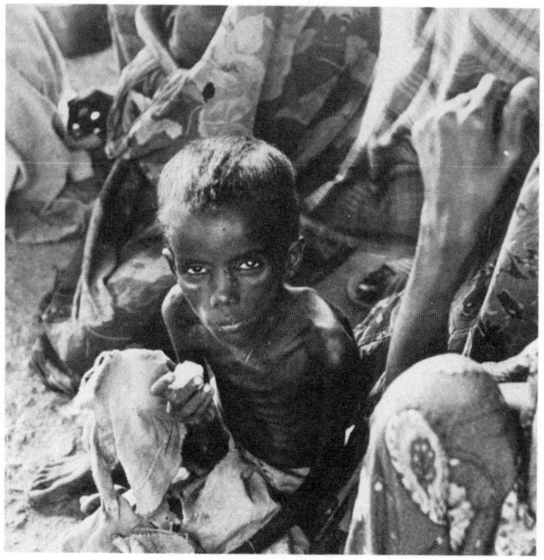

Fig. 9.3 *Food being distributed to refugees*

9.2 The basic structure of food

Every piece of food that we eat can be analysed to find the main chemical elements that it contains. These elements combine to form the main groups of chemical compounds found in food. They are:

(a) The *protein* group – food containing mainly the elements carbon, hydrogen, oxygen, nitrogen and sulphur.
(b) The *carbohydrate* group – food containing the elements carbon, hydrogen and oxygen.
(c) The *fat* group – food containing the elements carbon, hydrogen and oxygen, but in different quantities within the molecules compared with the carbohydrate group.

Any particular food may consist of just one of these groups, or it may contain them all. For instance sugar, which is sucrose extracted from sugar-cane or sugar-beet, is purely carbohydrate. However milk contains protein, fat and carbohydrate. Table 9.1 lists some common foods and the groups they contain.

Question 1 Using Table 9.1 write down which meat and two vegetables you would select to give the largest protein level, but the lowest possible fat and carbohydrate levels.

The groups that are present within a food determine its value to our bodies.

TABLE 9.1 **CHEMICAL GROUPS CONTAINED IN SOME COMMON FOODS**

Food	Protein (%)	Fat (%)	Carbohydrate (%)	Energy value (kilocalories/100 g)	(kJ/100 g)
Cereals/pastries					
Biscuits, plain, mixed	7	13	75	435	1827
Bread, wholemeal	8	2	47	228	958
Bread, white, large loaf	8	1	53	243	1021
Cornflakes	7	1	88	367	1541
Flour, white, household	11	2	78	350	1470
Rice, polished, boiled	2	Less than 0.5	30	122	512
Pastry, short, baked	8	33	55	548	2301
Fruit					
Apples, English eating	Less than 0.5	Traces only	12	45	189
Bananas	1	Traces only	19	77	323
Peaches, canned in syrup	Less than 0.5	Traces only	23	87	365
Oranges	1	Traces only	9	35	147
Peanuts	28	49	9	603	2533
Meat					
Beef, steak, stewed	31	9	0	206	865
Chicken, roast	30	7	0	187	794
Lamb leg, roast	25	20	0	292	1226
Sausages, pork, fried	12	25	13	326	1369
Turkey, roast	30	8	0	196	823
Bacon, streaky, fried	24	46	0	526	2209

TABLE 9.1 **continued**

Food	Protein (%)	Fat (%)	Carbohydrate (%)	Energy value (kilocalories/100 g)	(kJ/100 g)
Milk products					
Butter, slightly salted	Less than 0.5	85	Traces only	793	3331
Cheese, Cheddar	25	35	Traces only	425	1785
Milk, fresh, whole	3	4	5	66	277
Milk, dried, skimmed	3.5	Less than 0.5	49	326	1369
Yoghurt, low-fat	5	2	5	54	227
Egg, fresh, whole	12	12	Traces only	163	685
Lard	Traces only	99	0	920	3864
Margarine	Less than 0.5	85	0	795	3339
Fish					
Cod, fried	21	5	3	140	588
Herring, fried	22	15	2	235	987
Vegetables					
Beans, baked	6	Less than 0.5	17	93	391
Beans, runner, boiled	1	Traces only	1	7	29
Brussels sprouts, boiled	2	Traces only	2	16	67
Cabbage, spring, boiled	1	Traces only	1	8	34
Carrots, young, boiled	1	Traces only	5	21	88
Cauliflower, boiled	2	Traces only	1	11	46
Cucumber, raw	1	Traces only	2	9	38
Lettuce, raw	1	Traces only	2	11	46
Onions, raw	1	Traces only	5	23	97
Peas, fresh, boiled	5	Traces only	8	49	206
Potatoes, old, boiled	1	Traces only	20	80	336
Potatoes, old, chipped	4	9	37	239	1004
Potato crisps	6	38	49	559	2348
Tomatoes, raw	1	Traces only	3	14	59
Sugars/sweets					
Sugar, white	Traces only	0	100	394	1655
Boiled sweets	Traces only	Traces only	87	327	1373
Chocolate, milk	9	38	55	588	2470
Jam	1	0	69	261	1096
Drinks					
Coffee	13	15	29	301	1264
Tea, Indian	14	0	0	58	244

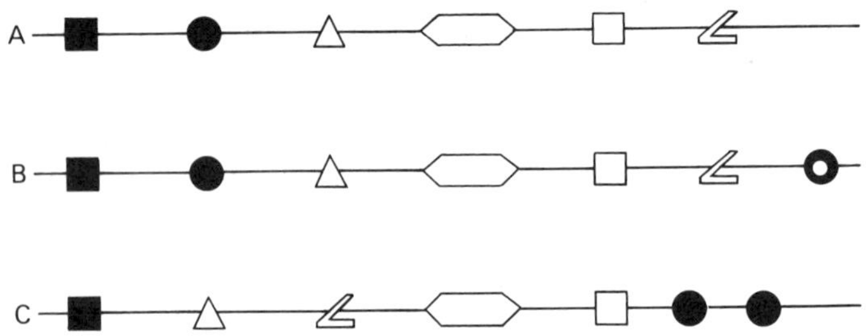

Fig. 9.4 *Amino-acids link together to form proteins*

9.3 **How proteins are used in the body**

We have already discussed in Section 9.1 why the body needs food. Can these needs be linked with the chemical groups found within food? Take growth, for instance: for the body to grow it must be able to manufacture its own cell material (protoplasm) and to do this it needs chemical substances called *amino-acids*. Amino-acids are the 'blocks' for the building of proteins. Just as there is a limit to the types of building block and brick available to the building trade, so there are only about twenty known amino-acids. In the same way that builders are able to build thousands of different houses from so few brick types, the linking together of a variety of amino-acids in different groups, ways and quantities allows millions of proteins to be produced (Fig. 9.4).

A tremendous variety of proteins are found within the tissues of the human body. For instance, human hair is just a chain of proteins. When hair is permanently waved the chemicals within the perm solution destroy some of the normal molecular links or bonds. New molecular bonds are then built up when a neutralized solution is used, and these create the curl in the hair.

The human body can make only about half the amino-acids it needs from simpler substances. The remainder, the *essential amino-acids*, must be supplied in the exact form that the body requires. *First-class protein* food is the richest source of these essential amino-acids. As you would expect, the proteins in first-class protein food closely resemble our own body proteins. Therefore most of our bodies' essential amino-acids are normally supplied through eating animal produce, for example eggs, fish, cheese, meat and milk (Fig. 9.5).

Most plant produce is a *second-class protein* food. A second-class protein does not contain all the essential amino-acids in the correct quantities. Therefore if a person wishes to live entirely on plant produce, he should ensure that it is a 'good mixture'. In this way one plant food will make up for the amino-acid shortage of another. *Vegans* are people who will not eat any food of animal origin. *Vegetarians* will not eat any food obtained by slaughtering an animal, but usually will accept other forms of animal produce in their diet.

Question 2 Can you name any foods from animals that a vegetarian could add to his diet to provide the essential amino-acids?

Amino-acids are used for growth, repair and replacement of tissues within the body. However, any extra which are not used in this way are carried in the bloodstream to the liver (page 160). Here they are changed chemically and can be used in respiration to give energy. *Kwashiorkor* is a disease of very young children and is caused by a lack of first-class protein (protein deficiency). It is common in countries where there is little food of animal origin, not even milk. The baby usually develops symptoms when it is weaned, and is no longer getting milk from its mother. The baby's skin becomes dry; it then cracks and

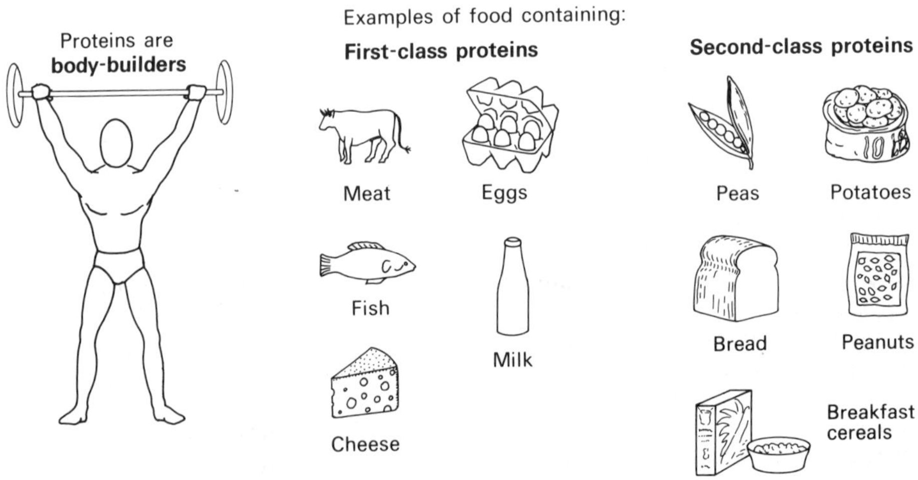

Fig. 9.5 *Protein foods*

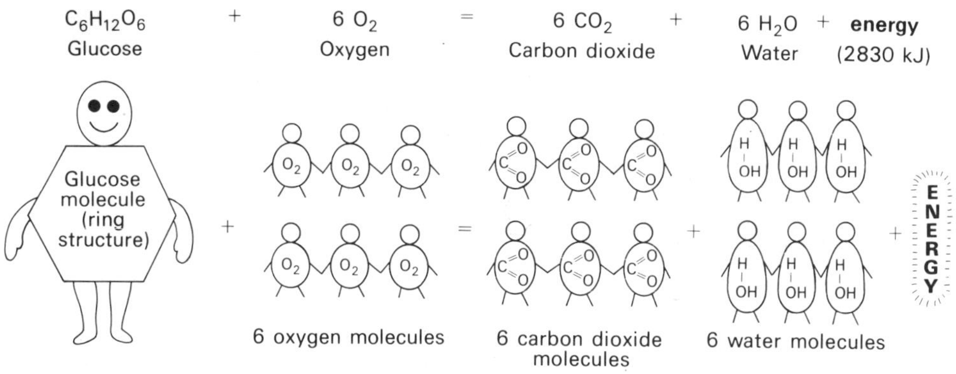

Fig. 9.6 *Internal (tissue) respiration*

scales. The child may even die if its protein supply is not increased through a special high-protein diet. Hyderabad mix and ettu paste (used in Uganda) are two food supplements for such malnourished children. They contain basically the local staple food plus high-protein foods such as soya beans, skimmed milk or some other form of animal protein.

9.4 How carbohydrates and fats are used in the body

Carbohydrates and fats are the energy-giving fuels of the body, The energy is released during tissue respiration, when the large carbohydrate or fat molecule is broken down into the smaller molecules of carbon dioxide and water (Fig. 9.6 and page 8). The breaking of the chemical bonds releases energy which can be used to build up the cell's energy store of ATP (page 41). Thus energy is readily available for the body's activities (Fig. 9.7). Some energy is lost as heat, which warms the body (page 186); 1 g of fat releases twice as much energy as 1 g of carbohydrate.

Carbohydrates can be found in food in three forms: *sugars, starches* and *cellulose*. Sugars dissolve, are sweet and are crystalline in nature. Starches are insoluble, tasteless and are powdery substances. Sugars and starches are energy givers. Cellulose is the fibrous material of the cell walls of plants. We are unable to digest cellulose (page 156) so for us it has no energy value. However, cellulose is a very important part of our diet as it forms the bulk of roughage (page 159).

Fig. 9.7 *Foods which provide energy*

Question 3 Name two foods in each case that would provide you with (a) starch, (b) sugar and (c) cellulose.

There are many varieties of sugar. This comes about because of the different arrangement of the atoms within the sugar molecules. *Glucose* (grape-sugar) and *fructose* (fruit-sugar) are simple sugars (*monosaccharides*) whose molecules have the formula $C_6H_{12}O_6$. More complex sugars,

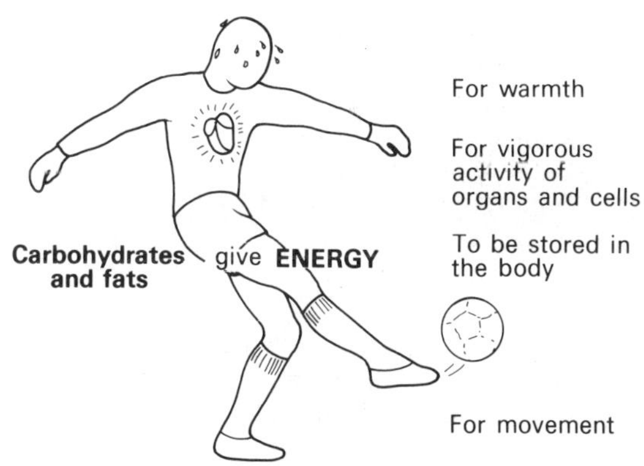

for instance *sucrose* (cane-sugar) and *lactose* (milk-sugar), are termed *disaccharides*. Their molecules are made up of simple sugar molecules.

Fats and oils fall into three groups: *saturated*, *monounsaturated* and *polyunsaturated*. Recent opinion suggest that eating large quantities of saturated fats increases the possibility of heart-disease or certain cancers, such as breast cancer. This is based on the high occurrence of these

Fig. 9.8 *Press advertisement from the Health Education Council's 'Look After Yourself' campaign*

diseases within communities which eat a lot of meat and fat, compared with the low occurrence within communities whose diets lack these two foods. It is thought that the culprit causing these diseases is the substance *cholesterol*, which is found in large quantities in saturated fats. Cholesterol is taken into the blood when saturated fats, basically animal fats, are eaten. The level of cholesterol in the blood can be reduced by eating less fatty meat and more fish, vegetables and polyunsaturated fats which are of plant origin (Fig. 9.8).

9.5 Vitamins and mineral elements

A good cook can tell you that the success of a casserole often depends upon the addition of small quantities of various herbs and seasoning. Similarly, if the body is to be maintained in full health it must receive, within its diet, *vitamins* and *mineral elements*. These do not need to be present in large quantities, but without them tissues cease to function correctly, and eventually the visible signs of *deficiency diseases* develop. Table 9.2 lists some of the vitamins that are important to the human body. Although their effects were known in previous centuries, as in the case of scurvy, vitamins were not thoroughly investigated until early this century. Since then many vitamins have been isolated and the contributions that some make to the healthy working of the body have been recorded (Table 9.2). They are named with letters of the alphabet, sometimes with an additional number to identify different types. Eating a mixed diet provides all the necessary vitamins. However at certain times, such as during pregnancy and early childhood, vitamins can be taken to prevent any possible deficiencies (page 19). Certainly the giving of cod-liver oil daily, or vitamin drops, is one of the reasons why rickets is now rare.

Rickets develop because without vitamin D the body cells cannot make use of calcium or phosphorus salts. These are just two of the *mineral salts* that are vital to the body tissues. Mineral salts contain the elements that the

TABLE 9.2 **VITAMINS**

Name	Influence upon the body	Deficiency diseases	Foods which are rich in this vitamin
A†	Growth in children. Healthy skin and mucous linings, e.g. nose, mouth, excretory passages. Healthy eyes and manufacture of visual purple (page 215)	Poor growth. Membranes dry up and become less resistant to 'invaders'. Difficulty in seeing in the dark. Eye's fluids dry up (page 212). The eye becomes inflamed – *xerophthalmia*	Halibut oil, cod-liver oil, liver, butter, milk, cheese eggs Body can manufacture it if supplied with *carotene*. This is found in vegetables, e.g. carrots, spinach, cabbage, tomatoes, watercress Not affected by cooking or storage
$\mathbf{B_1}$	Growth in children. Healthy nervous system. Seems to influence the building of proteins	Poor growth. Inflammation of nerves, eventually paralysis and death – *beri beri*. Loss of appetite, digestive disturbances. Muscles lose their tone	Wholemeal bread and flour, whole rice (not in polished, where husk and germ have been removed), potatoes, yeast extract, fish roe, peas, bacon Dissolves in water, but not if added to it when it is boiling Added to some foods, like flour, by law
$\mathbf{B_2}$ (riboflavin)	Growth in children. Healthy skin. Assists energy-giving foods	Poor growth. Skin cracks, particularly around the mouth. General feeling of weakness	Liver, milk, cheese, eggs, beef Dissolves in water. Deteriorates in hot and bright conditions
B group (nicotinic acid)	Normal activities of skin. Nervous and digestive systems	Redness of skin. Digestive and nervous disturbances – *pellagra*	Liver, bread, meat (lean), whole wheat, potatoes
B group (folic acid)	Formation of red blood cells	Anaemia	Liver,eggs, milk, meat, fish, yeast, spinach
C	Growth in children. Healthy gums. Natural healing processes. Convalescence after fevers	Poor growth. Anaemia. Loss of physical well-being. Aches and pains. Bleeding under the skin, in the mouth and gums, poor healing of wounds – *scurvy*	Fresh fruit and green vegetables, citrus fruits – oranges, lemons, limes. British sailors were called 'limeys' as they once ate these fruit on long voyages to avoid scurvy. Easily destroyed, therefore best to eat sources of the vitamin raw

† *Particularly required by children and expectant mothers*

Table 9.2 continued on page 120

TABLE 9.2 **continued**

Name	Influence upon the body	Deficiency diseases	Foods which are rich in this vitamin
D†	Formation of strong bones and teeth. Body needs it before it can use calcium and phosphorus	*Rickets* in children. Bones are soft and bend easily. Teeth easily decay	Cod-liver oil, fatty fish (e.g. herring), liver, eggs, milk, butter, sardines Sunlight acts on a substance under the skin. In ultraviolet light this forms vitamin D. Sunlight is essential for the young
K	Assists clotting of the blood	Blood capillaries easily damaged. Bleed a lot	Green vegetables, tomatoes, potatoes, carrots, liver

† *Particularly required by children and expectant mothers*

body cells need for particular functions. For instance, calcium and phosphorus are essential to bone structure and without them bones are softer than normal. Such bones easily bend under pressure. In rickets there is severe bending of the long leg bones. Some of the main elements and their importance to the body are listed in Table 9.3.

9.6 The energy value of food

In your physics lessons you may already have discussed various machines and the energy they use in order to do work. Different types of energy are obtained by converting one form to another in an energy chain. A car engine uses the chemical energy obtained when a fuel is burnt. If we want to know what work the engine can do, we must find out how much energy is released by the burning fuel. This is easily done by taking a known quantity of fuel, burning it and measuring how much heat energy is released. The energy value of a food may be measured in the same way.

When we measure the energy value of food, what units do we use? The older unit of heat energy is the *calorie* and it is likely that in everyday life you will see food energy values given in calories or kilocalories:

$$1 \text{ kilocalorie} = 1000 \text{ calories}$$

Kilocalories are often written in advertisements, slimming diets and popular literature as *Calories* (with a capital C).

However, as we discussed in Chapter 3 (page 00), the unit of energy used by scientists is the *joule* (J). Also

$$1000 \text{ joules} = 1 \text{ kilojoule (kJ)}$$
$$1\,000\,000 \text{ joules} = 1 \text{ megajoule (MJ)}$$

You may need to use calories as well as joules. The formulae below show how you can change from one to the other:

$$1 \text{ calorie} = 4.2 \text{ joules}$$
$$1 \text{ kilocalorie} = 4200 \text{ joules}$$

and

$$1 \text{ joule} = 0.24 \text{ calories}$$
$$1 \text{ kilojoule} = 240 \text{ calories} = 0.24 \text{ kilocalories}$$
$$1 \text{ megajoule} = 240 \text{ kilocalories}$$

The way in which food is oxidized in the body is a complex reaction yet it is similar to burning. It can be assumed that when tissue respiration takes place the food produces the same amount of energy as if it were burning. Table 9.1 lists some common foods and their energy value per 100 g.

TABLE 9.3 **MINERAL ELEMENTS AND SALTS**

Name	Some of the main needs for these elements and salts in the body	Deficiency disorders	Foods rich in these minerals
Calcium†	Bone and teeth formation – calcium phosphate or calcium carbonate. Clotting of blood. Correct muscle function	Poor bone and teeth structure. If an expectant mother's diet is deficient the calcium is removed from her own bones and teeth to supply the foetus. Rickets	Milk, cheese, eggs, fruit, watercress
Sodium†	In the composition of blood as sodium chloride (common salt) Blood composition is very close to that of sea-water: the 'internal sea' or 'internal environment'	Main mineral salt of the blood fluid. If the mineral strength of the blood varies, so will that of the tissue fluid. This can severely affect the activity of the cells (page 62). It is essential to maintain this internal fluid at a definite strength (concentration) if the body is to survive – *homoeostasis*. During exercise, when salt is lost from the body in sweating, muscle cramps can develop	In table salt or cooking salt
Phosphorus or **phosphates**†	Bone and teeth formation. Manufacture of ATP and cell membranes	Rickets. Dental decay. Energy requirements alter because of a shortage of ATP	Cheese, liver, egg yolk, milk, bread, oatmeal, cabbage, potatoes, other green vegetables
Iron†	Manufacture of haemoglobin within the red blood cells	Anaemia	Liver, sardines, bread, red meat, baked beans, cabbage and other green vegetables, oatmeal
Iodine†	Manufacture of thyroxine by the thyroid gland (page 179)	Swelling of the thyroid gland – *goitre*. Mental dullness. Can arise as a local deficiency because of environmental conditions (page 181)	Fish, water plants such as watercress Added to table salt in manufacture
Potassium	Correct growth – present in most cells. Found in the muscles and blood cells	Poor growth	Cereals
Magnesium	Important in the composition of bone	Poor bone structure	Vegetables and other plant produce

† *Those most likely to be deficient*

9.7 How much energy is required by the body?

When put away for the night in its garage, a car uses no fuel at all. This is not the same for us when we are in bed at night. Although we are not active all our internal engines, from the heart down to the smallest skin cell, are at work and using up fuel. Therefore a minimum quantity of energy is needed to keep us alive throughout the day and night. This energy is called *basal metabolism* (Fig. 9.9(a)).

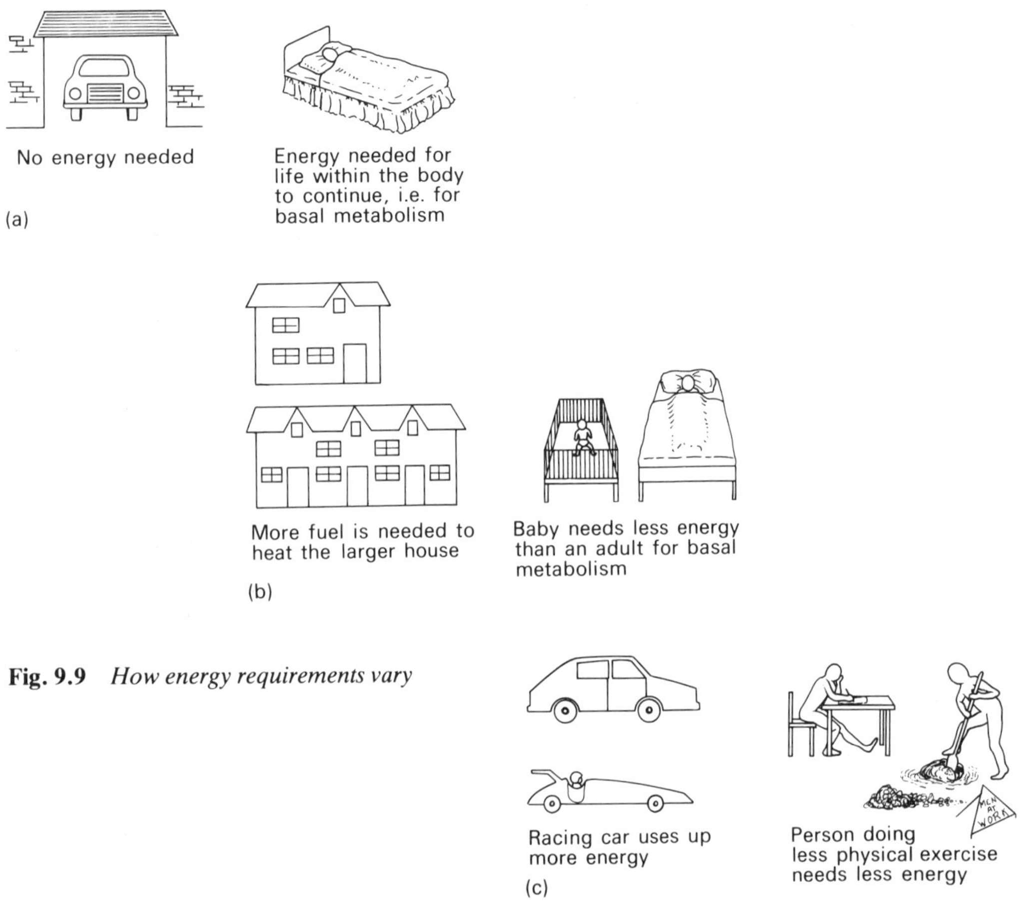

Fig. 9.9 *How energy requirements vary*

Does it take the same amount of fuel to warm a large house as a small house (Fig. 9.9(b))? Probably you will agree that the small house will need less fuel. Similarly, a person's basal metabolism depends upon his structure, i.e. height, weight, surface area, age and sex (Fig. 9.9(b)). Obviously this gives rise to a wide range of different basal metabolic rates. However, it is possible from the results of many experiments to produce tables of average figures (see Table 9.4). From such tables a person's individual basal metabolic rate can be estimated. Having worked out our basal metabolism we must then consider the extra demands for energy that are made on our bodies in our everyday lives. Once again we can make a comparison with a car. We know that if a car is driven at high speed the fuel consumption rises (Fig. 9.9(c)). Similarly, the human body's energy demands increase during very active periods. A person doing a job requiring heavy physical labour will need more energy than a person sitting at a desk writing (Fig. 9.9(c)). Fig. 9.10 shows the extra amounts of energy that would be needed per minute for a variety of activities.

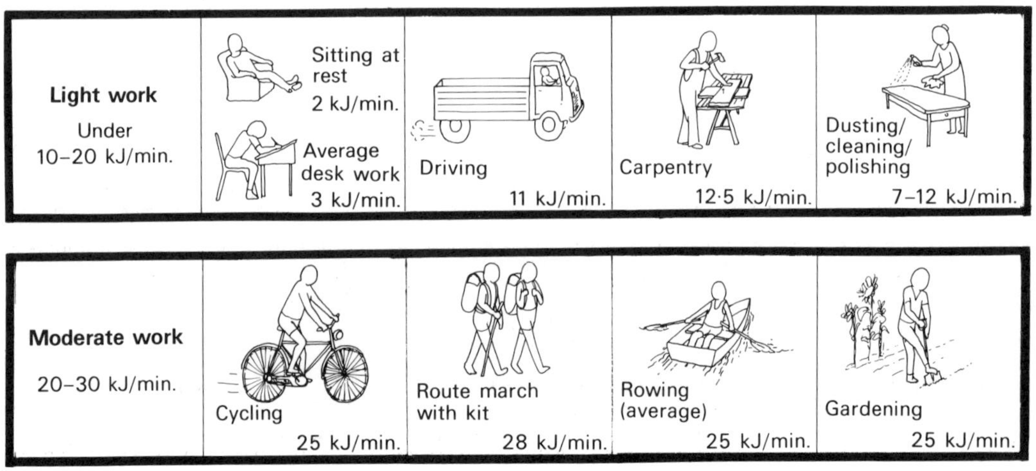

TABLE 9.4 STANDARDS FOR BASAL METABOLIC RATES

Age	Basal metabolic rate (kJ/m² hr) Male	Female
1	222	222
3	215	214
5	206	203
7	198	190
9	189	179
11	180	176
13	177	169
15	175	159
17	171	152
19	164	149
20	162	148
25	157	147
30	154	147
35	153	147
40	152	146
45	151	144
50	150	142
55	148	139
60	146	137
65	144	135
70	141	133
75	139	131
80	138	129

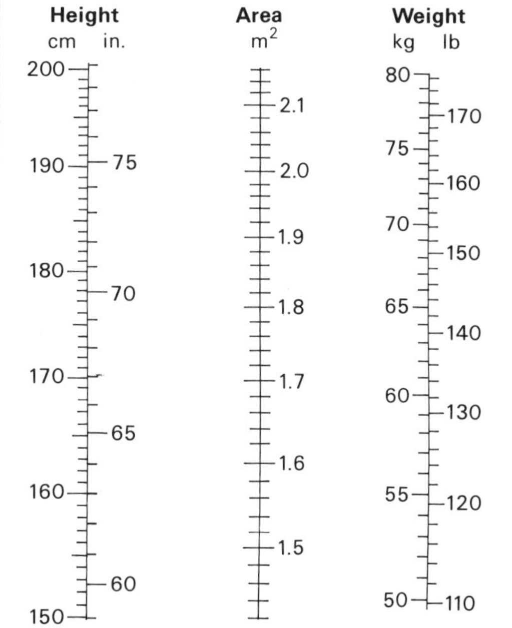

Place a ruler across from your height on the left-hand scale to your weight on the right-hand scale. The point at which the ruler crosses the centre scale will give your surface area

TABLE 9.5 A BOY'S ACTIVITIES DURING 24 HOURS

Activity	Time spent
Sleep	11 hours
Cycling to and from school	30 minutes
Class-room work	4½ hours
Football	1 hour
Cycling to and from swimming-baths	30 minutes
Swimming	1 hour session
Woodwork with his dad on a go-kart	1½ hours
Homework	1½ hours
Listening to records and watching T.V.	2½ hours

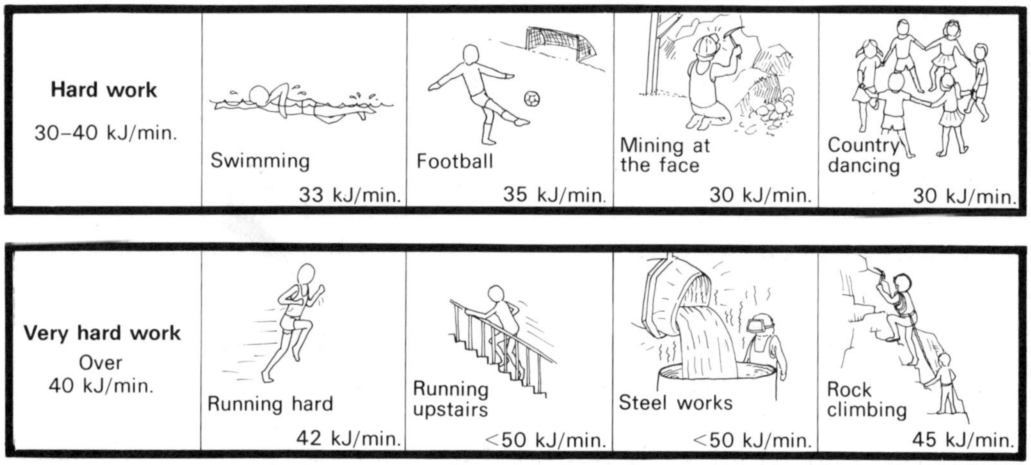

Fig. 9.10 *Extra energy needed for a variety of activities*

Question 4 Table 9.5 gives the record of the activities of a boy of fifteen for a period of 24 hours. Can you work out his energy demand for those activities alone? (Ignore basal metabolism.)

TABLE 9.6 **RECOMMENDED DAILY INTAKES OF ENERGY AND NUTRIENTS FOR THE U.K. 1969 (DEPARTMENT OF HEALTH AND SOCIAL SECURITY)**

Age range in years	Occupational category	Body weight (kg)	Energy (Cal)	Energy (MJ)	Protein (g)
Boys and Girls					
0 to 1	–	7.3	800	3.3	20
1 to 2	–	11.4	1200	5.0	20
2 to 3	–	13.5	1400	5.9	35
3 to 5	–	16.5	1600	6.7	40
5 to 7	–	20.5	1800	7.5	45
7 to 9	–	25.1	2100	8.8	53
Boys					
9 to 12	–	31.9	2500	10.5	63
12 to 15	–	45.5	2800	11.7	70
15 to 18	–	61.0	3000	12.6	75
Girls					
9 to 12	–	33.0	2300	9.6	58
12 to 15	–	48.6	2300	9.6	58
15 to 18	–	56.1	2300	9.6	58
Men					
18 to 35	Sedentary	65.0	2700	11.3	68
	Moderately active	–	3000	12.6	75
	Very active	–	3600	15.1	90
35 to 65	Sedentary	65.0	2600	10.9	65
	Moderately active	–	2900	12.1	73
	Very active	–	3600	15.1	90
65 to 75	Assuming a sedentary life	63.0	2350	9.8	59
75 and over		63.0	2100	8.8	55
Women					
18 to 55	Most occupations	55.0	2200	9.2	55
	Very active	–	2500	10.5	63
55 to 75	Assuming a sedentary life	53.0	2050	8.6	51
75 and over		53.0	1900	8.0	48
Pregnancy	–	–	2400	10.0	60
Nursing mother	–	–	2700	11.3	68

1 MJ = 10^6 joules

Table 9.6 gives figures for the recommended daily energy intake. Obviously an individual must assess his own requirements in relation to these figures.

9.8 How do we meet this demand for energy?

The body's demand for energy can be met only by a supply of food. By knowing a person's daily needs (Table 9.6) and the energy value of food (Table 9.1) you can calculate the amount and types of food that should be eaten.

The planning of specific diets requires the specialized knowledge of a *dietician*. People suffering from certain illnesses may have to observe strict diets worked out for them by the hospital dietician. For instance, Fig. 9.11 illustrates the personal diet of a person who has suffered total kidney failure. People who are overweight may wish to follow specialized diets. These may be given to them by their doctors, taken from articles in magazines or received at clubs set up to help people lose weight.

Most people eat the food which they like, and as we have already seen in Fig. 9.1 there is an extensive choice to whet their appetites. However, as we saw at the end of the last chapter, the change in the pattern of diseases has led many people to suggest that we must become far more aware of our bodies' need for a *balanced diet*. This does not mean going to a supermarket armed with a calculator to work out to the last decimal place the energy value of every food on the shelves! Our purchasing should be aimed at providing in our daily diet the following:

(a) Enough protein. Table 9.6 shows some of the recommended figures.
(b) Enough of the energy foods to supply each individual's energy requirements. (No more than 35 per cent of this should come from fat.)
(c) Water is essential for the maintenance of our internal body fluids and cellular activity. Seventy per cent of our body weight is water, and survival without

Your diet has been carefully planned to include all the foods that are important for health.

There are certain foods which have to be restricted because you are unable to get rid of excess amounts of:

Protein – as in meat, fish, cheese, eggs, milk, flour
Sodium – mainly as sodium chloride (ordinary salt)
Potassium – large amount found in fruit and vegetables

Because the protein content of this diet is restricted, you must take good-quality protein foods. The best types of protein for this diet are EGGS and MILK.

Daily allowances
One egg – preferably split between two meals
20 ml (1/3 pint) milk – measure this out each morning and keep it in a jug
180 g (6 oz.) boiled potatoes *or* 120 g (4 oz.) roast potatoes *or* 90 g (3 oz.) chips *or* 120 g (4 oz.) boiled rice
6–8 average helpings of fruit and vegetables
60 g (2 oz.) fresh *double* cream
One helping of porridge

To make the diet more interesting, you can exchange *half* of the egg on three days a week:

1/2 egg = 20 g (3/4 oz.) lean meat, liver, kidney or tripe
or 20 g (3/4 oz.) chicken, duck or turkey
or 30 g (1 oz.) fish

On exchange days you may have meat, chicken or fish in the amounts shown above but you must also have the other half egg *as egg*.

Salt
None at table
None in cooking

The total energy (calorie) content of your food intake tends to be low because many foods are restricted. It is very important to eat enough calories each day so that your protein allowance can be used efficiently, and to keep your weight constant. You must therefore eat as much of the following high-calorie foods as possible:

Fats and oils for frying
Butter – aim for at least one pound per week, used on bread, vegetables and in baking
Preserves, e.g. jam, honey, marmalade
Sugar
Protein-free bread – better toasted or fried
Protein-free biscuits and pasta
Protein-free flour – see recipes
Caloreen – a white powder like glucose but non-sweet, so it can be used in savoury foods, e.g. mashed potato as well as with fruit and in drinks
Hycal – drink in six flavours. Aim for one bottle daily

Both Hycal and Caloreen can be used in cooking (see recipes).

Cooking instructions for vegetables and potatoes
When boiling vegetables, use a large volume of water. All vegetables contain less potassium when boiled. The potassium leaches out into the cooking water. *DO NOT USE THIS WATER.*

Pressure-cooking is *not* suitable.

Boil potatoes for 10 minutes before roasting or chipping.

Fig. 9.11 *The personal diet of a person with kidney failure*

water is impossible after two or three days. Where there is a large loss of water from the body, such as during a fever or with vomiting and diarrhoea, more water than normal must be drunk to prevent *dehydration.* A loss equal to 1 per cent of body weight produces a sensation of thirst (page 175). A loss of 10 per cent would be fatal.

(d) Roughage to stimulate the removal of waste matter regularly from the body (page 159).
(e) Vitamins and mineral elements.

These requirements of our daily diet can be met by selecting a day's food from three groups:

Group 1 Meat and animal produce.
Group 2 Fruit and vegetables.
Group 3 Extras such as bread, cakes, biscuits, crisps, potatoes and cream.

Each day there should be a helping from Group 1. Thus we make sure that our body is being supplied with its first-class protein and fat. A selection from Group 2 will provide second-class protein, carbohydrate, protective vitamins and roughage. Group 3 will mainly provide additional carbohydrate. Foods within Group 1 tend to be more expensive than those in Group 3. However, Group 1 foods provide the necessary daily supply of first-class protein, so it is essential to use what money we have to buy our food wisely. This may need careful thought before going shopping, so that we shop for the right things with the money that is available. Elderly people particularly may need some help with this problem, especially if they find it difficult to get to the shops to obtain their provisions (Fig. 9.12). Care over our nutrition is essential at every stage in our lives to ensure good physical and mental health.

Fig. 9.12 *The 'Meals on Wheels' service*

9.9 **Stored energy within the body**

ATP and glycogen

If we burn oil or coal in our homes we always make sure that we have supplies stored away. Similarly, the human body is able to store energy within the cells. This is essential as most of our activity takes place between meals and not during them! In Chapter 4 we have discussed the energy stored in the form of ATP and glycogen. Glycogen is insoluble and therefore ideal for storage in cells. The liver is an excellent storehouse for glycogen, making it from the sugar *glucose* that reaches it in the blood (see page 160). Glycogen can be converted back to glucose as the body needs it. The glycogen in the liver can supply glucose for a period of up to 18 hours after a meal. If there are no further food supplies after this time the body starts to break down its fat store.

Fat storage

Excess glucose within the body is stored as droplets of fat within cells forming *adipose* or *fatty tissues* (see Fig. 1.4). It is found under the skin as well as in and around many organs. It can be of great value as a heat insulator (page 190). Generally it acts as a reserve fuel for energy; therefore regular

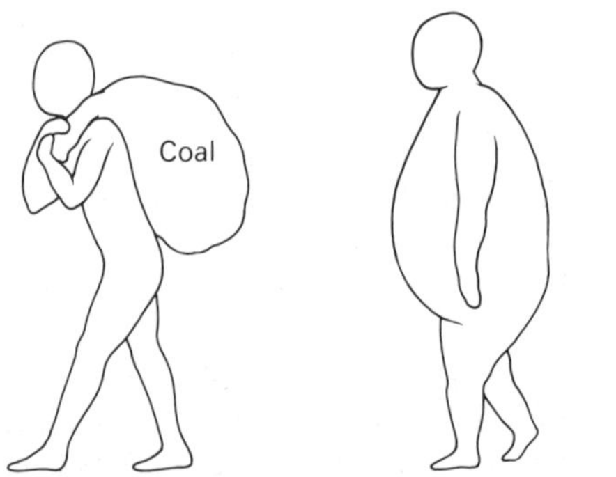

Would your body suffer if you always carried your coal-store around with you?

An increase in body weight puts a similar strain on the body

Fig. 9.13

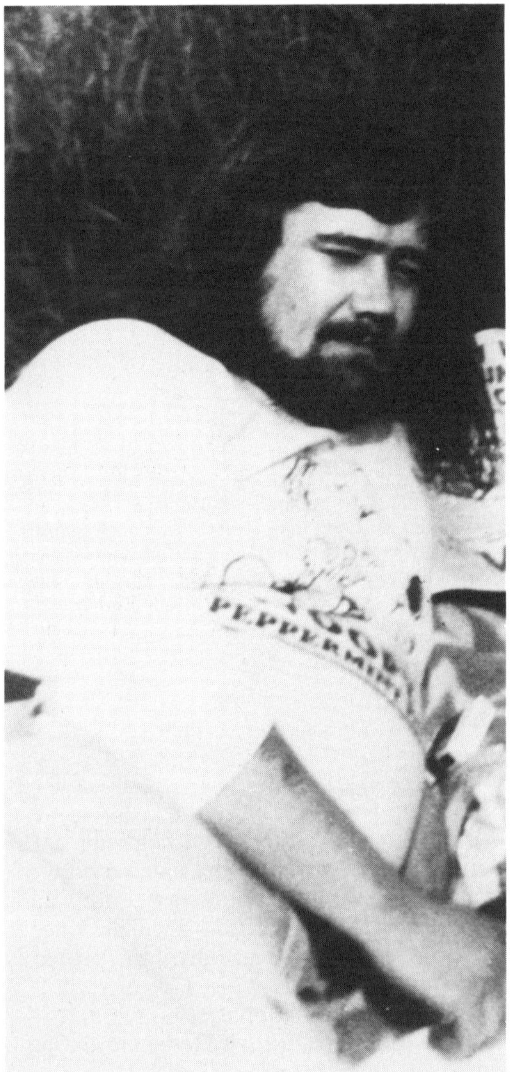

WeightWatchers

Mr Sam Simpson

Starting Weight: 19st. 9lb.
Goal Weight: 13st. 10lb.
Date attained: 2.8.78

Fig. 9.14

activity can 'burn off' fatty tissue. The amount of reserve fat laid down by the body is different for each individual. However, it is becoming clear that, if the fat or carbohydrate eaten exceeds the energy needs of the body, it will be laid down as fat. This can result in an increase in body weight, which makes extra demands on the body (Fig. 9.13). Becoming too fat or obese can present a person with other, more personal problems (Fig. 9.14).

Other physical problems can cause a person to become overweight, so it is best to seek medical advice before starting any drastic slimming campaign. However, for most people a decision taken at an early stage in life to eat the food the body needs, in the right quantities, may prevent a weight problem in later life.

Answers to questions in the text

1 White meat such as chicken, with spring cabbage and broccoli tops.

2 Eggs, milk, cheese.

3 (a) Most cereal crops, e.g. rice, barley, wheat and food made from them, such as flour, bread, potatoes and most breakfast foods.
(b) Honey, most sweet fruits, cane- or beet-sugar, milk sugar (lactose).
(c) Any two vegetables.

4 Cycling
25 × 30 + 25 × 30 = 1500 kJ
Football
35 × 60 = 2100 kJ
Swimming
33 × 60 = 1980 kJ
Classwork and homework
3 × 6 × 60 = 1080 kJ
Woodwork
12.5 × 90 = 1125 kJ
Leisure time, sitting at rest
2 × 2.5 × 60 = 300 kJ
Total = 8085 kJ.

THINGS TO DO

1 For a period of time record all that you eat and drink. It is necessary to record the weight of the food, so before you start construct a conversion table. For example take level or rounded tablespoons of various foods and weigh them. List your results as follows:

1½ rounded tablespoons of flour = 28 g
1 rounded tablespoon of sugar = 28 g

You should now be able to serve your food in tablespoons, and use your conversion table to estimate its weight.

Solid food such as bread can be weighed as a standard size, for instance, one slice. Record your daily intake in a table:

Day			
Main Course	**Name**		
	Weight		
Fruit and vegetables	**Name**		
	Weight		
Fillers	**Name**		
	Weight		
Water	**Volume**		

By using the chart on page 114 you can calculate your daily energy intake and protein consumption. Compare your results with the recommendations on page 124. Also analyse the mineral and vitamin content of your diet.

2 Use the results obtained above to calculate your *weekly* energy intake and protein consumption. Then calculate your *annual* demand. Select one or two foodstuffs and calculate what weight of these particular foods would be required to satisfy your demands; for example, how many kilograms of meat? You can find out the cost of this amount of food to the housewife.

3 Calculate how many joules of energy are required for a certain exercise (page 38). Using Table 9.1, which gives the energy values of various foods, calculate the weight of a particular food which would be needed to 'fuel' the exercise. Repeat this calculation for a variety of exercises.

4 In the class-room you will carry out various food tests:

Starch is identified through iodine solution changing in colour from light brown to blue.
Sugars (with the exception of beet-sugar) are identified by the formation of a red precipitate when the solution is heated to boiling-point with Benedict's reagent.
Protein is identified through the Biuret test, when a purple colour appears on the addition of dilute sodium hydroxide solution and blue copper sulphate solution.

These food tests require special solutions so you will not be able to carry them out at home. However, you can try the following:

(a) *Water content* Weigh the sample of food. Place it in a warm oven. At intervals remove and reweigh it. Any changes in weight will indicate the water content of the food.
(b) *Fat content* Squeeze small samples of food between sheets of blotting-paper (newspaper will do). Dry the blotting-

paper over a radiator. Any fat in the food will produce a greasy stain which can be seen when the paper is held up to the light.
(c) *Glucose content* You may be able to obtain a type of Benedict's reagent from the chemist. Its name is 'Clinitest'. These tablets are used mainly to trace glucose in urine samples. However, if a sample of food is crushed and made into a solution with water, a tablet may be added to reveal traces of glucose in the food.

5 Trace the origin of the food you eat. Draw a world map and pin-point where various foods are produced. Collect any newspaper literature about food and use it in a 'food montage'. Describe how this montage might differ in another country.

6 Collect a large variety of pre-packed and tinned foods. List the contents given on the label or packet. Identify any food analogues and determine the reason for other additives, for instance colouring or flavouring.

7 Make a chart, putting plant foods into columns according to the region of the plant which is eaten. (For example, when we eat a cauliflower we are eating the flower of the plant.) Name some plants or parts of plants which must *not* be eaten.

THINGS TO WORK OUT

In Questions 1–5 write down the *one* answer which you think is correct.

1 A person will suffer from scurvy when their diet is deficient in
A iron
B vitamin B
C vitamin E
D vitamin C
E protein

2 A food rich in calcium is
A bread
B milk
C fish
D liver
E table salt

3 Vitamin D is needed in a diet to ensure
A strong bones
B healthy skin
C clotting of the blood
D the formation of red blood cells
E healthy eyes

4 White bread contains a high percentage of
A polyunsaturated fat
B saturated fat
C first-class protein
D second-class protein
E carbohydrate

5 Food that gives a blue-black colour with iodine solution contains
A sugar
B fat
C starch
D protein
E water

6 Which of the following is the daily energy requirement most suited to (a) a four-year-old girl, (b) a middle-aged man working as a gardener, (c) an active housewife?
(i) 15 000 kJ
(ii) 6700 kJ
(iii) 10 500 kJ

Explain your answers with regard to basal metabolism and the body's additional demands for energy.

7 A four-year-old child's meals are regularly provided as follows:

Breakfast:	cereal, toast, tea
Mid-morning:	biscuits, orange squash
Lunch:	spaghetti on toast, ice cream
Early-evening meal:	chips, jam tart, tea

(a) What essentials are missing from this diet?
(b) What might be the long-term consequences of such a diet?
(c) What changes would you make to this diet and what would be your reasons?

8 (a) How does the body make use of (i) proteins, (ii) carbohydrates?
(b) Explain the difference between first-class and second-class protein foods. Give an example in each case.
(c) What elements are present in all amino-acids?
(d) What is meant by malnutrition and what causes it? Distinguish between malnutrition and undernourishment.
(e) Describe *one* disease that is caused by protein deficiency. Give one way in which it may be prevented.

10 Food Hygiene

10.1 The handling of food

When buying food the customer should carefully consider not only its nutritional value but also its quality. For instance, the food value of vegetables and fruit varies with their freshness.

Question 1 Can you suggest another reason why it is important that the food we buy should be fresh, particularly animal products such as meat, milk and eggs?

Manufactured and processed foods, or food that has been treated to prevent it perishing, must by law have a code stamped on it. This code enables the source to be traced and the date of manufacture to be confirmed. Some packs are date-stamped and this indicates the date by which they must be sold or eaten. Environmental Health Officers (E.H.O.) make spot checks by visiting food shops. During such a visit the E.H.O. will look for food that is not fresh and which could be a hazard to health. The E.H.O. or the Sampling Officer may take samples away for analysis, which will be done at one of the public health laboratories. The E.H.O. will also look for cases of *cross-contamination*. Cross-contamination occurs when a food becomes infected with possible harmful micro-organisms from another source. Cross-contamination happens when the organisms are passed from man to food, or from food to food. To prevent cross-contamination from man to food shop assistants must practise good personal hygiene and the shop should provide facilities for this. Cleanliness is also essential within the structure and fittings of the shop and every effort must be made to prevent pests. The storage of food could lead to food to food cross-contamination. This happens when cooked or ready-to-eat food comes into contact with uncooked food which may harbour harmful micro-organisms. Such contact can happen indirectly, for example if a knife which has been used to cut up raw meat is then used to cut cooked meats.

A customer with a complaint about contamination of food, or any other concern about its fitness for human consumption, can contact the local E.H.O. This is in the customer's interest and helps protect the health of others. It is not only shops and markets which are carefully inspected. Factories that deal with food processing and packaging receive visits from an E.H.O.

Fig. 10.1 *(a) A house mouse at the grocers, soiling food by body contact and droppings. (b) In this case cockroach droppings are soiling the wall, but they could just as easily be soiling food.*

10.2 Ways of preserving food

Yeast moulds and bacteria will spoil our food and make it unfit to eat. Some bacteria produce poisons (toxins) that enter the food, and these may give us the symptoms of food poisoning, such as diarrhoea, vomiting, headaches and fever. The most dangerous bacteria that can cause food poisoning are staphylococci, salmonellae and clostridia. To prevent the action of these harmful micro-organisms in our food we can either kill all the organisms present (sterilize) or remove some of the factors they need for growth. Table 10.1 (page 130) outlines some of the methods used in food preservation.

Once the food has been preserved it has to be stored in conditions that will maintain the preservation. For example, it is no good storing dried food in a damp cupboard. Also it must be kept well away from any form of pest, such as mice (Fig. 10.1(a)) or insects (Fig. 10.1(b)). Preservation, if carried out correctly, does not appear to alter drastically the nutritional value of the food. However there is one exception: vitamin C, which is destroyed by heating.

TABLE 10.1 **FOOD PRESERVATION**

Aim	Method		Storage time	Precautions that should be taken
	Name	**Process**		
To kill the micro-organisms (sterilize) and prevent others from entering	**Pasteurization**	See page 132	Several days	Must be kept sealed and cool
	Canning	Carried out commercially. Food is heated to 90 °C. Placed and sealed in tins while hot. No air in cans. Micro-organisms are all killed by the heat. (Higher temperatures may be needed to kill spores)	Several years	A swollen or 'blown' can shows that gas is being produced by the activity of bacteria
	Bottling	Can be carried out in the home safely, as long as high enough temperatures are used. Fruit heated up to 100 °C. Vegetables heated to higher temperatures using a pressure-cooker. Vegetables are more likely to contain dangerous bacteria as they are grown near soil	Between growing seasons	Temperatures must be high. Seals on bottles must remain intact
To remove the factors which encourage the growth and multiplication of micro-organisms	**Refrigeration**	Bacteria thrive and reproduce at room temperature. Their activity stops at low temperatures (below 7 °C)	A few days only	Once the temperature rises within the refrigerator, or when the food is removed from it, the bacteria become active. Therefore the food must be used immediately
	Freezing	(a) Quick freezing – used commercially. Food cooled to a temperature of −18 °C (b) Deep-freezing – used in the home. Food to be frozen must be in contact with the sides or bottom of the freezer. Temperature at least −18 °C	Generally a maximum of three months. Instructions may be given on the packaging according to 'star' rating	As for refrigeration. Once thawed, food must not be refrozen
	Drying (dehydration)	Bacteria need moisture for growth. In drying, all moisture is removed from the food. For example, herbs hung up in air to dry out. Commercial freeze-drying involves quick freezing, then lowering the air pressure to 'draw off' the water	Between growing seasons	If the food becomes damp, bacteria will become active

TABLE 10.1 **continued**

Aim	Method		Storage time	Precautions that should be taken
	Name	Process		
To remove the factors which encourage growth and multiplication of micro-organisms	**Smoking**	Food suspended in smoke from a hardwood fire. Temperature should not go above 50 °C. Dries the food out	Between growing seasons	Must be stored in a cool, dry place. Always look for signs of decay
	Preservatives Sugar	Sugar prevents growth of bacteria. Jam made from fruit must contain 65% sugar if it is to keep	Between growing seasons	Food must always be in the preservative. Always look for signs of decay
	Salt	Salt draws the moisture out of the food and even the bacteria by osmosis, 9 (page 63)		
	Acid (vinegar)	Acid solution prevents growth of bacteria. Food is salted for a few days, then rinsed and put in vinegar (known as pickling)		

10.3 The cooking of food

Some foods can be eaten raw, while other foods must be cooked. The main reasons for cooking food are: to soften or tenderize it, to improve its flavour, to make it easier to digest, and to kill harmful bacteria. Cleaning and preparing is as important for foods which are to be cooked as it is for those, like fruit, which we eat raw. Any general cookery book will give the methods of preparation as well as the times and temperatures that should be used in cooking. It is essential to follow such guide-lines for two reasons.

First, the temperatures and cooking times that are given will ensure that the food is thoroughly cooked and sterilized.

Second, the nutritional value of the food will not have been destroyed. For example, the vitamin C content in fruit and vegetables is reduced by approximately 50 per cent when they are cooked. Should cooking of these foods be necessary it must be done as quickly and in as little water as possible.

Personal cleanliness, and the cleanliness of equipment and food preparation areas, is essential in home kitchens. Unlike factories, shops and restaurants we are not subject to visits from the E.H.O.! However, it does no harm to review and inspect our own standards from time to time. Remember that pathogenic micro-organisms flourish particularly well where there is moisture and in those temperatures where we feel most comfortable, i.e. in a warm room. For this reason cooked food which is not to be eaten immediately should be cooled as *rapidly* as possible after being removed from the oven. It should then be put into the refrigerator or a cool storage area (Question 2, page 132). Good standards within our kitchens are particularly vital when there are babies or young children in the home. A 'tummy upset' in a very young baby can be serious. Milk, the only food for a very young baby, is unfortunately an ideal medium for the growth and multiplication of bacteria. A natural uncontaminated source is the mother's breast (see page 256) but for a mother who bottle-feeds her baby there are certain aids which help her to provide sterile milk at every feed. A bottle-feed is prepared by mixing a measured amount of boiled (sterile) water with a measured quantitiy of powdered evaporated milk and pouring it into a sterilized bottle. The special powdered milks available for babies are prepared from cow's milk. This closely resembles human breast milk but contains more protein and less sugar: 30 cm^3 of cow's milk (either skimmed or full-cream) contains approximately 1 g of protein. Cow's milk is also a valuable source of protein, vitamins and calcium for a mother who is breast-feeding.

Fig. 10.2 *The unhealthy kitchen*

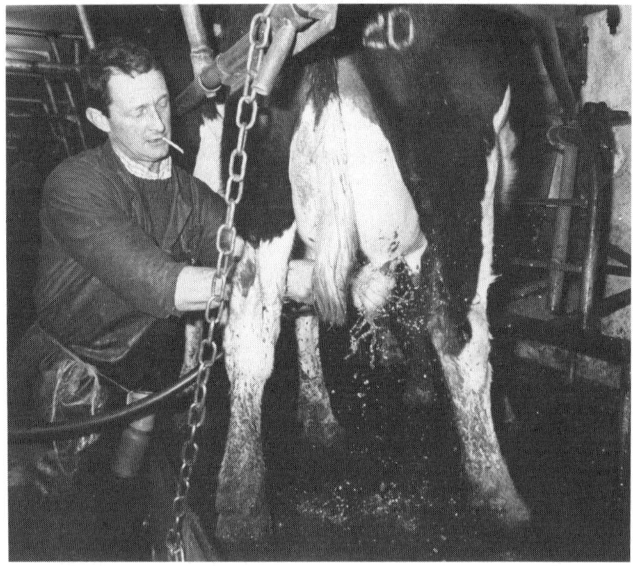

Fig. 10.3 *Cow's udders being washed before milking*

Question 2

(a) Why should the hot food not be put directly into the refrigerator from the oven?

(b) Fig. 10.2 shows a kitchen. What health hazards can you spot?

10.4 Milk

After infancy, children and adults are able to drink cow's milk – the familiar 'pinta'. Milk taken directly from a cow is called raw milk. It can be drunk fresh and warm immediately after leaving the cow but we would be taking a risk, because of the number of micro-organisms it could contain. Although the area where the cows are milked and the milking equipment are kept very clean, as seen in Fig. 10.3, it is still possible for bacteria to enter the milk. This is particularly true if the milk is exposed to air. Other micro-organisms, such as those that cause bovine T.B. (see Section 10.5), can enter from within the cow's udders. This type of contamination is not prevented when the cow's udder is washed before the milking equipment is attached.

It is some hours before the milk from the cow reaches our doorsteps. In raw milk the micro-organisms would have multiplied to a point where such milk could be dangerous to drink. So the milk is cooled to 5 °C after it is taken from the cows, then it is collected from the farm in tankers and taken to the dairy for the process of *pasteurization* (Fig. 10.4). This involves heating the milk to destroy the micro-organisms, but at a temperature and for a period of time that will not seriously destroy its food value and flavour. The two methods used are:

(a) heating the milk to 80 °C, holding it at that temperature for 30 seconds, then cooling it rapidly to below 10 °C;

(b) heating the milk to 63 °C, holding it at that temperature for 30 minutes, then cooling it to below 10 °C.

Question 3 Where do you think the word pasteurization comes from?

Samples are taken from each tanker-load of milk, and these are tested for micro-organisms and quality before this milk is pumped into the processing plant of the dairy. Similar testing is carried out throughout the process of pasteurization and on the milk when it is about the leave the dairy. Dairy staff are required to observe strict hygiene regulations, and all equipment and work areas are regularly washed and

Fig. 10.4 *Pasteurizing milk in a dairy and* (right)
Fig. 10.5 *as it is delivered to our door.*

sterilized. As in other food-processing industries the E.H.O. makes spot checks to see that every precaution is being taken to ensure that our milk is safe to drink.

10.5 Quality and grades of milk

Pasteurized milk can be kept for several days if it remains sealed and cool. If it is opened and left at room temperature micro-organisms will enter and multiply, rapidly turning it sour.

Cows are subject to *bovine tuberculosis* (not found in goats) and this disease can be passed on to us in cow's milk. Pasteurization kills the bacteria which cause this disease. However, farmers are required by the Ministry of Agriculture, Fisheries and Food to build up a dairy herd which is free of T.B. A veterinary surgeon must test every cow for T.B. once a year. Any diseased animal is removed from the herd. If the herd is free from infection the farmer is issued with a licence for a T.T. (tuberculin-tested) herd.

Brucellosis (undulant fever) is another disease which can be passed on through milk from both cows and goats. Again constant checks are kept on cows and milk for evidence of the disease, although it is possible for animals to be vaccinated against it.

Milk is tested for content (such as percentage of cream) and divided into different grades. We can identify these grades by different-coloured bottle caps, and purchase them at different prices (Fig. 10.5). Two other forms of milk are *homogenized* and *sterilized*. In homogenized milk the cream is mixed into the milk and does not separate out. If this milk is bottled and heated to very high temperatures we obtain sterilized milk. The milk lasts much longer than pasteurized milk. However it tastes very different and may be recognized by the different bottle and tightly-sealed metal cap.

Other forms of milk which can be stored are *powdered* (dried) milk, *condensed* milk and *evaporated* milk (canned). Evaporated milk is made by evaporating off 40 per cent of the water content of full-cream milk. Condensed milk is made by a similar process, but has sugar added. The care that is taken in supplying uncontaminated milk is also taken in supplying water to our homes.

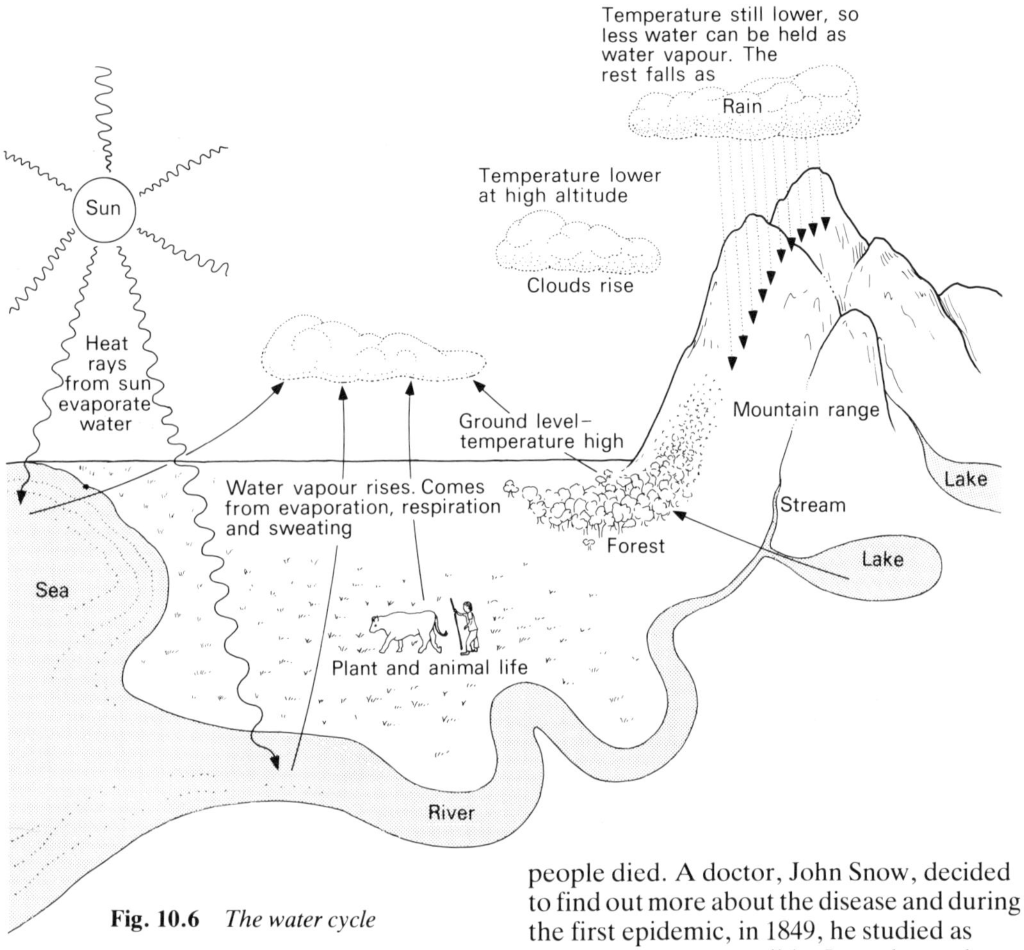

Fig. 10.6 *The water cycle*

10.6 The need for clean drinking-water

Fig. 10.6 shows what you have probably learned in other science lessons about the water and rain cycle on earth. In early times natural springs, wells, lakes and rivers were man's source of drinking-water. There were even schemes for piped water-supplies, for example Roman aqueducts. But gradually, as the population increased, as villages become towns and as town became cities, the demand for water grew, and so did contamination. In the middle of the nineteenth century, London suffered from three great epidemics of cholera and thousands of people died. A doctor, John Snow, decided to find out more about the disease and during the first epidemic, in 1849, he studied as many patients as possible. In each case he found that the disease started with pains in the stomach and intestines. Thus he became convinced that the source of the disease was something which the people had eaten or drunk.

Question 4 What sort of evidence to prove your theory would you have collected if you had been John Snow? How would this have indicated the source of the disease?

In 1854 a second epidemic broke out. John Snow studied cases in the area of London around Broad Street. In every case of cholera he found that the person had obtained water from the common pump in Broad Street itself. John Snow was able to prove his theory. In September 1854, he persuaded the Board of Guardians for that area to remove the handle from the Broad Street pump. Within three or four days everyone was convinced that Snow was right, for no new cases of cholera were reported. The epidemic had dramatically ceased.

It did not take long for the news of this discovery to spread to other areas of the country, and there was at last public awareness of the need to provide an uncontaminated water-supply. In fact, the problem with the Broad Street pump was that it was taking water from a shallow well (Fig. 10.7). Water flowing into the well passed through the material in the surrounding soil. This material could have been buried rubbish, sewage, decaying matter or even corpses in the cemetary, all producing dangerous micro-organisms which then entered the water and well.

10.7 Our present water-supply

In England today most surface water (Fig. 10.8) from rivers, streams, lakes and shallow wells is considered unsafe for drinking. Water from deeper wells (Fig. 10.7) and from very high natural lakes is considered to be a little safer but, even so, before it is pumped to our homes it is treated with chlorine gas.

The demand for water is so great nowadays that contaminated water must be taken and purified so that it is fit to drink. This means taking water from contaminated natural sources such as rivers and low-lying lakes. Fig. 10.9 illustrates how this water is collected, purified and piped to the customer. Throughout the process samples of water are taken to a laboratory and the quality checked. Some settlement of particles will have taken place in the reservoir, and other solid matter is removed in the water treatment works. If the water entering the plant is fairly clean it will need only to be strained, by being passed through screens of

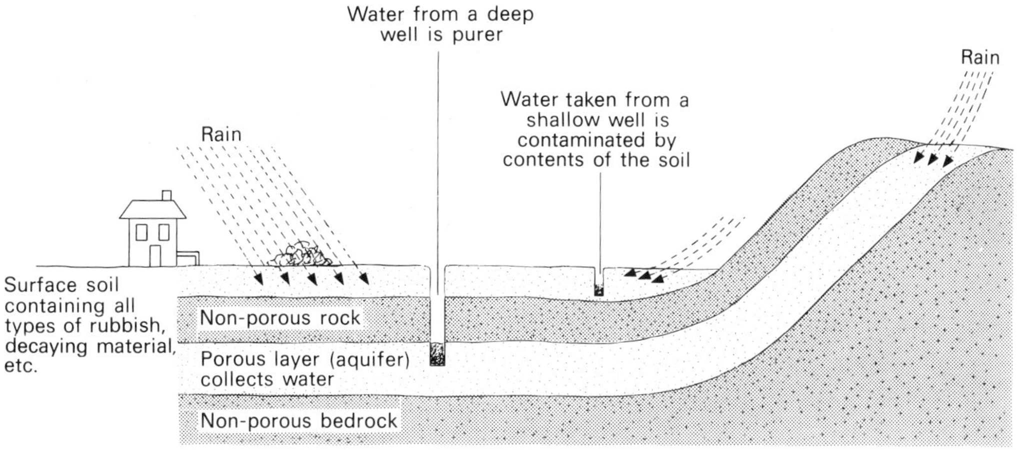

Fig. 10.7 *Water from deep and shallow wells*

Fig. 10.8 *Chemical foam on the River Mersey in March 1976. Now there is no foam but the river is still heavily polluted.*

fine wire gauze or allowed to trickle slowly through beds of fine sand. If the water is heavily polluted it is made to flow upwards through a *sedimentation tank*. It then trickles through a *gravity filter-bed* made up of layers of sand and gravel (see page 141). Particles and some bacteria are removed by this form of filtering because micro-organisms build up on the filters. These particular micro-organisms are scavengers of bacteria.

Even though the water leaving the filtration process is clear, it still contains some bacteria. Therefore this water, together with that taken from deep boreholes or from high lakes, is passed through a *disinfectant process*. In Chapter 7 we saw the use of chlorine in solution as a disinfectant. Chlorine gas, when added to water (usually 1 part of chlorine to 2 000 000 parts of water), will kill any germs it contains within 30 minutes. In such small quantities it remains virtually tasteless. The water then flows or is pumped to a *service reservoir*. This reservoir is near to the area it is to serve. It is covered and is often hidden underground. A water-tower may be used instead of a reservoir. Both of these can supply the daily needs of the customer. During the night they are topped up by pumps from the main source.

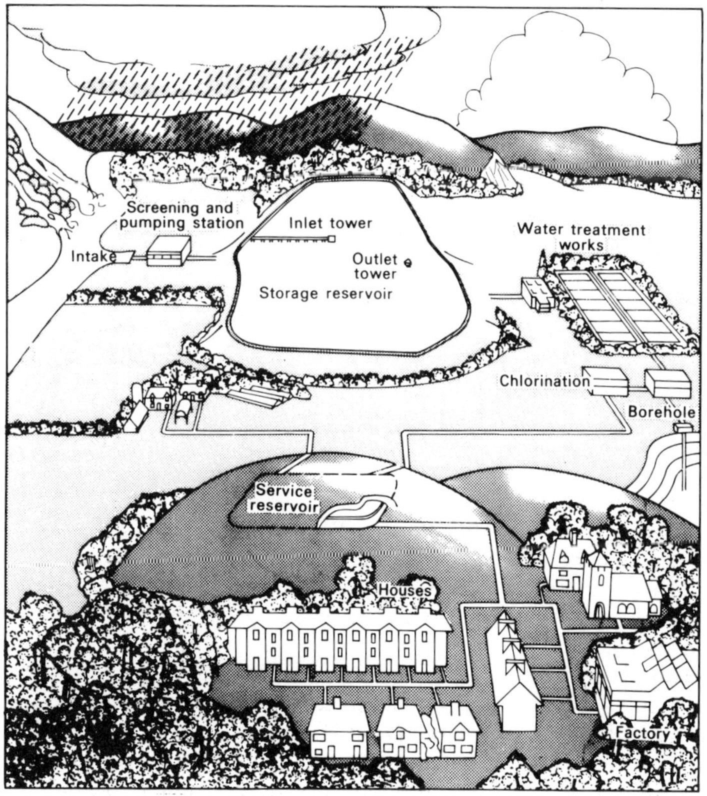

Fig. 10.9 *Water-supply and treatment of water*

Fig. 10.10 *Pump and overland pipeline drawing water from what is left of a reservoir (Sept. 1976)*

10.8 How much water do we consume?

In 1973 the Water Act formed ten water authorities and made them responsible for water-supply, sewerage, sewage disposal, land drainage, river management, control of water pollution, and the use of water for recreation such as fishing and boating. Payment for water and sewage disposal is made to the local water authority. Unlike energy we tend to treat water in our homes as if it were free and unlimited. It takes a drought like that experienced in 1976 (Fig. 10.10) to make us realize that we should be more careful in our demands. The North-West is the wettest part of England. Fig. 10.11 gives some of the figures provided by this authority, indicating the estimated demands made on the water-supply.

10.9 Parasites found in food and water

Chapter 7 discussed the occurrence of bacteria in food and water, and the diseases that they can cause in man. However, there are also larger organisms that can infect man through contact with water and food. The largest of these organisms are tapeworms.

Fig. 10.11 *Water-supply within the North-West Water Authority*

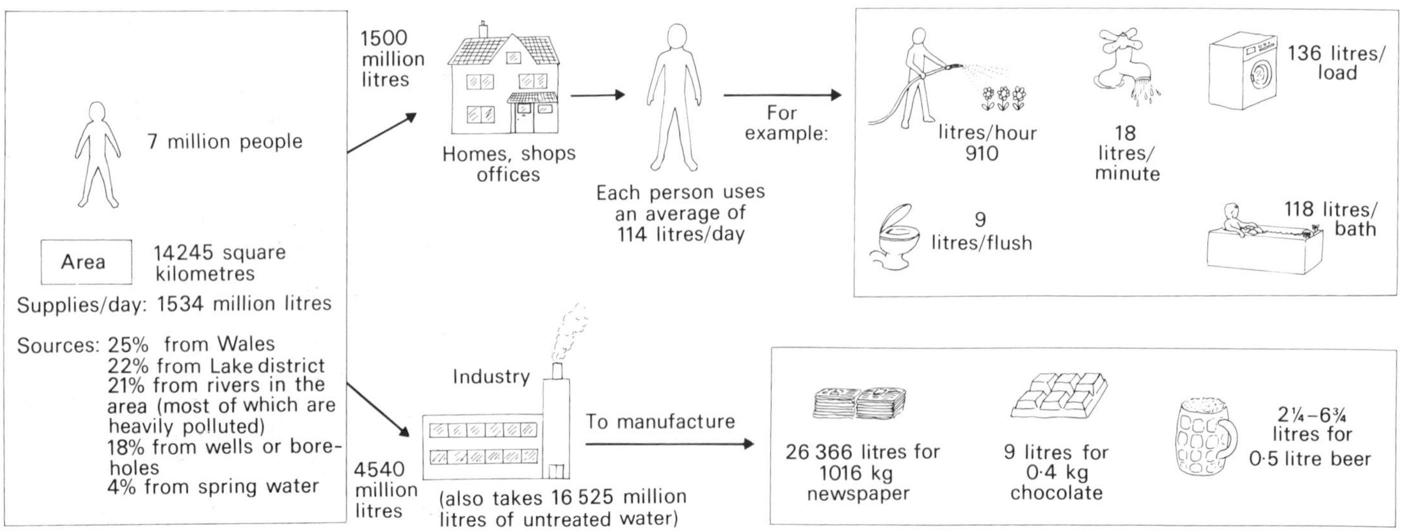

Tapeworms

Tapeworms, like many other parasites, need two hosts in which to complete their life cycle (Fig. 10.12). The larva of the tapeworm has hooks or suckers round its head region which enable it to attach itself to the lining of the intestine. It takes in food through its body wall from the material present in the intestine. As the tapeworm grows its end segments become sexually mature. These segments contain both male and female reproductive tissue: the male and female *gametes* fuse and fertilize within each segment. The tapeworm is said to be *hermaphrodite*, i.e. it is able to fertilize itself and needs no partner for reproduction. The segments break off as the eggs ripen and are deposited outside the host in the host's excreta.

These eggs, depending upon the type of tapeworm, will hatch if they are eaten by either cattle, fish or pigs. They develop in the intestine of the creature and then penetrate through to the fleshy areas of the body, such as muscle. Here they encyst, forming *bladder worms*. This larval stage of the parasite can be killed by very high temperatures. The cycle continues when this infested flesh is eaten (Fig. 10.12).

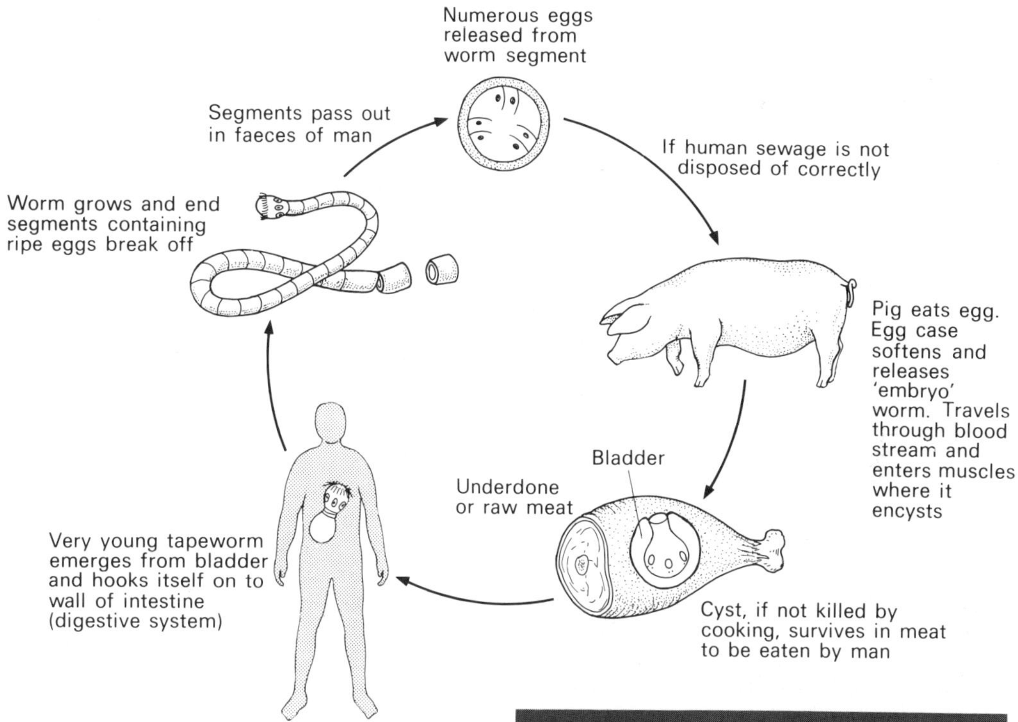

Fig. 10.12 *Life cycle of the pork tapeworm (*Taenia solium*)*

To break this cycle we can:

(a) ensure that human excreta and sewage is disposed of properly (page 164);
(b) use specialist inspectors at abattoirs to check carcasses for any signs of infection;
(c) if infection is diagnosed, give drugs to remove the tapeworm.

Question 5 Why do you think it is unwise to eat raw or undercooked meat?

Tapeworms that infest other creatures can develop in man in different ways. For example, if the eggs of a particular dog tapeworm are swallowed by man, they will hatch and the larvae will embed themselves in such organs as mucles, the liver and the brain. This infestation produces a cyst which can become very large and produce certain ill effects. There is evidence to show that children have suffered blindness owing to this type of infection. Owners can look out for worm segments in their pets' faeces and administer worming pills regularly.

Fig. 10.13 *Adult roundworm*

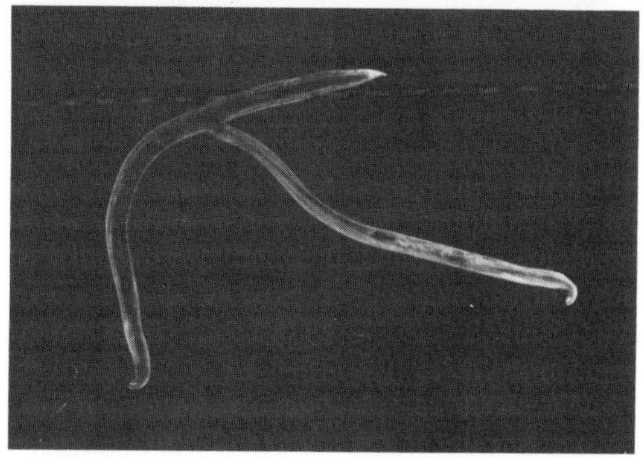

Fig. 10.14 *Hookworm*

Roundworms

Eggs are taken in through contaminated food or water. They develop into mature adults (Fig. 10.13) after burrowing through the intestine, entering the blood to reach the heart and lungs, then returning to the intestine. Hookworms (Fig. 10.14) are found in contaminated water and infect a person by burrowing through the skin. Severe infection can damage and block the organs through which the worms burrow.

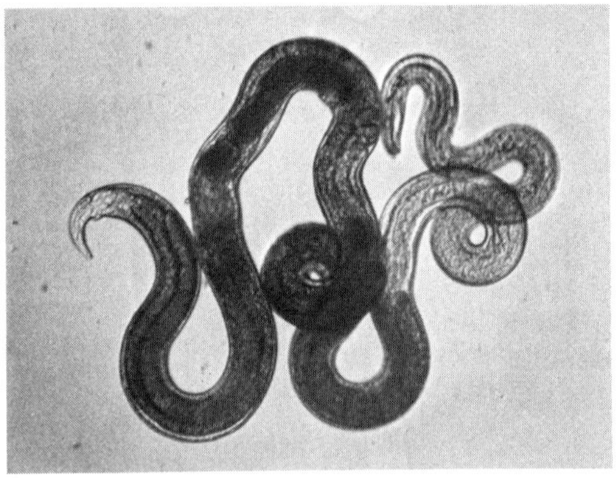

Fig. 10.15 *Threadworms – parasites commonly found in children*

Threadworms (pin-worm)
As seen in Fig. 10.15 these are very small parasites about 1 cm long, and they are among the most successful of all parasites. They only infect man, and often remain unnoticed for they have little effect on health. They are most commonly found in children and are spotted when a child wakes up during the night and complains of an 'itchy bottom'. This is caused by the worm leaving the back passage to lay eggs. Treatment consists of taking a course of a drug. It is usually prescribed for the whole family as infection can come only from another person. The itching makes the person scratch and eggs are transferred by fingers to clothing, bedding and other articles and even into the air. However the eggs cannot survive for very long outside the body unless they find another host.

10.10 **Food pests**

We often find ourselves in competition with other life forms for our supplies of food. When a food sample is ruined by these life forms we regard them as pests and usually try to find ways of preventing their action in future. Any gardener knows the value of nets to protect his fruit from birds but for smaller creatures, such as greenfly, he usually turns to 'chemical warfare'. *Insecticides* are invaluable, but intensive use of particular ones can have serious drawbacks. For instance, D.D.T. remains chemically active for years, and thus gradually accumulates in the soil, rivers and seas. It can reach such concentrations that it kills other life forms. It also accumulates in live tissues. If such chemicals are not seriously to affect our natural environment their use must be carefully regulated. In this way the disadvantages will not outweigh the gains in health and crop production.

Another solution is to develop and use alternative insecticides. For instance, *pyrethrum* has already taken the place of D.D.T. in our homes. The great advantage of pyrethrum is that it does not remain chemically active after use. When using any insecticide it is vital to obey the instructions on the can or wrapper.

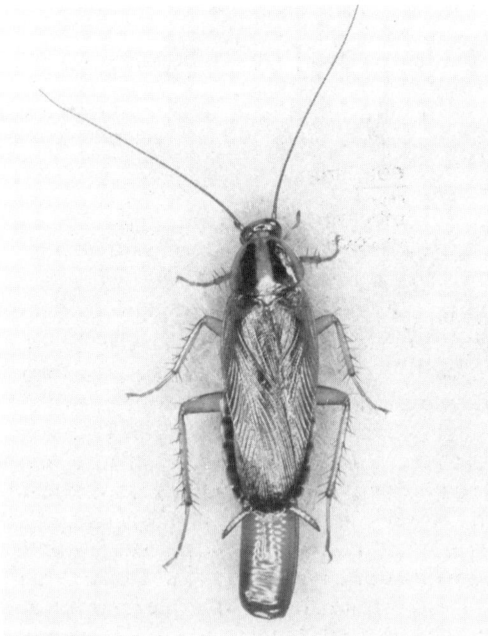

Fig. 10.16 *A German cockroach, commonly found in restaurants, hotels and hospitals. This one is carrying its egg capsule around with it.*

Fig. 10.17 *A house-fly*

Ants, cockroaches, houseflies, mice and rats (Figs 10.1 and 10.16) are some of the pests that are found in and around places where food is stored, cooked or disposed of. This includes home kitchens, food warehouses, food manufacturing premises, breweries, food shops, and canteens in factories and hospitals. The danger of any infestation of this kind is cross-contamination. For example, cockroaches live during the day in warm, dark places such as drains, sewers and rubbish heaps. At night they emerge in search of food. As they crawl over our food supplies they foul them with countless micro-organisms from their bodies and excreta. Tests have revealed that cockroaches can carry pathogenic bacteria such as salmonella, viruses such as polio and pathogenic protozoa. Therefore the presence of cockroaches should be viewed with as much concern as that of the house-fly (Fig. 10.17). Professional advice and pest control services are available to anyone who has a problem of pest infestation (Fig. 10.18).

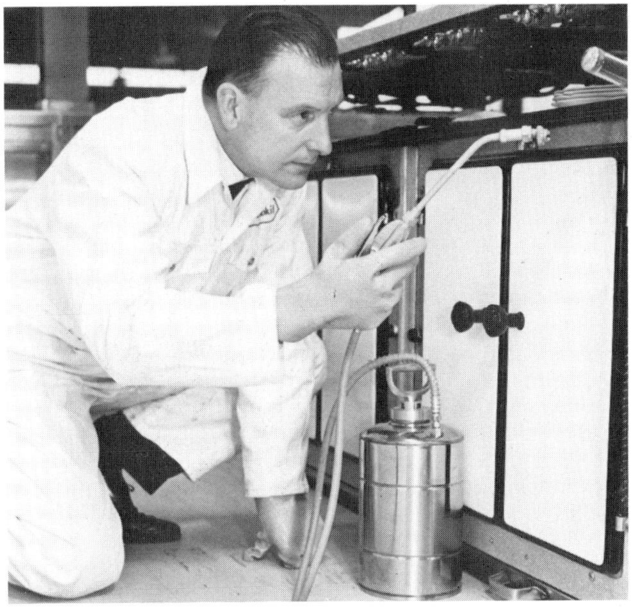

Fig. 10.18 *Spraying a canteen kitchen to rid it of insect pests*

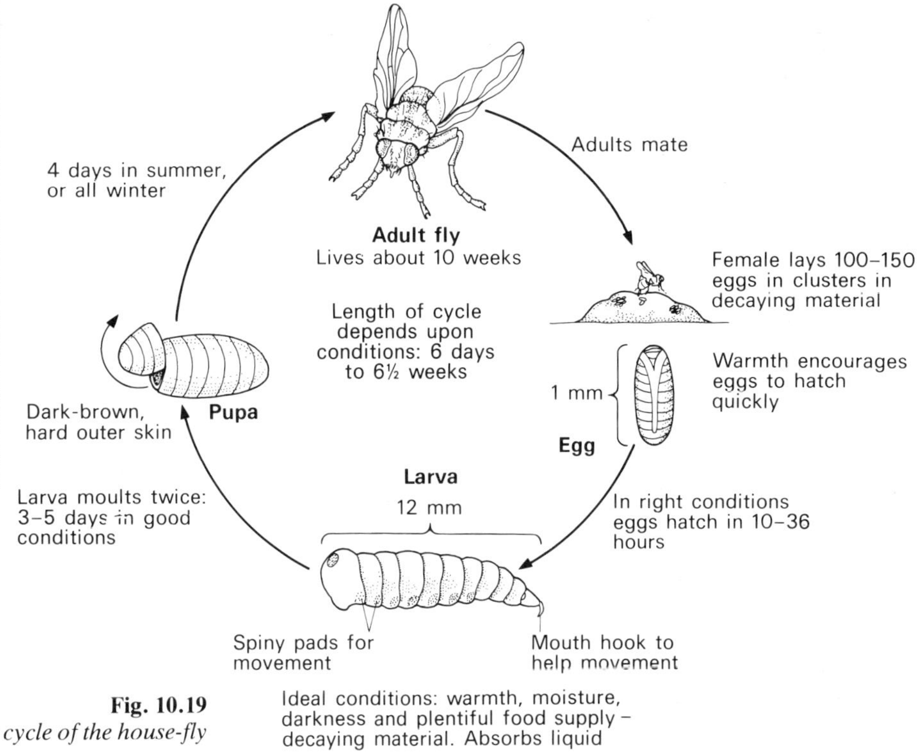

Fig. 10.19 *Life cycle of the house-fly*

10.11 The house-fly

Fig. 10.19 illustrates the life cycle of the house-fly. You may notice that, unlike other insects we have studied so far, this does not live for any part of its cycle in or on the human body. Nevertheless it is just as important a vector of disease. The reason for this is that it feeds both on decaying material and on human food (see page 132). The fly lands on its food and places the end of its mouth parts (the *proboscis*) on the food's surface. It then produces saliva, and brings up some of its stomach contents, both of which are poured on to the food through the proboscis. By this action the food is dissolved and can be sucked back up the proboscis into the mouth and the digestive system. There seems to be little harm in this until we consider where that fly may have been before it landed on our food! The female needs to lay its eggs in a warm, dark place where the hatched larvae can obtain food. This can be within any decaying vegetable or animal matter. Both sexes will also feed on such material. Thus the fly is a creature of dirty habits, crawling around on such things as decaying rubbish piles and heaps of animal excreta. With decay there are microbes, some of which may be pathogenic. These microbes are picked up on the hairy body of the fly, or they may enter its salivary glands and intestine. The fly transfers these microbes to the area where it next lands, and that could be the food you are just about to eat. In fact, individual flies have been found to have on or within their bodies as many as 533 000 000 organisms. These could easily include micro-organisms causing the diseases of dysentery, typhoid, polio and, in some areas of the world, cholera.

Protection against diseases carried by the house-fly involves (a) controlling the numbers of flies, (b) wrapping or covering food, and (c) organizing safe disposal of refuse such as sewage (page 164) and household waste.

Question 6

(a) How can we prevent dustbins becoming breeding places for flies?

(b) Why should dustbins be positioned as far as possible from the kitchen?

10 POINT CODES FOR HOUSEWIVES

1 Buy only from clean places. Get the food home clean.

2 Use clean containers in your home. Use your refrigerator properly. If in doubt, find out.

3 Keep family foods away from food for pets. Use separate utensils and crockery.

4 Wash your hands always before preparing food, always after using the W.C. See your children do too.

5 Cover cuts and sores with waterproof dressings. If you are not well with no one to take your place in the kitchen then be extra careful about personal cleanliness.

6 Keep food clean, covered and either cool or piping hot.

7 Reheated leftovers must be made really hot right through. Do the same with ready packed foods intended to be eaten hot.

8 Keep working surfaces clean. Use really hot, soapy water. A wipe with a dish cloth is not enough.

9 Stack washed and rinsed crockery and pans to drain. If you use drying cloths be sure they are clean.

10 Keep the lid on the dustbin.

The Health Education Council
78 New Oxford Street, London WC1A 1AH

Fig. 10.20 *'Food Code'*

The Health Education Council issues a leaflet which gives a ten-point food code for housewives and food trade workers. This can be seen in Fig. 10.20. If we wish to prevent the contamination of food and its dangerous consequences, we must observe this code.

Anwers to questions in the text

1 Harmful micro-organisms can invade and multiply within food. This is more likely the longer the food is kept.

2 (a) The temperature in the refrigerator would rise because of the hot food, so the other contents would no longer be cool.
(b) Fig. 10.21 outlines some of the health hazards.

3 It is taken from the name of a man, Louis Pasteur. Pasteur developed the process which we now call pasteurization when he was trying to prevent wine going sour.

4 Every patient should have been asked what they had eaten or drunk. If they had all eaten or drunk one thing in common this would have indicated a possible source of the disease.

5 Cooking meat at the right temperature and for long enough will kill the encysted tapeworm.

6 (a) Clean and disinfect the dustbin regularly. Wrap all waste in newspaper or bin liners.
(b) Fewer flies will enter the kitchen this way. The further away the bins are, the fewer flies will get into the kitchen.

Fig. 10.21 *Answers to Question 2(b)*

THINGS TO DO

Your teacher may be able to arrange for your class to visit a dairy, a food factory, a bakery or a water-treatment plant. Also an E.H.O. could visit you to talk about his work.

The action of micro-organisms in contaminating food can be studied through cultivating colonies on agar. It has already been stressed on page 84 that cultures of bacteria and moulds can, if mishandled, present a hazard to health. This section of your work must therefore be done with the guidance of your teacher and with strict observation of the safety precautions.

1 Visit a food shop. Make a list of everything which had been done by the management to ensure good food hygiene.

2 Imagine that a national competition has been launched to find a design for a poster which will be sent to every housewife in the country. The aim of the poster will be to point out to the housewife the possibilities of food-to-food cross-contamination in her kitchen. Draw your entry.

3 Peel and core an apple. Cut it into thin slices and thread them on to a string. Hang the slices in a warm place. This drying of the apple slices is a way of preserving them. Try drying other foods such as herbs and mushrooms. A good cookery book will describe other methods of preserving food.

4 Construct a montage to show how water is recycled. Include yourself and the town water-supply.

5 Put into a scrap-book any newspaper articles which deal with cases of (a) food poisoning and (b) prosecutions when food hygiene regulations have been broken. Use these articles to study how the health of the community is protected by the work of the local health authority.

6 Carry out a survey of commercially preserved foods. List the foods and describe any special recommendations that are issued by the manufacture concerning the method of storage, length of storage and cooking.

7 Take a sample of muddy water. Into a clean plant pot put a layer of large pebbles, followed by layers of gravel, coarse sand and finally fine sand. Filter your sample of muddy water. Find out how you should purify contaminated water in an emergency.

8 Find out how cheese is made commercially. Cheese can be made safely in the home by 'setting' pasteurized milk with rennet. *It must not be made from pasteurized milk which has been allowed to go sour*, as this could be a source of dangerous bacteria.

9 Draw a map of your area. Indicate the position of natural watercourses, water-towers, reservoirs, and pumping stations. Survey the conditions of the natural watercourses and use a colour code to indicate the level of pollution on your map. Pin-point any areas of water that are used for recreation.

THINGS TO WORK OUT

In Questions 1–4 write down the *one* answer which you think is correct.

1 The conditions that encourage the rapid multiplication of bacteria in food are
- **A** plenty of moisture and temperatures above 80 °C
- **B** no moisture and temperatures below 3 °C
- **C** plenty of moisture and temperatures below 3 °C
- **D** no moisture and temperatures between 15 and 35 °C
- **E** plenty of moisture and temperatures between 15 and 35 °C

2 The chemical which is added to water to purify it is
- **A** chlorine
- **B** ozone
- **C** oxygen
- **D** fluorine
- **E** carbon dioxide

3 People who drink raw milk run the risk of contracting
- **A** bronchitis
- **B** brucellosis
- **C** asthma
- **D** stomach ulcers
- **E** typhoid

4 Tapeworms are contracted by
- **A** breathing in air which contains the parasite's eggs
- **B** drinking contaminated water
- **C** bathing in water contaminated with human sewage
- **D** eating infected and undercooked meat
- **E** touching infected bedding or clothing

5 (a) What is meant by the term cross-contamination?
(b) List *two* measures that should be taken in the kitchen of a restaurant to prevent cross-contamination by
(i) human-to-food contact
(ii) food-to-food contact
(iii) animal-to-food contact
(c) Who is responsible for checking hygiene standards in all food establishments?

6 Salting, canning, pickling, deep-freezing and drying are all methods of food preservation.
(a) Describe *three* of these methods in detail.
(b) State how each method deals with the harmful micro-organisms in food.
(c) Name a food that can be preserved by each method.
(d) Select *two* methods and describe the precautions that must be taken during storage to maintain preservation.

7 (a) Give *three* contaminants that are present in rivers and lakes.
(b) Describe the processes that are used to purify such water before it is supplied to our homes.
(c) State *three* responsibilities of the local water authorities (do not include water purification).
(d) Name *three* diseases that are transmitted through water.

8 (a) Distinguish between the processes of sterilization and pasteurization of milk.
(b) State *two* ways in which human milk is different from cow's milk.
(c) What form of cow's milk is used in a bottle-feed for a baby? Give *three* precautions that should be taken in making up a bottle-feed.
(d) Name *two* other forms of milk that can be stored. How are they produced?

9 (a) Describe with diagrams the life cycle of a house-fly.
(b) How does an adult fly feed?
(c) Name *two* diseases that can be transmitted by the house-fly, and describe how the transmission takes place.
(d) Give *four* ways of preventing the transmission of disease by house-flies.

11 The Food Chain and Teeth

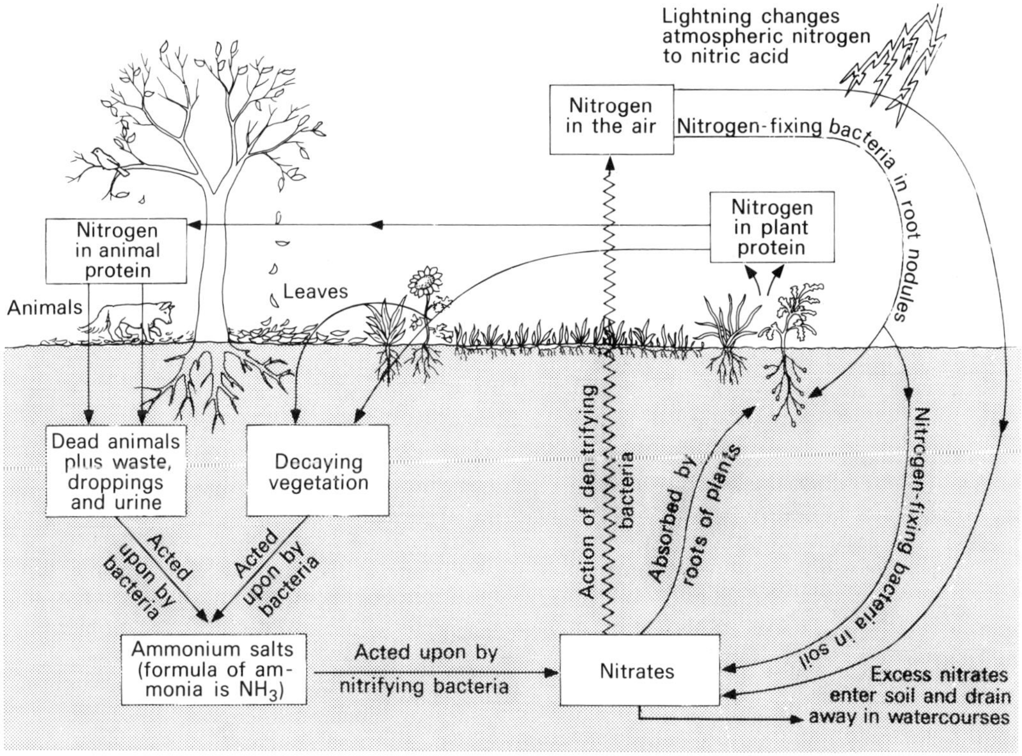

Fig. 11.1 *The nitrogen cycle*

11.1 **Energy and the carbon cycle**

Nowadays we are becoming more aware of the need to reuse many of the things that we would normally throw away. For example, many organizations collect waste paper. This waste paper is sent to a factory where it is mechanically and chemically treated to produce raw material from which new paper can be made. You may have used paper bags made from *recycled* paper. Recycling prevents accumulation of waste and the loss of valuable materials or energy. Nature has never been wasteful. The energy and food materials that are available on this planet are part of these natural recycling processes (see carbon cycle, page 44).

Energy is governed by a physical law known as the *conservation of energy*. This states that energy cannot be created or destroyed, but can be converted from one form to another (see Fig. 9.10). The living processes within organisms depend upon one source of energy, the sun. However, only plants can harness this energy directly. They use the energy of sunlight for the chemical process of *photosynthesis*: the food of both plants and animals is manufactured by this process. All animals depend upon plants for their food, either directly or indirectly. For instance, as we have seen in Chapter 9, we are able to obtain some proteins from plants, but other essential proteins must come from eating the flesh or produce of animals that have lived on plants. One of the essential elements in a protein molecule is nitrogen.

11.2 **The nitrogen cycle** (Fig. 11.1)

Most plants absorb their nitrogen from the soil in the form of compounds known as nitrates. These nitrates are present in the soil through the action of bacteria which we call *nitrifying bacteria*. Their existence depends upon the energy that they obtain by oxidizing the ammonia present in decaying organic material within the soil. In this way there is a natural recycling of nitrogen. You can also see in Fig. 11.1 that other bacteria help in this cycle: these are *nitrogen-fixing bacteria*. Unlike other life forms these bacteria can capture the free nitrogen that is present in air and use it to manufacture amino-acids.

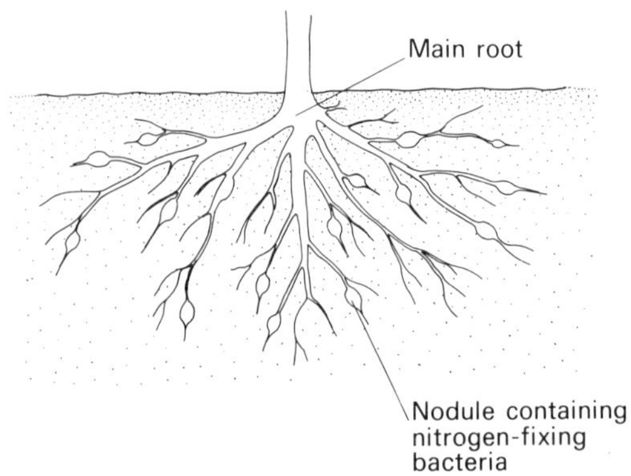

Fig. 11.2 *Nodules on the root of a leguminous plant*

Plants which are members of the pea family (Leguminosae) have these bacteria in swellings, called nodules, on their roots (Fig. 11.2). A farmer who grows a crop of clover and then ploughs it into the soil will be putting back the essential nitrogen which other crops have removed. This is one of the reasons for the practice of *crop rotation*. Fig. 11.3 illustrates a form of crop rotation that can be used in a small vegetable garden.

1st year	2nd year	3rd year	
Natural manure, Onions, Peas, Beans, Leeks	**Root crops**: Carrots, Turnips, Parsnips	**Greens**: Cabbage, Lettuce, Spinach	⅓ of plot
Greens: Cabbage, Lettuce, Spinach	Natural manure, Onions, Peas, Beans, Leeks	**Root crops**: Carrots, Turnips, Parsnips	⅓ of plot
Root crops: Carrots, Turnips, Parsnips	**Greens**: Cabbage, Lettuce, Spinach	Natural manure, Onions, Peas, Beans, Leeks	⅓ of plot

Fig. 11.3 *One form of crop rotation for a vegetable garden*

11.3 Fertility of the soil

A system of crop rotation helps the soil to keep its 'natural goodness', i.e. all the elements which are needed for the healthy growth of plants. Any person growing crops, whether a farmer owning many hectares or a gardener with a small vegetable plot, depends upon the condition, texture and fertility of the soil. Digging in rotting organic material from a compost heap, or animal manure, will help to separate the soil particles and so break up the soil. It will also provide the elements needed by plants, particularly nitrogen, phosphorus, potassium and calcium. The simple rule to remember is that the soil feeds the plants, plants feed animals, animals manure the soil, and manure feeds the soil.

Question 1 Most small gardens do not have their own source of animal manure. If the gardener wants to make a compost heap of rotting vegetation, what can he use from the garden and from the kitchen?

If a farmer or gardener finds it impossible to maintain the fertility of the soil by natural manuring, he may decide to use *artificial fertilizers*. The use of chemical fertilizers does give rise to problems: for example, natural manure adds fibrous material called *humus* to the soil. Humus helps to keep the soil particles apart, producing good air spaces for ventilation and drainage. Chemical fertilizers, if used for long periods, can destroy this soil structure through loss of natural humus. Excessive use of fertilizers also leads to surplus nitrates draining through the soil into natural waterways and lakes. Microscopic organisms and certain plants in the water grow and multiply rapidly on this rich nitrate food supply. When these organisms and plants die they are broken down by the action of bacteria. These bacteria are *aerobic*, i.e. they need oxygen for their activities. With such a lot of material to work upon they use up all the oxygen that is available in the water. Thus the water becomes fouled with an excess of rotting and decaying material, and other life forms die through lack of oxygen. The drainage of other nitrates from sewage-works (page 162) can add to this problem in our natural waterways. Water authorities constantly monitor the nitrate level in our water-supplies. Very high levels can be harmful, especially to the very young.

11.4 Food-chains

Recycling of essential elements such as carbon, nitrogen and oxygen means that all forms of life on earth are interdependent (page 44). We are dependent upon other life forms for our food supplies. By tracing where a food comes from we can construct a *food-chain* for any animal. Fig. 11.4 shows different ways of illustrating such a food-chain. All these diagrams simplify the natural process that takes place and each illustrates a different feature of the food-chain. The *food-pyramid* (Fig. 11.4(b)) illustrates that energy is lost each time it is transferred from one living form to another. This means that the animal at the top of the pyramid must eat a larger number of the organisms beneath it in the food-chain in order to get its full energy requirements. At each level the total weight of the organisms will be less than that on the lower levels. The food-web (Fig. 11.4(c)) illustrates the true complex nature of a food-chain, i.e. how an organism usually feeds on a great variety of other organisms.

11.5 Population and the demand for food

Our survival depends upon there being enough food for us to eat. It is estimated that two-thirds of the world's population (which is at present approximately 3552 million) is undernourished; people who are undernourished lack either the correct quantity of food or enough first-class protein. We have

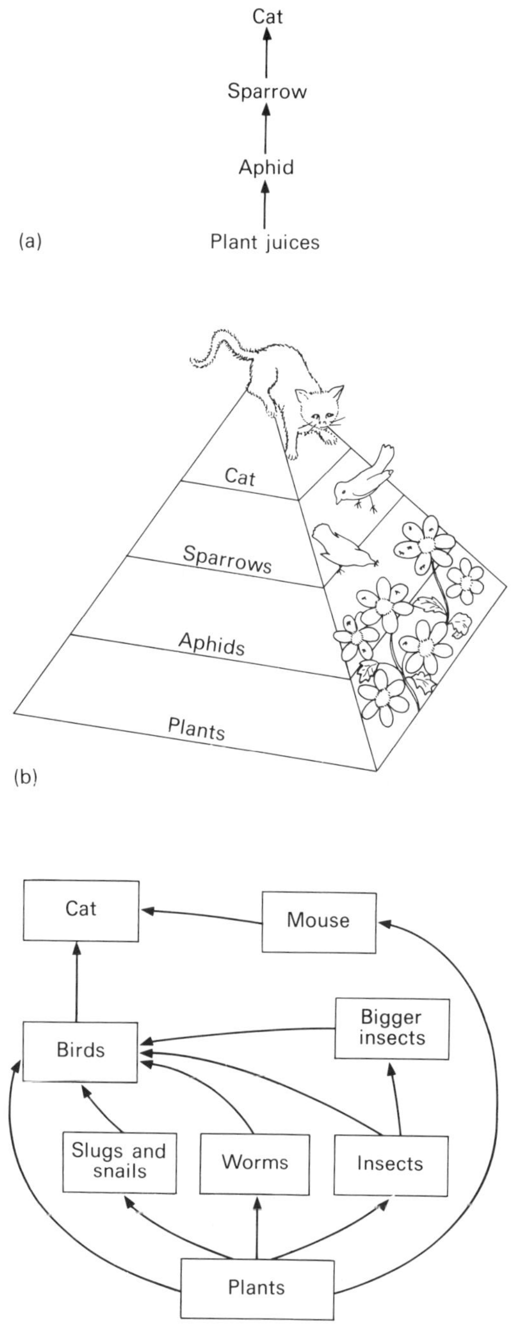

Fig. 11.4 *(a) Simple food-chain. (b) Food-pyramid. (c) Food-web*

already seen in Chapter 9 some of the specific diseases that may arise from certain food deficiencies. To these we should add the fact that children whose diets lack first-class protein in the very early stages of life are generally slower mentally than children who have no such food deficiency. Our brains undergo rapid growth in the first three years of life. In fact by the age of three the brain has grown to 80 per cent of its adult size. This development relies upon the vital amino-acids that are present in first-class protein food. Without these amino-acids the growth of brain cells is below normal growth levels. Therefore, to feed the world's population in the future, the provision of first-class protein will be essential. However, as we have already observed in our study of food-chains, the 'conversion of plants to meat' is a very inefficient process in terms of energy.

It is estimated that by the year 2000 the world's population will have doubled (Fig. 11.5). Twenty per cent of this figure will be the population of developed countries. Any increase in population means that there is a greater demand for vital foodstuffs. A country may not itself be able to meet this demand as its agricultural land may have been used up for housing and industry. Thus food has to be bought from the world market, and as demand increases, so supplies become short and the prices rise steeply. At present much work is being done on finding alternative food supplies and ways of increasing our production of food by new growing methods.

One of the richest sources of plant protein is the soya bean. The plant that yields this bean looks very like a French-bean plant. It is grown in warmer climates such as central America and central Europe. The production of soya-bean protein is particularly efficient in terms of land use. For example, an area of land put down to soya-bean production would yield approximately five times more protein by weight than the beef which would be produced if the same area of land was used to graze cattle.

Soya bean is used in the production of

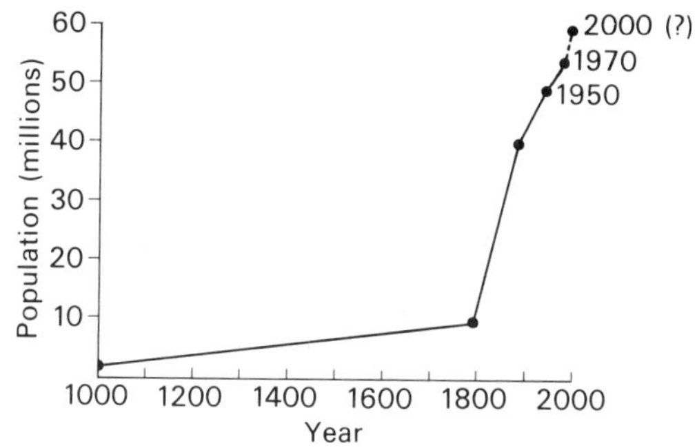

Fig. 11.5 *The growth of the population of Britain*

animal feeds. It is a valuable source of protein concentrate that can be added to other foods that humans eat. It is also becoming a valuable food in its own right when used as *textured vegetable protein* (Colour plate 3). The term 'textured' indicates the problem of the acceptance of a new source of food by the public. The type of food we eat is often part of the country's tradition, for example the 'roast beef of England'. So no matter how nutritious a new food source may be it is often rejected, even by undernourished people, because it does not resemble the traditional diet. As new foods (food analogues) are developed from alternative sources, scientists have to find ways of giving them an acceptable flavour, colour and texture. You might like to discuss our attitude to food in this respect.

Research is being carried out by producing food analogues by such methods as culturing yeast on the unwanted by-products of the oil industry, and culturing micro-organisms on carbohydrates or cellulose. The culture formed by these methods is removed at intervals and then processed to yield a white powder.

Other methods of coping with the increasing demand for food involve cutting down waste in our present methods of farming, using every area of land by reclaiming waste land, and storing food so that it is not attacked by the weather or pests. New varieties of wheat and vegetables which will resist disease and grow more strongly in the fields are under development. Some plants, for instance lettuces, are already grown intensively without soil. The are supported so that their roots dip into water that runs through plastic channels in the greenhouse. The water contains all the elements that the plants need for growth. This method is called *hydroponic culture*.

The sea contains a wealth of food. It may be possible to 'farm' its microscopic plankton and use this in the manufacture of animal feeds or food analogues. Some fish-farms are in operation in Britain as a result of research into the breeding, hatching and growing of fish in large quantities.

11.6 **Teeth**

The survival of an organism depends not only upon its ability to obtain food but also on its ability to avoid being eaten.

One part of an animal that is closely linked with its ability to obtain and devour food is its teeth. Fig. 11.6(a) illustrates the teeth of a *carnivore*: an animal such as a wild dog that preys on other animals and eats flesh alone. Fig. 11.6(b) shows the teeth of a *herbivore*: an animal such as a rabbit that only eats plants. Fig. 11.6(c) illustrates the teeth of an *omnivore*: an animal that eats both meat and plants. The differences that we can see in these three sets of teeth have close links with the animal's diet. For instance, the function of human teeth is to bite and slice the food, then to grind it down into smaller particles.

To carry out these actions the teeth must apply a mechanical force. This force is supplied by muscles which move the lower jaw up against the upper jaw. You only have to place your finger in your mouth and close your jaws to discover the extent of that force! It is an example of the lever principle. However, other mechanical devices are at work, and this is seen more clearly if we consider the shape and positioning of the teeth within the jaw (Fig. 11.7).

If you feel each tooth in your mouth you will find that there are three basic shapes. The wedge-shaped teeth at the front of your mouth are *incisors*. Like an axe, incisors have a sharp, narrow edge which is perfect for cutting and slicing through food. *Canines* are found at the side of the incisors. They are more rounded and end in a definite point. This shape makes them most suitable for puncturing flesh. As you can see in Fig. 11.8, no vampire can operate without his canines! Once the canines have penetrated the flesh they become anchored among its fibres, and with a tugging action they can tear it apart. You have probably used this technique to tear flesh from a chicken leg.

Question 2 Is the force applied at the back of the jaw greater or smaller than that applied at the front?

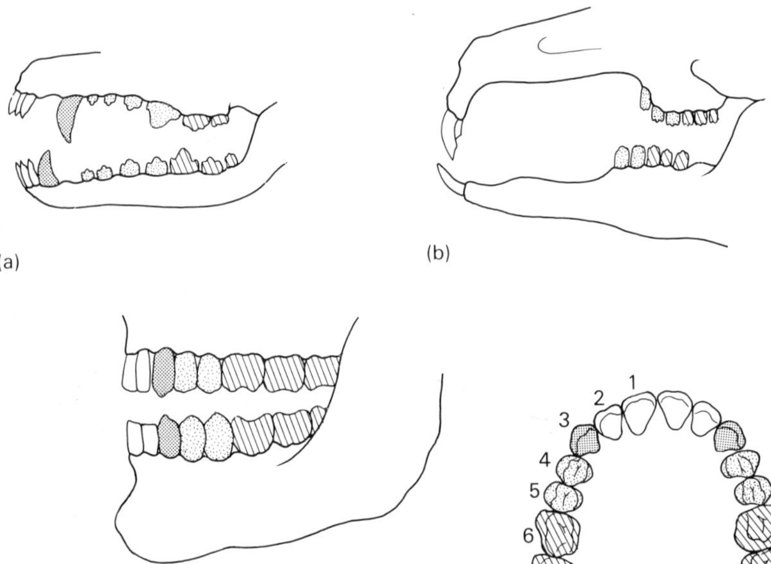

Fig. 11.6 *(a) Teeth of a carnivore (flesh-eater): wild dog (e.g. dingo in Australia). (b) Teeth of a herbivore (plant-eater): rabbit. (c) Teeth of an omnivore (plant- and flesh-eater): man*

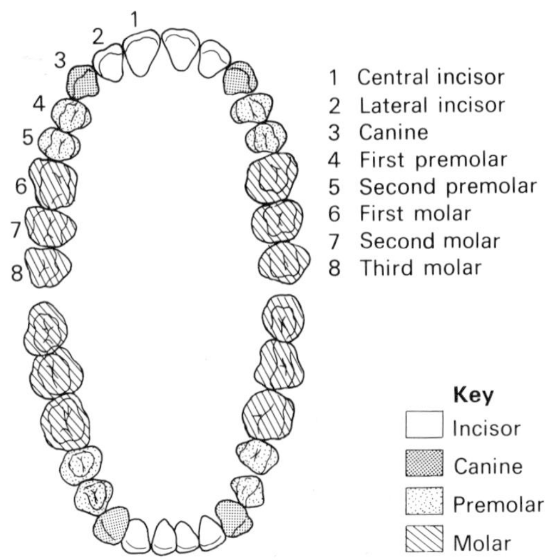

Fig. 11.7 (Right) *Shape and position of teeth within the human jaw*

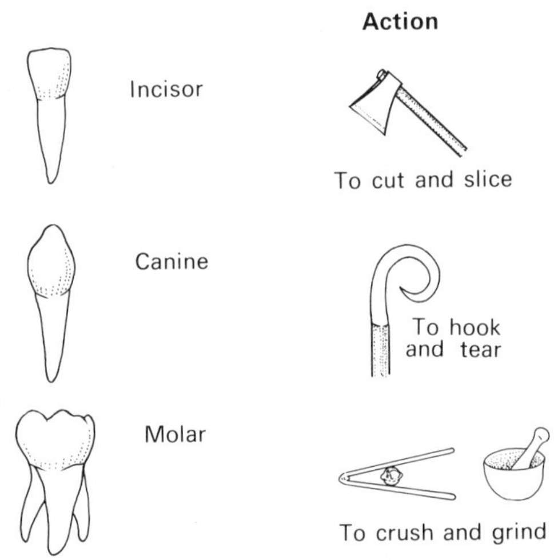

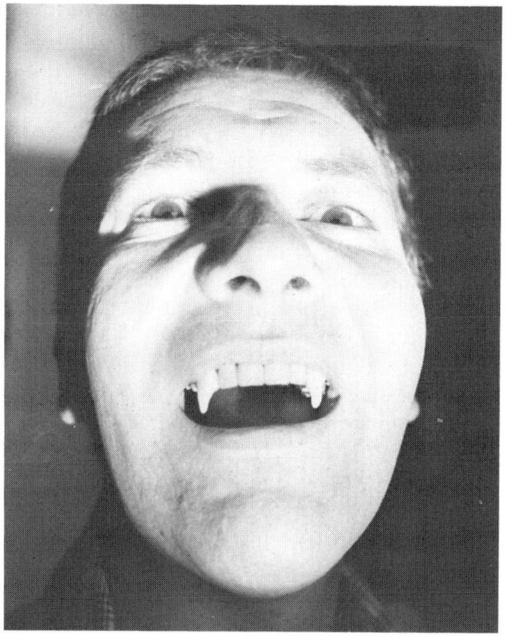

Fig. 11.8 *This actor was provided with his vampire teeth by the local dental school!*

11.7 The development and growth of teeth

Teeth first start to develop in the human embryo at about the sixth week as *tooth buds*. This first set of teeth, often referred to as *milk-teeth* or *deciduous teeth*, is present beneath the *gums* when the baby is born. After a few months the baby's second set of teeth, known as *permanent teeth*, begin their development out of sight within the jaw. You can see that the foundations of both milk-teeth and permanent teeth are laid down long before they appear. Therefore the expectant mother must make sure that her diet in early pregnancy includes the elements essential to dental tissue. These elements, calcium and phosphorus, are found in milk and cheese. Vitamins D and C are also essential. Similarly, all of these materials should be contained in the diet of a very young child to ensure the formation of strong permanent teeth.

Like other developments within the body, when and how the first tooth appears will vary among individuals. Some babies will cut their teeth with great distress, others will not. On average a baby's first tooth erupts at approximately seven months. Within the next seventeen months the remaining nineteen teeth will appear (Fig. 11.9). Fig. 11.10 is an X-ray photograph of the jaws of a six-year-old child. You can see the permanent teeth in the gums under the milk-teeth.

The *molars* are positioned at the back of the mouth. They are much wider than all the other teeth, and their flattened surfaces together with their position in the jaw makes them most suitable for crushing food. In fact the surface of molars are not completely flat. They have ridges with rounded mounds called *cusps* on either side of the lower area. *Premolars* possess two cusps and are positioned next to the canines. The remaining group of molars are much larger, extend right to the back of the mouth and possess up to four cusps. When the jaws are closed the surfaces of the molars opposite to one another are able to interlock. Further movement of the jaws allows the surface of the molars to slide across one another, thus grinding the food.

Question 3 Fig. 11.6(b) shows the teeth of a herbivore. Why do you think it has a long row of molars with deep ridges?

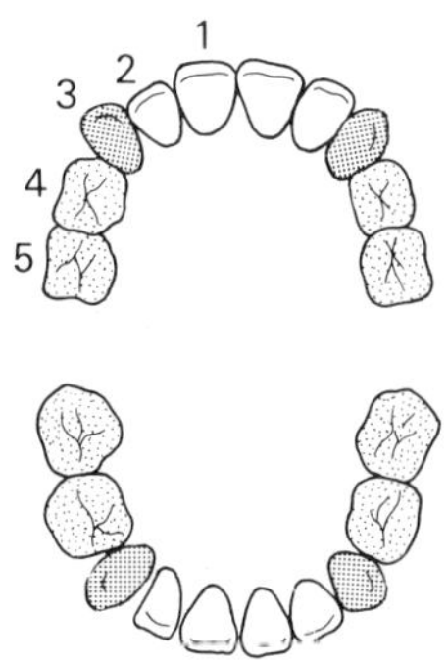

		Eruption	Lost
1	Central incisors	6–8 months	6–8 years
2	Lateral incisors	7–9 months	7–8 years
3	Canines	16–18 months	9–12 years
4	First molars	12–14 months	10–11 years
5	Second molars	20–24 months	10–11 years

Fig. 11.9 *Growth of milk-teeth*

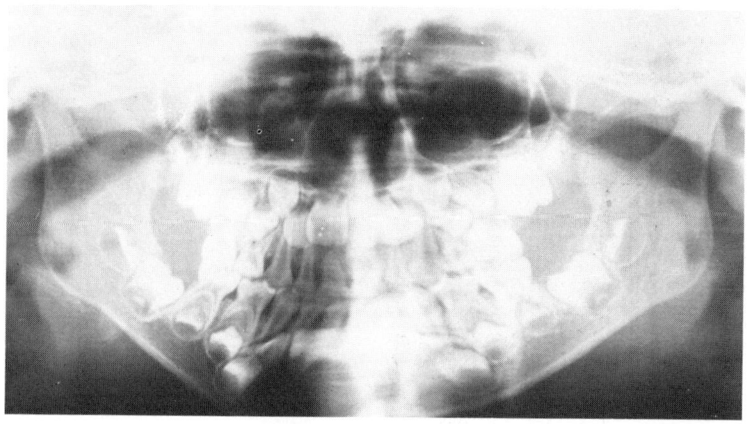

Fig. 11.10 *X-ray of the jaw of a six-year-old child. The permanent teeth are visible under the gum*

11.8 The structure of a tooth

Only a small fraction of a tooth is visible above the gum tissue covering the jaw-bone (look again at Fig. 11.10). This visible portion is the *crown* of the tooth. As this is the surface which bites, tears or grinds the food it needs to be made of an extremely hard material. *Enamel*, a hard non-living material consisting mainly of calcium phosphate, is ideal. Fig. 11.11 shows a molar tooth in cross-section. You can see that the enamel layer covers and protects a layer of *dentine*. Dentine is a material very similar to bone, although it is not as hard as enamel. The dentine extends downwards into the socket in the jaw-bone to form the *root* of the tooth. A layer of *dental cement* surrounds the root. Embedded in the cement are the ends of fibrous strands which stretch across to the jaw-bone like anchor chains. Thus the root of the tooth is held firmly in position but at the same time is cushioned against quite severe impacts. Even so a really hard bang that might occur in an accident or on the games field can push a tooth out of place. On such an occasion an X-ray can be used to see what damage has been done to the root. It is at the bottom of the root that blood-vessels enter the tooth through a narrow channel. If the tooth is displaced so that these blood-vessels are severely damaged, the tooth will blacken and die.

Question 4 Why do you think the tooth dies under these circumstances?

The blood-vessels form capillaries within the central *pulp cavity* of the tooth. This cavity also contains the sensitive end of the nerve (page 223), which enters the tooth by the same channel as the blood-vessels. Very few of us have not experienced the pain that is produced when the enamel of the tooth ceases to protect these sensitive areas!

11.9 **Permanent teeth**

When a child is about six years old the twenty milk-teeth begin to be replaced. Most children are very proud of their first 'wobbly' tooth, displaying it to everyone who shows an interest. What has happened to the firm root anchor? If you study Fig. 11.10 again you will be able to see that some milk-teeth have hardly any root left. Their roots have been gradually reabsorbed by the natural activity of the cells. Without a firm root the crown begins to wobble and usually falls out as the permanent tooth grows through. If you have ever needed to have a tooth pulled out by a dentist, you may have noticed its long root compared with that of a tooth which has fallen out naturally.

Gradually all the milk-teeth are lost, to be replaced by permanent teeth in the same order that the first teeth grew. By twelve years the replacement is complete and there have been several additions (Fig. 11.12). For instance, at about six years the first of the new molars erupts behind a milk-molar. The milk-molars are then gradually replaced by premolars. The second molars are last to erupt, behind the first molars, at about

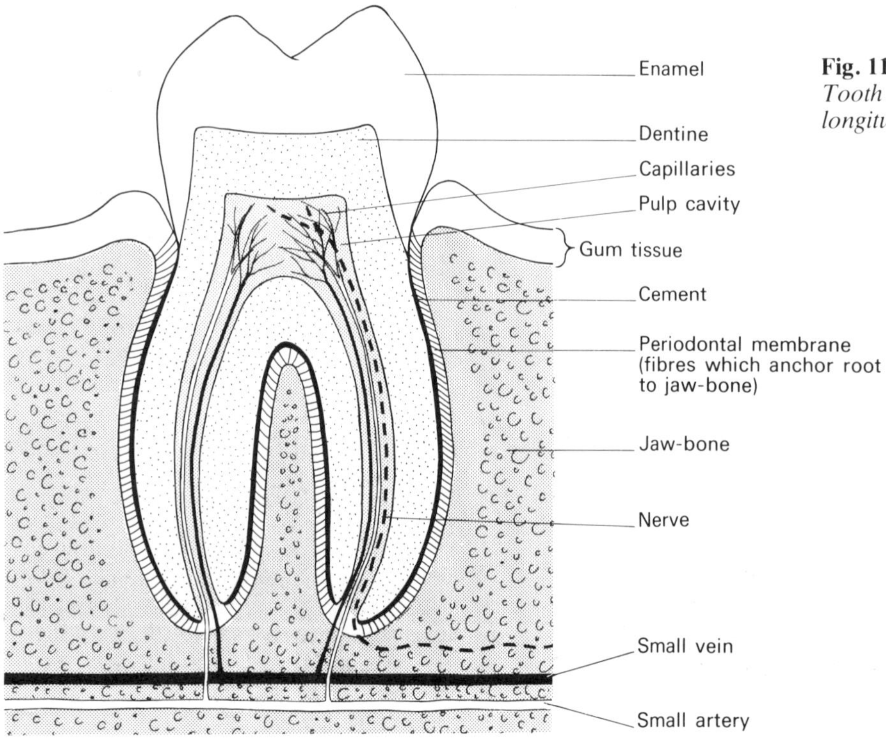

Fig. 11.11 (Left) *Tooth in longitudinal section*

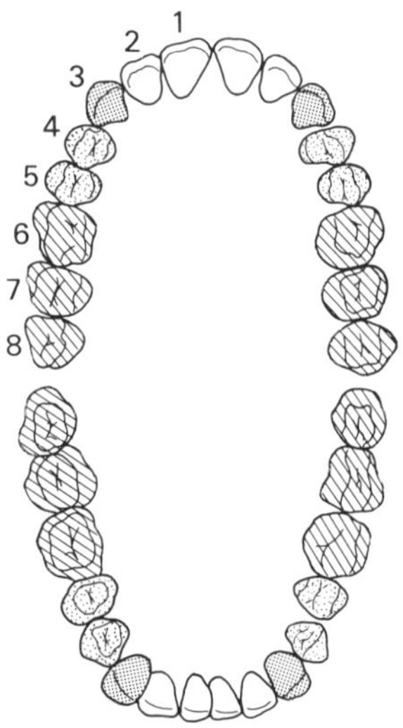

		Eruption
1	Central incisors	6–8 years
2	Lateral incisors	7–9 years
3	Canines	9–12 years
4	First premolars	10–12 years
5	Second premolars	10–12 years
6	First molars	6–7 years
7	Second molars	11–13 years
8	Third molars (wisdom teeth)	17–21 years

Fig. 11.12 (Right) *Growth of permanent teeth*

twelve years. A third set of molars, often called *wisdom teeth*, erupts at about eighteen to complete the full set of thirty-two teeth in the adult mouth. Some people never get their wisdom teeth, in others, their appearance may be painful or may overcrowd the jaw so that they have to be removed. Fig. 11.13 is an X-ray photograph of the jaw of an adult. You can see that the wisdom teeth have not erupted through the gum in this instance.

There is a shorthand way of writing down the number of incisors, canines and molars that make up our full set of thirty-two teeth. It is written for one side of the jaw as follows:

$$\mathrm{i}\,\frac{2}{2}\quad \mathrm{c}\,\frac{1}{1}\quad \mathrm{p}\,\frac{2}{2}\quad \mathrm{m}\,\frac{3}{3}$$

Total number of teeth = 16 × 2 = 32.

This is known as a *dental formula*. The upper row of figures gives the number of teeth within that particular group in half the upper jaw. The lower row of figures represents half the lower jaw. This method of writing a dental formula can be applied to any other sets of teeth.

Question 5 What is the dental formula of the carnivore illustrated in Fig. 11.6(a)?

There is no replacement for our permanent teeth other than artificial teeth in the form of *crowns* or *dentures*. When human teeth are fully grown the root channel becomes much smaller. There is enough blood to keep the tooth alive but not enough for further growth, so once the permanent teeth have grown the body cannot replace any piece that is broken off or worn away. In a herbivore growth continues as the root channel remains open, and the surfaces of a herbivore's molars are continually replaced. On the other hand, in the wild an elderly carnivore whose teeth are worn or lost cannot survive, for the animal can neither catch its food nor defend itself. How does the loss of teeth affect our lives?

11.10 Tooth decay and gum disease

As with many other body structures that we have discussed, the foundations of healthy teeth and gums are laid down before birth and during early childhood. It requires a very long-term plan if we wish to keep our teeth and gums healthy for a lifetime. Colour plate 4 shows some of the results of not adopting such a plan. Unlike the carnivore it is not our survival that is threatened, but how we look and how we feel. Rotten and decaying teeth are very painful (Colour plate 5). If left untreated an abscess can develop at the root, and poisons may even spread into the bloodstream. Inflamed gums can produce similar problems. Thus bad teeth and gums are physically unpleasant. They can also be an embarrassment: for example, bad teeth do nothing to improve appearances, and breath becomes offensive. In Chapter 7 we discussed decay and the bacteria that cause it. *Tooth decay* is no different: it too is caused by the action of bacteria.

Fig. 11.14 *Plaque, the bacteria that cause dental decay, as it appears under an electron microscope. You can see that different types of bacteria are present*

Fig. 11.14 is a highly magnified photograph of *plaque*. Plaque contains thousands of bacteria, and grows as a sticky film over the teeth. At first it is an invisible enemy, but if it is not removed it builds up, and a much harder substance called *tartar* is laid down on the teeth. Tartar appears as a yellowish film near the gums and has to be chipped and cleaned off by a dentist.

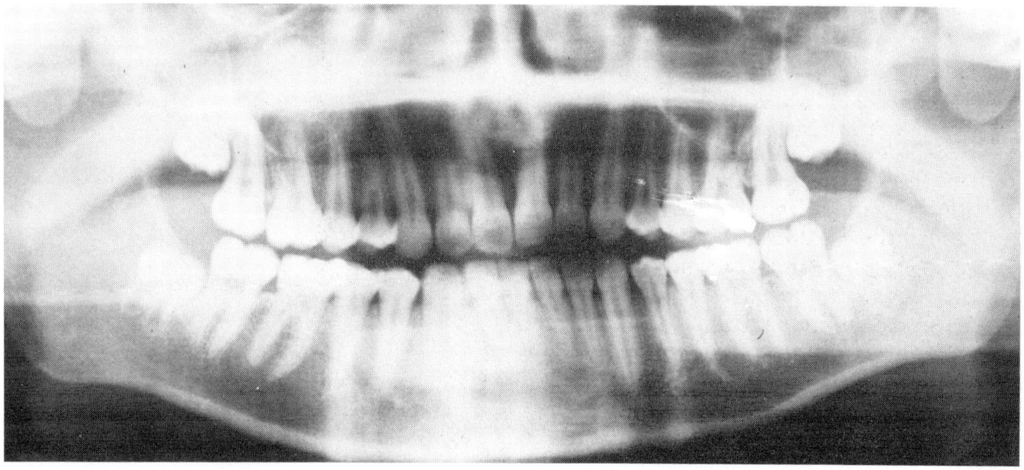

Fig. 11.13 *X-ray of the jaw of an adult whose wisdom teeth have not erupted*

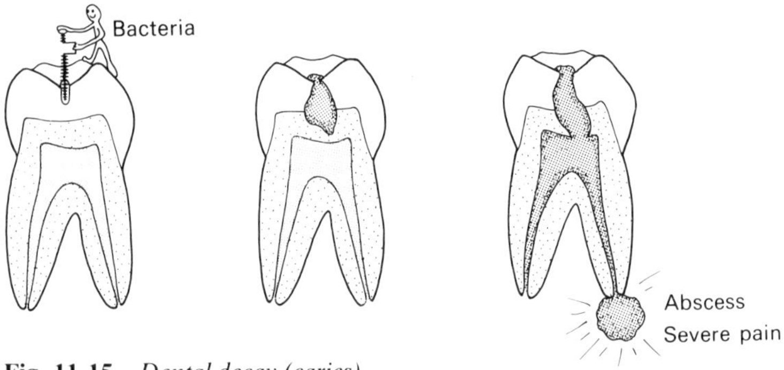

Fig. 11.15 *Dental decay (caries)*

The bacteria within plaque will work on sugary bits of food left in the mouth after a meal. This action creates an acid which is strong enough to attack the enamel surface of the tooth. If this erosion of the enamel is not treated, the decay (dental caries) spreads even further, reaching the dentine layer (Fig. 11.15). At this point the enamel surface will probably collapse. Decay spreads very quickly once it reaches the softer dentine layer. Eventually it will reach the soft pulp region, the whole tooth will become infected, and an abscess may form within the gum. As a result the tooth may be too badly infected to be saved.

Plaque is also the cause of gum disease, for it accumulates most rapidly in the region where gum and teeth meet. The poisons produced by the bacteria affect the gums so that they become swollen, inflamed and bleed easily (Colour plate 6). Also the gums may recede, allowing pockets of decay to reach the dental cement and destroy the tooth's support, so that it becomes loose in the jawbone.

11.11 **Prevention of dental decay and gum disease**

All efforts should be made to prevent plaque building up on the surface of teeth. This can be done by the simple process of brushing. Dentists can provide a special tablet which contains a harmless dye which stains plaque red. If such a disclosing tablet is chewed after brushing your teeth you might get the result shown in Colour plate 7(a). The plaque that remains is clearly visible. By further brushing you can find the best way of removing all traces of plaque on or between the teeth (Colour plate 7(b)). The right technique and the time needed to brush the teeth effectively to remove all plaque can therefore be worked out.

It has been estimated that 50 per cent of the population never visit a dentist. Obviously regular visits to the dentist for check-ups on the condition of our teeth and gums are essential. Children should have check-ups every four months. An adult should visit the dentist about twice a year. In this way any condition which might give rise to a serious problem will be recognized at a very early stage and difficulties probably prevented.

Nowadays it is generally accepted that fluorine helps to prevent dental decay. Calcium fluoride is found in tooth enamel and contributes to the strength of this material. In some areas the water-supply is fluorinated, so that fluoride is absorbed into the bloodstream and reaches the enamel this way. Scientific trials have proved that children brought up in areas where there is approximately one part of fluoride to one million parts of water have 60 per cent less tooth decay than those drinking water containing little or no fluoride. A dentist may advise the taking of fluoride tablets, or the coating of the teeth with a special fluoride preparation. Up to the age of twelve, perhaps even later, fluoride is absorbed into the enamel from fluids in the mouth. Many toothpastes contain fluoride, which when brushed on to the teeth helps to harden the exposed enamel. It is important to adjust the amount of fluoride in water to the correct amount, for excessive quantities can discolour the enamel.

Question 6 What types of food will encourage dental decay?

In Britain one in every four people will possess some form of artificial teeth before they reach their twenty-first birthday! There is plenty of evidence to link our high consumption of sugar and sweets with this large incidence of tooth decay. For instance, during and immediately after the Second World War, when there was rationing and sweets were scarce, tooth decay decreased. Sweets every day brings a smile to a child's face, but will that child feel able to smile at twenty-one? If we wish to preserve good teeth we have to accept a decrease in our intake of sweet, sugary foods and liquids.

Answers to questions in the text

1 Dead leaves and cuttings from the garden as well as the scraps left over from vegetable preparation in the kitchen can be used to produce a very effective compost heap.

2 Larger, because this point is a short distance away from the fulcrum of the jaws.

3 Plant material needs a great deal of chewing and grinding to break it down, so a herbivore needs a large grinding tooth area. The elephant is the proud owner of the biggest molars in the world.

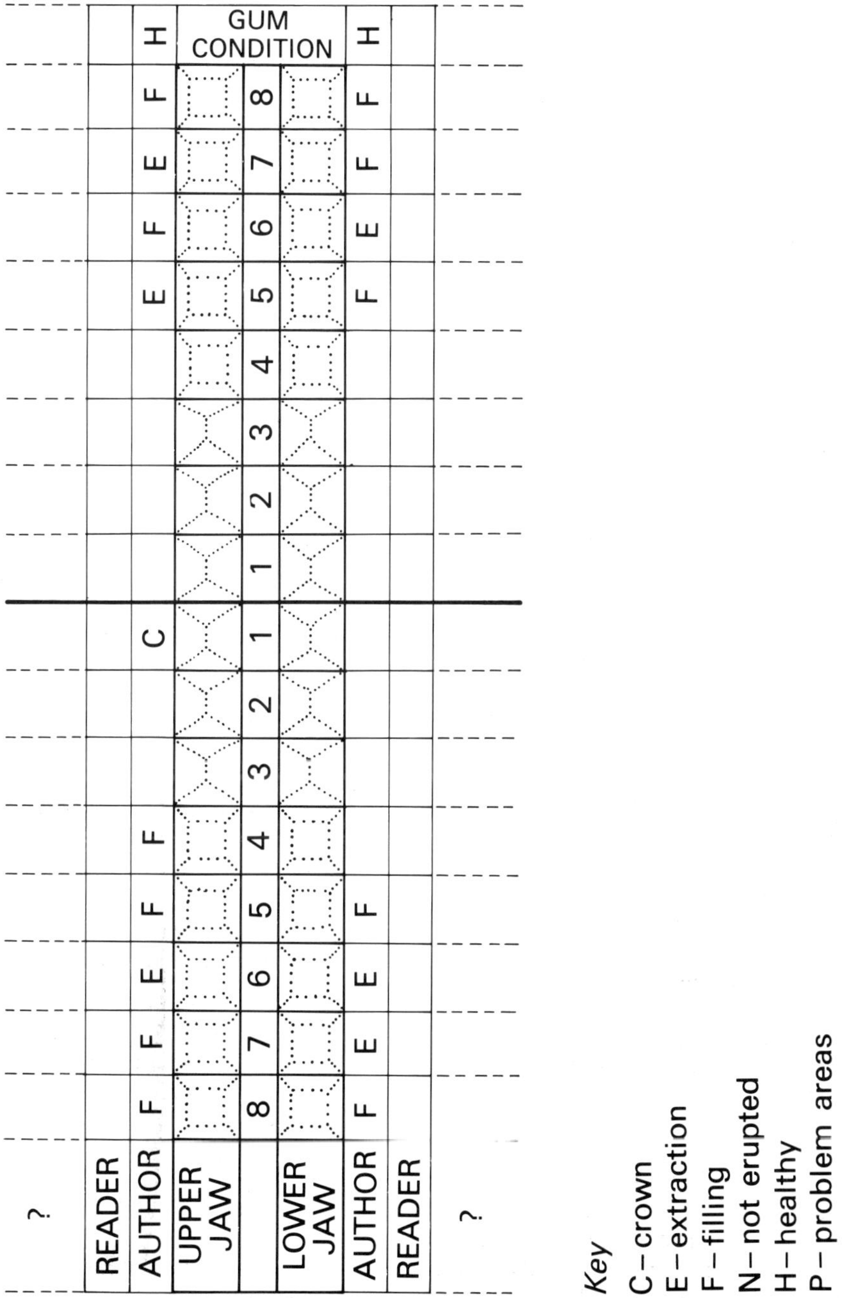

Fig. 11.16 *Record of author's teeth*

4 The blood-vessels would no longer be able to deliver an adequate blood supply to the tooth. Thus living tissue within the tooth would be deprived of food and oxygen.

5 $i\frac{3}{3}\quad c\frac{1}{1}\quad p\frac{4}{4}\quad m\frac{2}{3}$

6 The bacteria in plaque need sugar: without it their activity ceases. Thus all sugary drinks or solids encourage dental decay.

THINGS TO DO

1 Draw a map of the world. Pin-point the areas where there is famine and starvation. Assemble material and information about the work of an organization which is active in these areas (for example Oxfam).

2 Dig up a clover plant and observe the nodules on the roots. By using newspaper advertisements or old packet labels construct a montage of artificial fertilizers and aids to natural manuring.

3 Choose a site near your home or school and study the organisms that live within that area. Try to construct a food-web for the area.

4 Fig. 11.16 shows part of a dental record chart of the author's treatment. Construct a similar chart and record the condition of your teeth. Make up a dental questionnaire which asks such questions as: How many sweets do you eat? How often do you brush your teeth? When do you visit a dentist?

5 Record which teeth you use and how you use them to eat (a) an apple, (b) a chicken leg (without using a knife and fork), and (c) a hard ginger biscuit.

6 Get together with a group of friends after you have all eaten a meal. Each person should select a different method for cleaning their teeth, for example:

(a) swilling water round the mouth
(b) eating an apple or carrot
(c) using a tooth-brush
(d) using a tooth-brush with toothpaste
(e) using a tooth-brush and dental floss.

Each person should then use a disclosing tablet (page 150) to reveal the effectiveness of their method.

7 If possible visit a natural history museum where skulls of a large variety of creatures are on display. Select three of the skulls to draw. Compare their teeth and discuss how they are suited to each creature's way of life.

THINGS TO WORK OUT

In Questions 1–3 write down the *one* answer which you think is correct.

1 The chemical added to water to help prevent dental decay is
A chlorine
B fluorine
C iron salts
D iodine
E ozone

2 The dental formula of an adult woman should be

A $i\,\frac{3}{3}\quad c\,\frac{1}{1}\quad p\,\frac{4}{4}\quad m\,\frac{2}{3}$

B $i\,\frac{2}{2}\quad c\,\frac{1}{1}\quad p\,\frac{2}{2}\quad m\,\frac{3}{3}$

C $i\,\frac{2}{1}\quad c\,\frac{0}{0}\quad p\,\frac{3}{2}\quad m\,\frac{3}{3}$

D $i\,\frac{3}{3}\quad c\,\frac{1}{1}\quad p\,\frac{3}{3}\quad m\,\frac{3}{3}$

E $1\,\frac{3}{2}\quad c\,\frac{1}{1}\quad p\,\frac{3}{2}\quad m\,\frac{3}{3}$

3 Nitrogen-fixing bacteria are found in
A the stems of a pea plant
B the roots of a dandelion
C the roots of clover
D the leaves of clover
E the roots of grass

4 Which of the following food-chains is incorrect? (Write it out correctly.)
(a) plankton – fish – seal – polar bear
(b) plant juices – aphid – sparrow – sparrow-hawk
(c) grass – cow – man
(d) decaying plants – earthworms – chicken – fox
(e) decaying organic matter – spider – housefly – bird

5 (a) Draw the basic shapes of canine, incisor and molar teeth. State the function of each type of tooth
(b) Compare the number of teeth of a five-year-old with that of an adult.
(c) Name the vitamin and an element that are essential in the formation of healthy teeth.

6 (a) Explain what is meant by a food pyramid.
(b) Give *two* ways in which a continual increase in population can threaten man's survival.
(c) Describe how food supplies may be increased by
(i) new farming techniques
(ii) better storage
(iii) developing alternative food supplies

7 (a) Draw a simple diagram of a human tooth. Label all its parts.
(b) State *three* advantages of the hard enamel surface.
(c) Why does a diet containing large quantities of sugary food encourage dental decay?
(d) What is pyorrhoea?

8 Describe, using a diagram, the nitrogen cycle. How may a farmer make use of this cycle in a system of crop rotation in which *no* artificial fertilizers are used?

12 Digestion

12.1 Why does food need to be broken down?

Throughout the previous chapters we have seen how the cells of the body need food for their various activities. However, for food to enter a cell it must be able to pass through the cell membrane. To do this its molecules need to be of the right size. Digestion is the process by which foods made up of large molecules are broken down and converted into molecules small enough for the cells to accept. As we saw in Chapter 7, the body regards foreign matter that enters it as an invader. For instance, when a thorn punctures the skin the body will usually bring its defences into action to reject the foreign material. Food will be regarded as foreign matter too unless it is broken down into a form that the body is able to accept. The breaking-down process is carried out in the *digestive system*. The digestive tract acts like a conveyor belt, carrying the food along within the body while it undergoes treatment (Fig. 12.1). Once the food is of the right form and size it is taken into the bloodstream. It is *absorbed* from the digestive tract into the body.

12.2 The breaking-down processes and enzymes

The breaking down of the food is carried out by two methods: mechanical and chemical.

Question 1 Where does the first mechanical process take place and what carries it out?

The chewing of our food is under our conscious control and we should do it thoroughly if further digestion is going to be fully effective. You may have noticed that as food is chewed it becomes much wetter. This 'wetting' of the food happens automatically, and is caused by a liquid called *saliva* being poured into the mouth from three pairs of glands. Fig. 12.2 shows where these glands are to be found. If you have ever suffered from mumps, you will have become very conscious of these glands, as they swell up with the infection. We only have to think of our favourite food, smell it or see it on the other side of the table to increase the flow of saliva to the mouth (see Chapter 18).

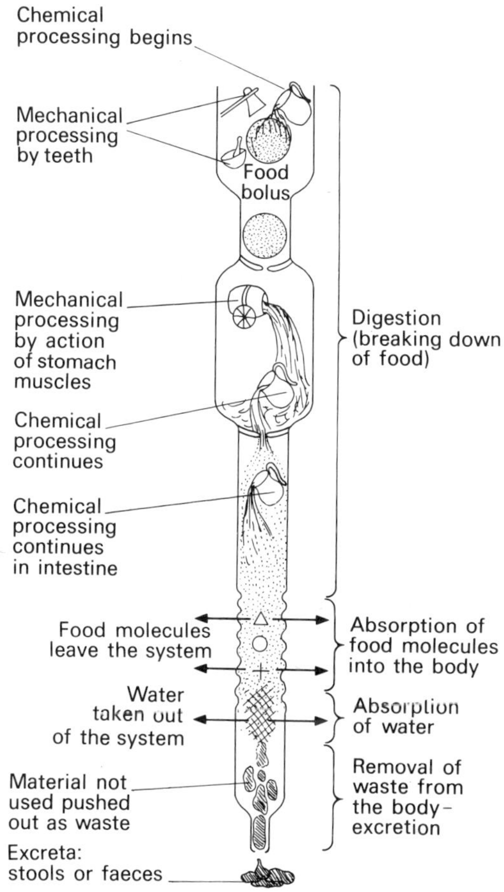

Fig 12.1 *The processes that take place within the digestive tract*

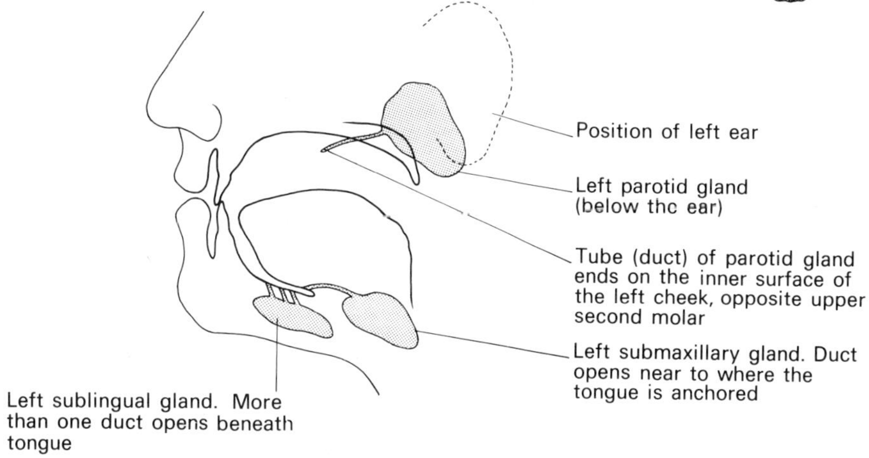

Fig. 12.2 *Position of the glands that produce saliva (not to scale*

Because the food is wet it is easier to swallow. However saliva does have another purpose: it contains an *enzyme*, therefore it begins the chemical processing of the food. In other science lessons you may already have studied substances called *catalysts*. A catalyst speeds up a chemical reaction. The chemical reactions that take place inside the body would in a laboratory need very high temperatures, concentrated solutions and several hours to complete. In the presence of the body's catalysts, i.e. enzymes, they take place at body temperature (39.9 °C) and in a matter of minutes. Each enzyme is suited to assisting one particular reaction within the body. Most cells in the body can make up to 200 different enzymes! They work best at temperatures between 35 and 40 °C; in fact, extremely high temperatures will destroy enzymes.

The other condition which determines the action of an enzyme is whether the surroundings are *acid*, *neutral* or *alkaline*. When a liquid is neutral, it is said to have a *pH value* of 7. A solution that is alkaline has a value on the pH scale higher than 7. An acid solution has a pH value lower than 7. Each body tissue and fluid has its own pH value. A particular enzyme will work only in conditions that have the required pH value. Table 12.1 summarizes some of the properties of enzymes.

TABLE 12.1 **SOME FACTS ABOUT ENZYMES**

Where made	**Type of substance**	**Function**	**Favourable conditions for action**
In cells	Protein	To speed up chemical reactions of living cells. However, an enzyme is unchanged by a reaction. Enzymes are found working both inside and outside cells	Temperature between 35 ° and 40 °C Each enzyme is suited to working at a particular pH value

12.3 **The digestive tract** (or alimentary canal)

The digestive system is in the part of the body known as the *abdomen*. Like the thorax, the abdomen has a lining, a membrane called the *peritoneum*. Folds within this membrane support the digestive organs, keeping them in position yet allowing them to move against one another without friction. Parts of this membrane are also used for the storage of fat, which becomes quite obvious when we put on weight! The diagram alongside Table 12.2 shows the digestive system laid out so that its total length may be seen. In an adult it would measure about 9 m. We shall consider the mechanical and chemical processing that takes place by looking at certain sections of the system. These are described in Table 12.2 (see pages 156/7).

12.4 **The oesophagus**

Once food is thoroughly wetted and chewed it is swallowed as a *bolus*. The action of swallowing (page 231) is also under our control. However, once the food enters the oesophagus we lose all control over its movement. This now takes place automatically (Fig. 12.3). The automatic action which moves the food along is *peristalsis*. It is brought about by the action of muscles within the walls of the digestive system. The easiest way to understand this action is to imagine a long tube like a stocking. If we put a ball into the top of the stocking, how can it be moved along to the toe end? We simply squeeze our hands around the stocking and push the ball along from behind (Fig. 12.4(a)). Similarly, once the walls of the oesophagus are stretched by the entry of food, the muscles in these walls respond by contracting inwards behind the food and relaxing in front of it (Fig. 12.4(b)). This circle of muscular pressure travels down the length of the digestive system, pushing the food along. At the same time, other muscular activity mixes the food with the secretions that are 'fed' into the system. This mixing is strongest in the widest region of the digestive system, the *stomach*.

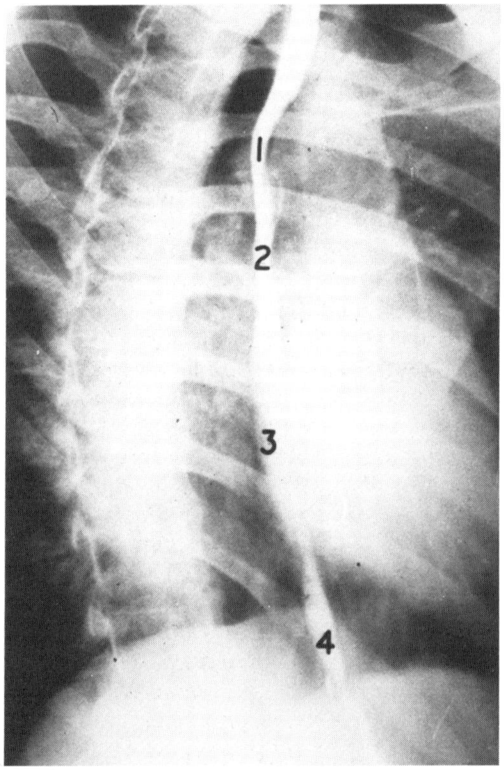

Fig. 12.3 *X-ray showing how food passes along the oesophagus*

12.5 **The stomach**

This J-shaped organ has a capacity of approximately 1½ litres and is 30 cm long when full of food. It is supported by a membrane attached to the liver and is

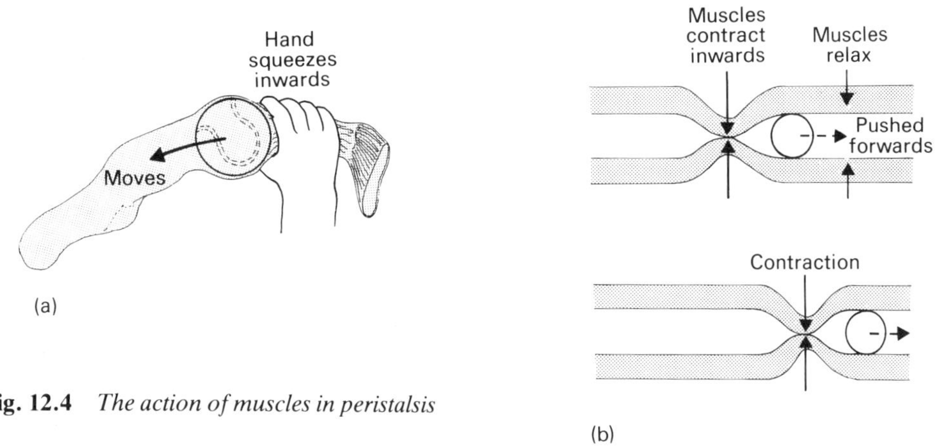

Fig. 12.4 *The action of muscles in peristalsis*

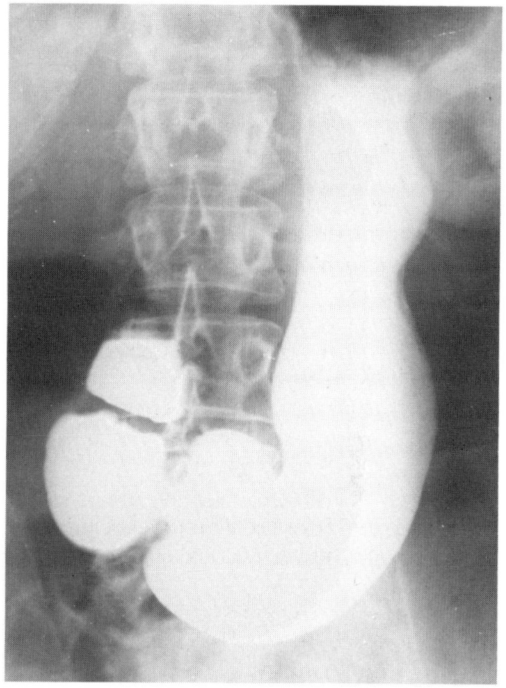

Fig. 12.5 *X-ray of stomach containing a 'barium meal'*

positioned in the left-hand side of the abdomen, just below the diaphragm. Two muscular rings, *sphincters*, control the movement of food in and out of the stomach. The *cardiac sphincter* (A in the diagram shown in Table 12.2) lies between the oesophagus and the stomach; the *pyloric sphincter* (B in the diagram shown in Table 12.2) lies between the stomach and the duodenum. The purpose of any sphincter is to act as a valve in the 'pipeline': it keeps the food in a section until all the processing is complete. Fig. 12.5 shows an X-ray photograph of the stomach. Under normal circumstances the stomach would not show up on an X-ray (page 26), but in this case the person has swallowed a barium meal. Barium in the form of barium sulphate will not dissolve in water; a suspension of barium sulphate particles in water does not allow X-rays to pass through it. Thus any organ containing barium sulphate will show up on an X-ray. In fact the passage of the barium meal along the digestive tract can be watched on an X-ray machine's screen. This is used to check that the digestive tract is functioning correctly and is an aid to doctors in their diagnosis.

The stomach acts mechanically on the food, tenderizing and liquidizing it rather like a cement mixer. Water is added to cement in a mixer; in a similar way, liquid known as *gastric juice* is added to food in the stomach. Gastric juice is secreted by glands in the mucous membrane (*mucosa*), which lies in folds, forming an inner lining of the stomach (Fig. 12.6).

Question 2 Why do you think the inner membrane of the stomach lies in so many folds?

Gastric juice is always present in small amounts in the stomach, just as saliva is always present to moisten the mouth. However, once food is present these secretions are automatically increased.

The mixing action of the stomach is produced by muscles lying in three directions in the stomach wall: along its length, around its width, and across it diagonally. By the action of these muscles the solid food is converted into a semifluid called *chyme*.

Chemical processing also takes place within the stomach, because gastric juice contains hydrochloric acid and two enzymes. The action of these chemicals is described in Table 12.2 on page 156/7.

Fig. 12.6 (Below) *The inner membrane of the stomach, and the gastric glands*

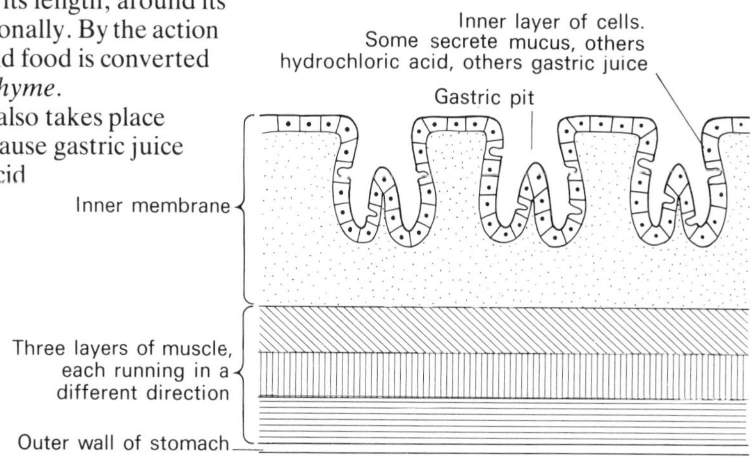

TABLE 12.2 DIGESTION

Parts of the alimentary canal	Mechanical processing that takes place	Secretion or juice (chemical processing)	Acid or alkaline	Where the digested food passes into the body tissues	Time taken for materials to pass through each section
Mouth	Food mechanically broken down by chewing. Bolus formed, which is swallowed	Saliva	Alkaline	Some drugs may be absorbed here	Under our own control until swallowed
Oesophagus	None	None	—	None	Liquids move rapidly, solid food more slowly
Stomach	Muscular action churns food like a cement mixer. This tenderizes and liquidizes the food	Gastric juice	Strong acid	A little absorption takes place, of such materials as alcohol, water, sugar (glucose) and drugs	Depends upon type of food. Average meal 5 hours. Chyme released at intervals during this time by action of the pyloric sphincter
Duodenum		Bile (from liver) Pancreatic juice (from pancreas)	Alkaline Alkaline	None	3–4 hours Duodenum about 30 cm long
Ileum		Intestinal juice	Alkaline	Main region for absorptionof food. Food molecules pass through the intestinal walls and into the blood	Ileum about 6 m long
Colon		Mucus – assists passage of material		Main region where water passes from the undigested material back into the blood	12–24 hours Colon about 2 m long
Rectum				This very muscular region pushes all the undigested food, mucus and dead cells out of the body	

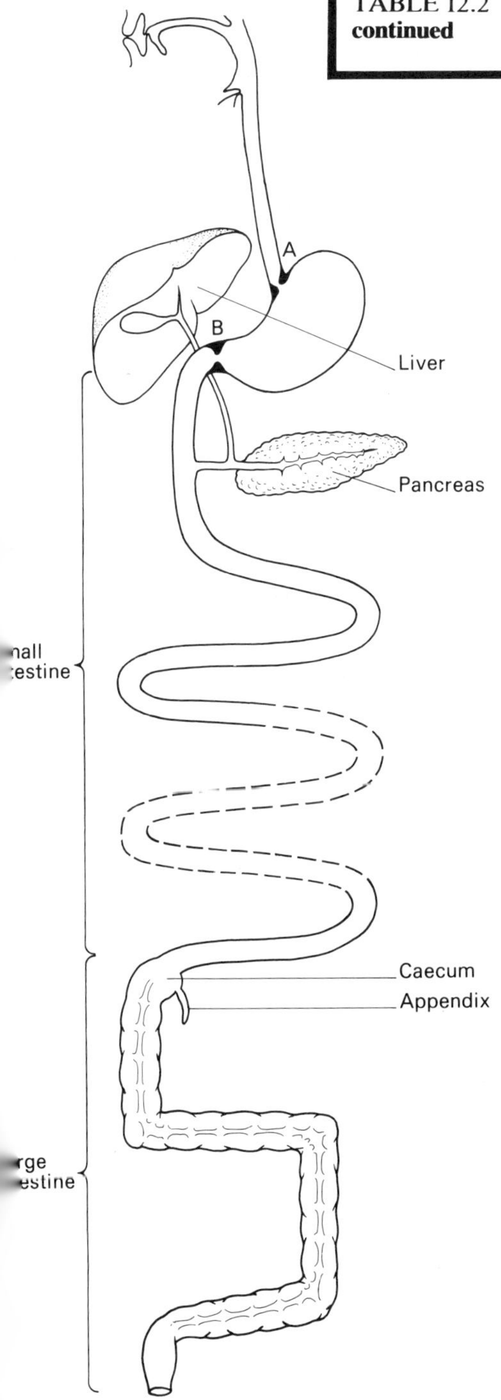

TABLE 12.2 continued

Digestive juice	Enzymes contained in digestive juices	Changes that the secretions cause in food: Carbohydrates	Fats	Proteins
Saliva	Salivary amylase	A small amount of cooked starches to sugar (e.g. maltose)		
Gastric juice	Dilute hydrochloric acid (not an enzyme)	Kills bacteria, therefore sterilizes. Neutralizes alkaline saliva and stops action of enzyme in saliva		
	Rennin – present in the very young			Clots or coagulates milk protein
	Pepsin			Broken down to peptones
Bile			Emulsifies. Acts like a detergent, breaking fat into droplets	
Pancreatic juice	Amylase	Starch to complex sugars (e.g. maltose)		
	Lipase		Become fatty acids and glycerol	
	Trypsinogen			(Acts when it meets entrokinase)
Intestinal juice	Maltase and lactase	Complex sugars to simple sugars (e.g. maltose to glucose)		
	Lipase		Become fatty acids and glycerol	
	Invertase	Complex sugars to simple sugars		
	Entrokinase releases Trypsin			Peptones to peptides
	Erepsin			Peptides to amino-acids

Gastric juice is very corrosive. It would burn a hole in many materials and you may have felt its burning effect at the back of your throat when some of it is 'burped' back in an attack of wind, or when you are sick. The cells of the stomach are normally resistant to it as they are protected by a thick layer of mucus. Also they are replaced every two to three days. However, if the mucous layer breaks down the stomach lining can become irritated and inflamed. When this happens a *stomach ulcer* may form. Drugs such as aspirin (page 90) seem to irritate the stomach lining and destroy the natural resistance of the cells. Excessive alcohol can produce severe inflammation and loss of appetite. Infections such as influenza and gastro-enteritis can produce similar inflammation. This may result in vomiting, which is the stomach's automatic reaction when it needs to rid itself quickly of its contents. Most of us will at some time suffer from *indigestion*: a feeling of discomfort in the upper abdomen. This can have many causes, such as nervous tension, overeating or eating food which is too rich. Persistent attacks may indicate a physical ailment. People often claim that their indigestion is due to 'over-acidity'. Sometimes excessive hydrochloric acid is produced, but only a medical diagnosis can prove it. As with other forms of medicine it is not wise constantly to take doses of anti-acid pills to cure persistent indigestion.

Food in the stomach keeps us going between meals, as quantities of chyme are released at intervals into the small intestine. If there is nothing for the stomach to work on we get symptoms of *hunger*.

12.6 **The small intestine** (duodenum and ileum)

Bile is produced in the liver, stored in the gall-bladder and released into the duodenum when food is present. Bile neutralizes the acid chyme and within an alkaline solution the enzymes of the *pancreas* are able to process the food (see Table 12.2). Chemical processing continues with the addition of *intestinal juice* from the glands of the mucous lining of the *ileum*.

By this stage the food has been broken down into the size and form that the body will accept. So it is within the ileum that *absorption* of food takes place. The mucous lining of this region also lies in folds, producing a large surface area (about 10 m^2). As shown in Fig. 12.7 it is covered with small finger-like extensions called *villi*. Each villus contains blood capillaries and a small lymph vessel, a *lacteal*. It also contains a muscle which can cause a change in volume and movement. The surface movement of villi resemble seaweed moving with the sea.

Amino-acids and simple sugars such as glucose enter the blood through the walls of the villi. They are then transported through the hepatic portal vein to the *liver* before they enter the main blood system. The reasons for this will be given in Section 12.8. The fatty acids and glycerol bypass the liver because they are absorbed into the cells lining the villi. Here they become tiny droplets of fat which find their way into the lymph. These droplets are then transported round the lymphatic system. They finally enter the bloodstream in the neck region (page 160). Absorption of salts, minerals and vitamins also take place, so all that remains within the digestive system is material which is of no use to the body. This enters the large intestine via another sphincter.

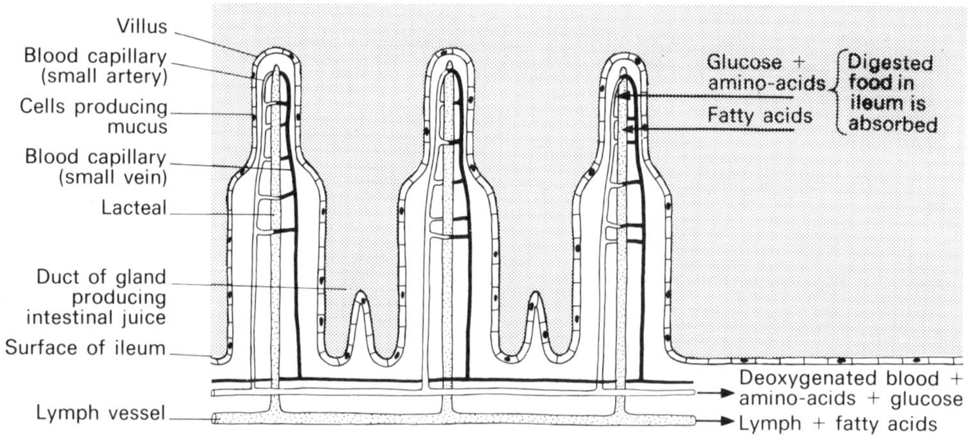

Fig. 12.7 *Villi in the ileum*

12.7 **Large intestine (colon and rectum) or bowels**

The sphincter and the short section of the colon known as the *caecum* prevents any backflow of material. Attached to the caecum is a small tube 6–10 cm long, the *appendix*.

Question 3 Can you suggest what happens to this organ when a person suffers from *appendicitis*?

The colon is wide and 'crimped' in external appearance. However its internal membrane is straight and smooth with a constant secretion of mucus. There are a large number of bacteria present in this region and they act

Fig. 12.8 (Above) *The arrows indicate the points at which samples were taken to study contamination of faecal bacteria.* (Below) *A bacterium which is found in faeces (*Escherischia coli*)*

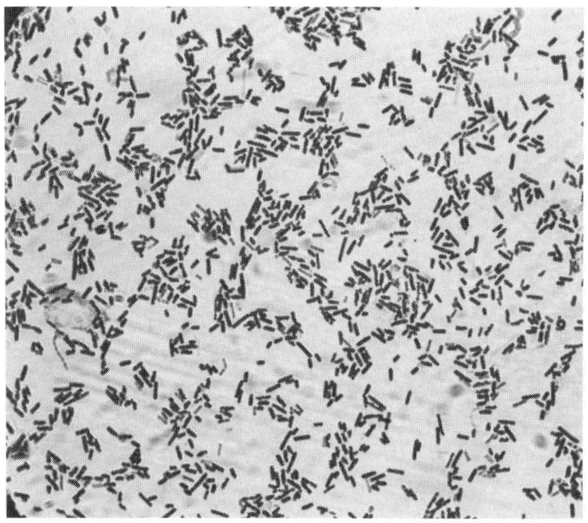

TABLE 12.3 **FREQUENCY WITH WHICH SURFACES WERE CONTAMINATED BY FAECAL BACTERIA (%)**

Sampling point	Male	Female
Cubicle door lock	5	4
Cubicle door handles	5	6
Flush handle	6	5
Entrance door handle, outside	12	8
Entrance door handle, inside	23	20
Tap handles	25	29
Water in W.C. pedestal	38	24
Floor in cubicle	39	46
Under flushing rim	53	38
Wash-basin overflow	55	59
W.C. seat	68	58
Urinals	89	–

upon material to decompose it. The presence of these intestinal bacteria in a water-supply can indicate that the supply is being contaminated by human waste. Also the number that are present on surfaces in and around lavatories can indicate to us the risk that there is to health. The greater the contamination the greater is the risk of cross-infection.

The results of a survey into this problem are shown in Fig. 12.8 and Table 12.3. You might consider how this information can help us to decide our actions on such things as personal hygiene and the daily use of disinfectants (page 86). It also indicates how the maintenance and considerate use of public washrooms and lavatories by everyone is essential if these facilities are to be an asset to the health of the community and not a hazard.

The material within the colon contains a great deal of water. This moisture is needed by the body and is therefore absorbed into the blood. Some of the remaining salts and sugars may also be absorbed at this stage.

We have already seen that peristalsis occurs when the digestive system is stretched by food entering it. Peristalsis is thus a response to bulk. If we eat enough fibrous foods, such as vegetables, raw fruit and whole wheat, there will be sufficient undigested cellulose to provide this bulk. The material will be moved along naturally. It is better for the body to remove this waste material than to leave it in the colon for long periods, because toxic materials can be reabsorbed into the bloodstream.

Waste material leaves the colon and enters the rectum. As it accumulates it stretches the walls of the rectum; this stretching urges the body to pass a motion. The *anus* is opened by relaxation of the *anal sphincter*. The muscles of the rectum then contract; also pressure inside the abdomen is increased by the downward movement of the diaphragm and contraction of the abdominal walls. As children grow up they learn to control this action. However, it is better not to suppress this natural urge for too long because *faeces* or *stools* left in the rectum become hard owing to reabsorption of water into the bloodstream. They are then more difficult to move. Difficulty in ridding the body of faeces is termed *constipation*. This condition is less likely to occur if plenty of fibrous foods and water are added to a person's diet (page 117). A *laxative* irritates the colon, causing it to expel its contents. An *enema*, i.e. an injection of a volume of water into the colon through a tube placed in the rectum, produces the same effect. If barium sulphate is added to this water then an X-ray picture of the bowels can be taken just as with a barium meal.

Inflammation of the colon's lining membrane owing to a physical ailment or infection can cause *diarrhoea*. The stools become very liquid because they are not kept long enough in the colon for water to be removed. Cases of severe diarrheoa, particularly in young children, can cause the body to lose so much water that the tissues become dehydrated. Thus a fluid (non-milk) diet is essential for a person suffering from this ailment.

Human digestion considered as a process is very efficient: 99 per cent of the carbohydrate, 92 per cent of the fat and 93 per cent of the protein is absorbed from the material available within the system. Any inefficiency of the system is due to its inability to digest cellulose.

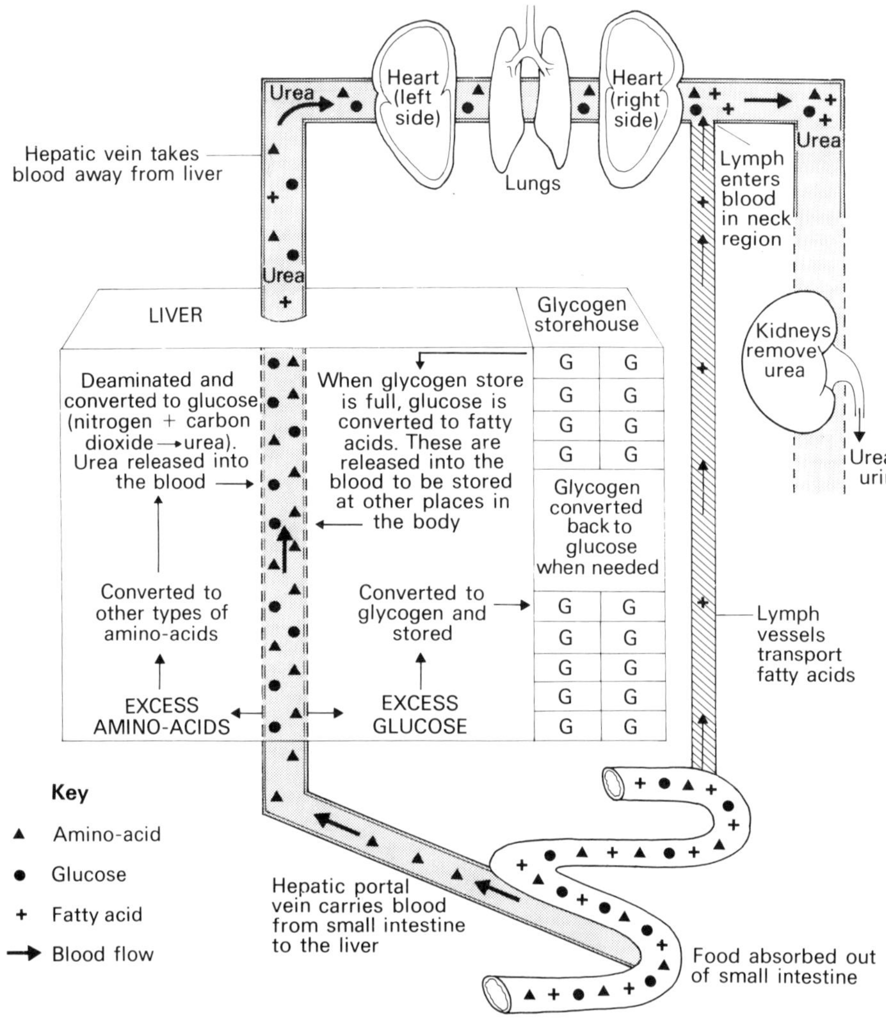

Fig. 12.9 *How the liver regulates the supply of amino-acids and glucose in the blood*

12.8 **The chemical factory of the body – the liver and homoeostasis**

The liver is a chemical-processing plant concerned with keeping the body tissues supplied with adequate amounts of amino-acids and glucose. Fig. 12.9 illustrates how the liver carries out this function, and how it acts as a storage depot for excess glucose. The activities of the liver are so numerous that it accounts for 25 per cent of the energy demand of the body. Fig. 12.10 summarizes some of them. The liver also releases a lot of heat energy which can be distributed all over the body by the blood. The liver's action (deamination) on excess amino-acids releases a highly toxic waste, *urea*. This has to be despatched via the bloodstream to the kidneys for disposal.

The liver is one of the largest organs of the body and it requires a high concentration of oxygen, which reaches it through the *hepatic artery*. Blood which comes to the liver through the *hepatic portal vein* has little oxygen but a good supply of food molecules. These may be used by the liver cells, processed, stored or released once more into the bloodstream as described in Fig. 12.9.

Think of a central-heating system. Do you have to keep running to the boiler and switching it on or off every time the room temperature gets too high or too low? It is much more likely that you will have a thermostat in the sytem. This automatically switches the system on when the room temperature falls below the temperature you

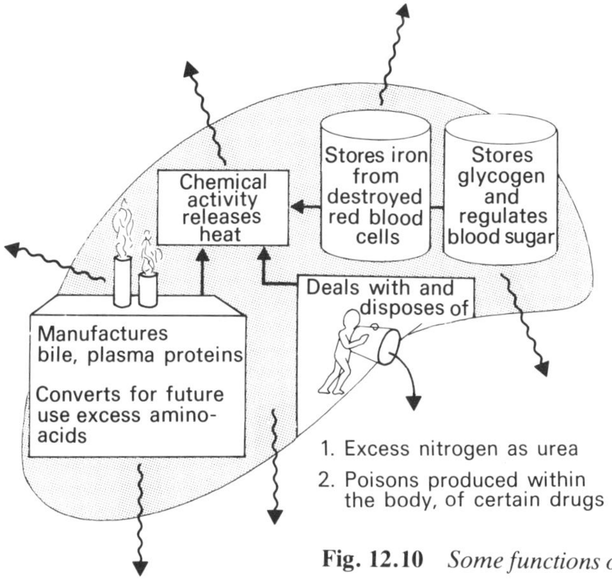

Fig. 12.10 *Some functions of the liver*

Fig. 12.11 *Inside the London sewers*

have selected. It switches it off when the temperature reaches that which is required. Thus the thermostat regulates the system for us. Our own body has a number of mechanisms which are regulators which, like a thermostat, function to keep the conditions inside the body within certain limits. This regulation of the body's *internal environment* is called *homoeostasis*. If homoeostasis ceases the survival of the whole body is threatened (page 188).

Question 4 Can you name two fluids in the body whose composition must be kept within certain limits?

All of the liver's functions help to regulate the concentration and composition of the blood and the body fluids. The liver and its functions are vital to homoeostasis.

12.9 Sewage

Human faeces have an unpleasant odour, contain harmful bacteria and are an ideal breeding ground for vectors of disease such as flies. It is essential that there is safe disposal of this human waste if we are to prevent disease within a community.

Sewerage systems were introduced into this country by the Romans, but they fell into disuse after the Romans left. No organized system for the whole community was developed for about 1400 years, until 1850.

In the Middle Ages the dwellings of the wealthy had *privies* or *garde robes*. These were basically an opening over a narrow tunnel which passed through the wall of the house and ended outside, over the moat or in a hole in the ground (a *cesspit*). Where this type of facility did not exist sewage was simply thrown out of doors, possibly into a *midden* or a channel which ran down the centre of a street. The first water-closet was invented during the reign of Elizabeth I but only the wealthy were to benefit from this development during the next 200 years. Even by the early nineteenth century such facilities were available to only a small proportion of the total population. Although this type of disposal of sewage from the home might be seen as a welcome development, we should remember that the contents of such closets were still discharged into natural water-courses or open cesspits. No thought was given to treatment of the waste and the rivers and streams in towns became public sewers (see Fig. 7.2). The problems that this situation created can be seen in the story of the River Thames.

12.10 'The river that died and came alive again'

The population of London in 1800 was 1 million, and by 1870 it was 4 million. Human and industrial waste poured into the tidal basin of the Thames within central London. There was persistent epidemics of cholera and in 1858 M.P.s were made very aware of another problem: they were driven from the House of Commons by the smell coming off the Thames! Sheets soaked with disinfectant were hung at the windows to try and dispel the awful odours. A parliamentary bill was rapidly passed and Joseph Bazalgette was put in charge of cleaning up London. He designed over 100 miles of sewers (Fig. 12.11). There were two sets of interconnect-

Fig. 12.12 *Raw sewage*

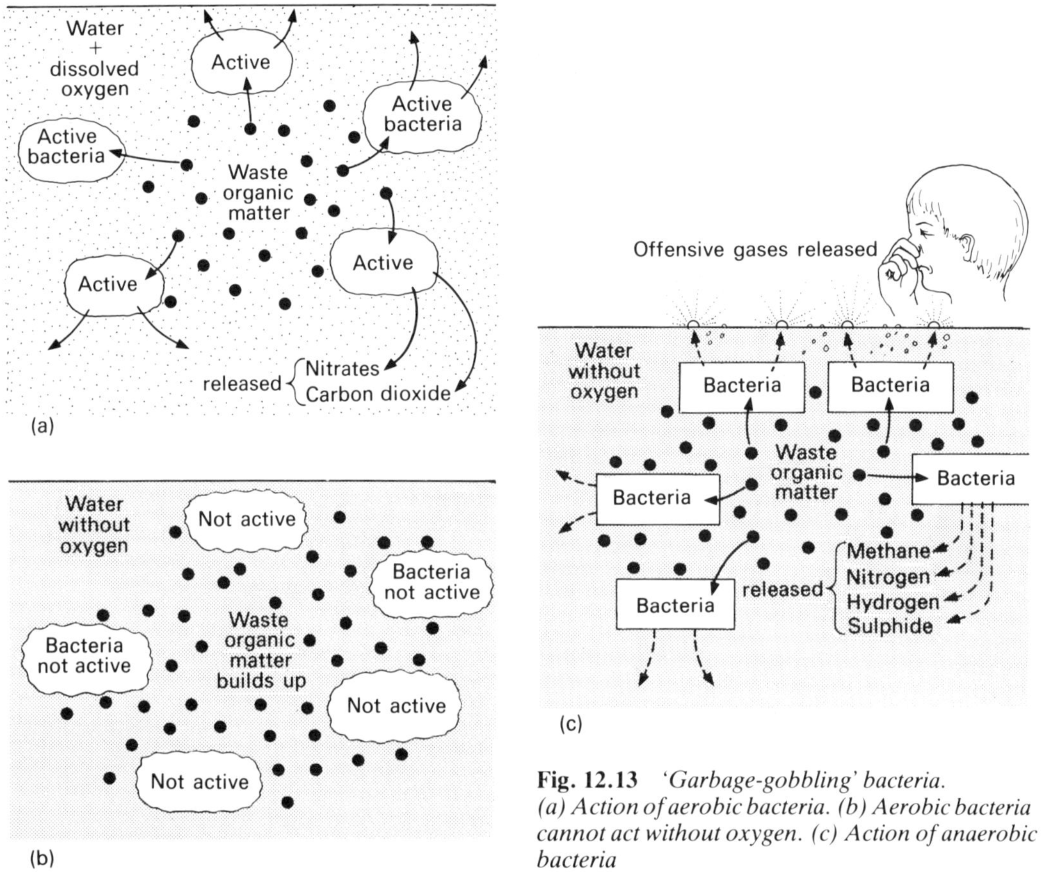

Fig. 12.13 *'Garbage-gobbling' bacteria. (a) Action of aerobic bacteria. (b) Aerobic bacteria cannot act without oxygen. (c) Action of anaerobic bacteria*

ing trunk sewers – one to serve London north of the Thames, the other south of the Thames. These took sewage away from central London and it was stored in large reservoirs further down river at Barking in the north and Crossness in the south. At ebb-tides the untreated (*raw*) sewage (Fig. 12.12) was released into the river, in the hope that it would flow out to sea. However, the peculiar action of the tides within this stretch of the river meant that a good deal of the sewage was carried back towards central London! Bazalgette's sewerage scheme was completed in 1875. Cholera and typhoid epidemics ceased but the Thames still remained dirty and foul, because treatment of the sewage consisted only of separating solids from liquids. The river could not cope naturally with the mass of organic waste that was poured into it every day.

As you can see in Fig. 12.13(a) certain groups of bacteria (pages 80 and 144) work on waste organic matter, destroying its complex molecules. Carbon dioxide and nitrates are released. In order to carry out this particular process these bacteria require oxygen: they are *aerobic*. When the bacteria are presented with a large quantity of waste organic material, they become so active that they begin to use up all the oxygen that is dissolved in the water (Fig. 12.13(b)). Without dissolved oxygen other forms of life within the water die, adding to the decay. In 1947 the Thames was declared chemically dead – it contained no dissolved oxygen.

When there is no oxygen other types of bacteria, which are *anaerobic*, begin their activities (Fig. 12.13(c)). The result of their activities is a blackened stretch of water that gives off bubbles of evil-smelling gases. The gases are ammonia, methane (commonly known as marsh gas) and hydrogen sulphide, which smells like rotten eggs. In 1950 the Thames was black for a stretch of 50 km within central London. The hydrogen sulphide given off from the river turned copper and brassware black within twenty minutes of being cleaned. In the 1920s biological treatment (see Section 12.11) of sewage was begun at Crossness and Beckton (Barking), but still pollution of the Thames was severe. Progress was hampered by the Second World War, and during air raids many sewers were damaged. In 1950 the Port of London

Authority mounted two surveys of the Thames: one looked into the source of pollution and the other into the tidal flow of the river. The pollution was found to be due to 79 per cent sewage, 12 per cent industrial waste and 9 per cent other sources. Two aims were adopted: to remove the smell, and to raise the oxygen level in the water so that fish might return. Every effort was made to improve sewage treatment. Industries co-operated in the clean-up; for example, together with other authorities, agreement was reached with detergent manufacturers that they should produce *biodegradable* products which could be destroyed by sewage treatment. This reduced the amounts of foam and toxic chemicals in the river. Gradually the Thames lost its smell and the recording of the return of fish and birds was begun (Fig. 12.14). The greatest excitement has been over the return of the salmon, several of which have been found.

Fig. 12.14 *Recording fish that are drawn into the screens of an electricity station that uses Thames water for cooling*

Question 5 Why are fish a good indication of the river becoming cleaner?

Within twenty years the Thames has become alive again (Fig. 12.15), levels of 94.8 per cent dissolved oxygen being recorded in 1977. All of this shows what it is possible to achieve by the correct disposal of man's waste and an anti-pollution policy. The clean-up of the Thames is now being used as a model for other waterways here and abroad.

In 1974 the *Water Act* established ten *Area Water Authorities*. This has given power to each authority to create overall standards of sewage disposal and water-supply. We are beginning to see the results of such co-ordinated effort as our rivers gradually suffer less from pollution and become areas for recreation. Sewage-treatment plants vary in construction but the scientific principles behind their operation are the same. Basically they depend upon the 'garbage-gobbling' activities of those greedy aerobic bacteria.

Fig. 12.15 *The Thames Water Authority has a floating laboratory, a launch called* Thameswater. *Scientists aboard the launch take samples from the Thames every fortnight and test them*

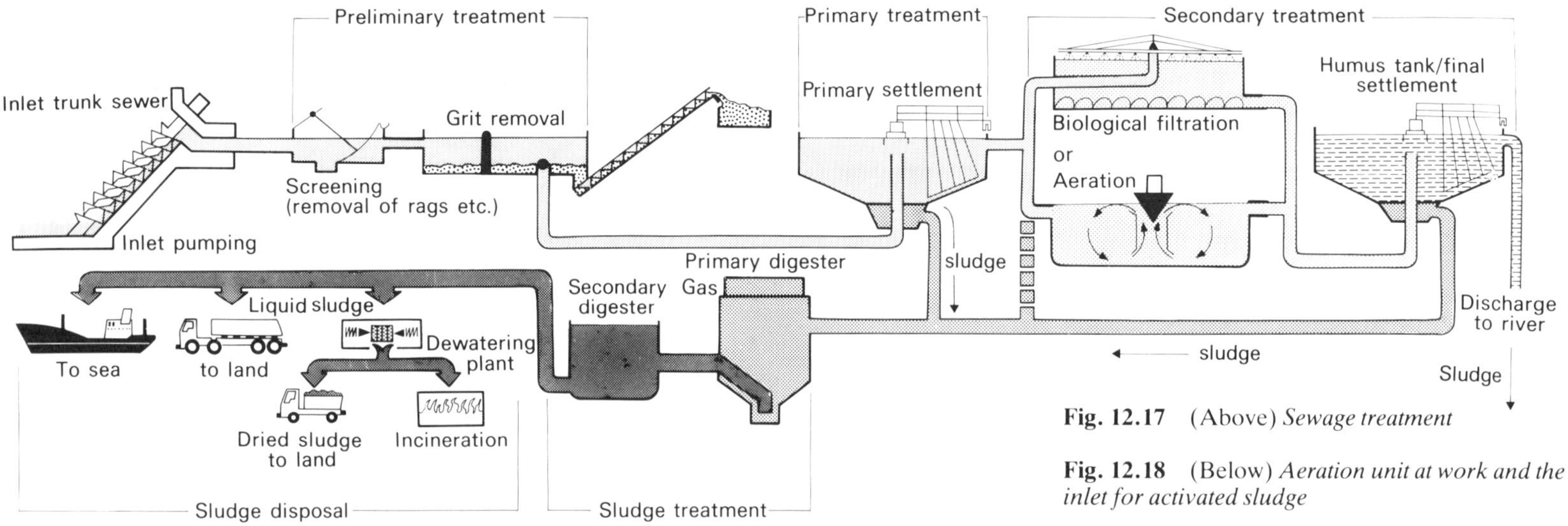

Fig. 12.17 (Above) *Sewage treatment*

Fig. 12.18 (Below) *Aeration unit at work and the inlet for activated sludge*

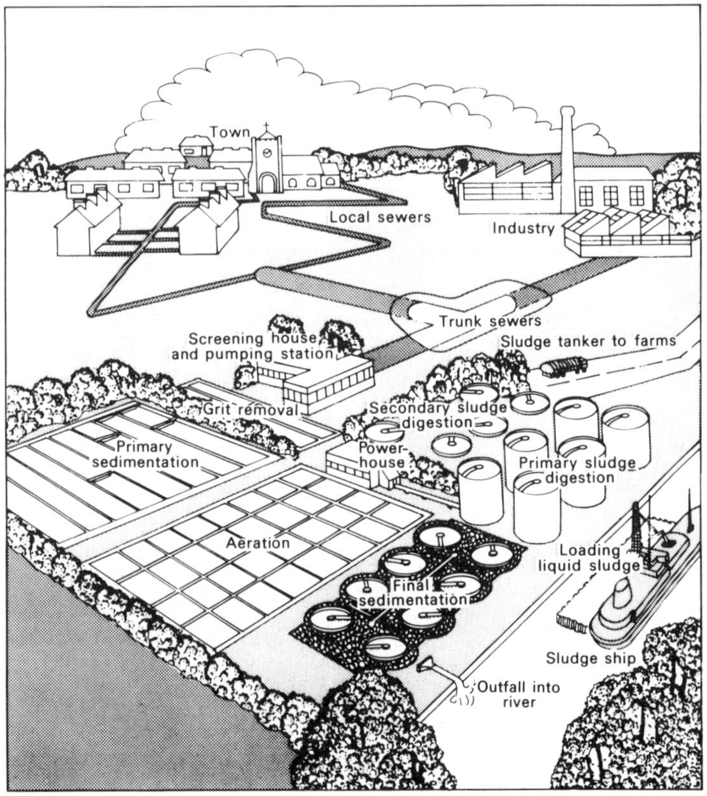

Fig. 12.16 *Treatment of sewage and waste water*

12.11 **Treatment of sewage**

Waste water that enters the sewage-treatment plant comes from homes and factories (Fig. 12.16). It contains both human and industrial waste. The description of the process given here is based on a sewage plant at Davyhulme in the North-West Water Authority Area (Fig. 12.17). This is one of the largest sewage plants in Europe, dealing with more than 341 million litres of sewage every day.

Screening and grit removal
Fine screens remove any solid material, for example paper and fabric, as the sewage passes through. The sewage then moves slowly along a series of tanks or channels where other solid material such as sand and grit settles out.

Settlement tanks
The sewage that enters these tanks still contains a suspension of fine particles and micro-organisms. These light particles take much longer to settle than the heavier particles of sand and grit, so the sewage is kept in the settlement tanks for up to six hours. The material that settles at the bottom of these tanks is called *sludge*. This is removed once a day for further treatment.

Biological filtration or activated sludge
Although a lot of organic material has been removed by settlement there still remain many dissolved impurities. This is where the aerobic 'garbage gobblers' are put to work. There are two ways of doing this.

The *biological* or *percolating filter* is the older method. This involves building a bed of rough material, such as clinker, which provides a large surface area for colonies of bacteria to grow on. Also there are large air spaces between the pieces of clinker which allow the sewage to trickle through and mix with air. Further mixing with air is achieved by sprinkling the sewage from a rotating arm on to the top surface of the clinker bed. This process involves building large areas of filter-beds and filtration takes some time.

A quicker and more compact method is the recent technique of *activated sludge*. This process was first developed at Davyhulme in 1914 and is now used in sewage plants all over the world. A culture of bacteria (activated sludge) is put into the liquid leaving the settlement tanks. Air is bubbled through the liquid under high pressure (Fig. 12.18). This thoroughly mixes the sludge and liquid. It also produces a very high concentration of oxygen and the bacteria are able to work vigorously at breaking down the impurities.

Final settlement tank
The activity of the bacteria purifies the liquid sewage in either of the above methods. It is then taken to a final settlement tank where the solid sludge can settle and be extracted. The liquid leaving these tanks is termed *final effluent* and is discharged into nearby water-courses. Laboratory checks are carried out on samples throughout the whole of the process.

Sludge treatment
Raw sludge that leaves the settlement tanks contains such things as pathogenic micro-organisms, which are potentially dangerous to health. To break it down even further, and to destroy nearly all the harmful micro-organisms, the sludge is *digested*. This is done by the action of the other group of 'garbage gobblers', anaerobic bacteria. The black evil-smelling sludge is placed in a tank and heated to 30 °C. All air is kept out of the tank and anaerobic bacteria are encouraged to decompose 50 per cent of the sludge over the period of three to four weeks. The resulting sludge has no smell and is safe to use as a fertilizer (Fig. 12.19). If there is no demand for it as a fertilizer the sludge is deposited well out of harm's way at sea. The gas given off from the digester is mainly a mixture of 25 to 35 per cent carbon dioxide and 65 to 75 per cent methane. This 'bio-gas' can be used as a heating fuel or to generate electricity. It is often used in this way in the sewage-works to power some of the plant machinery. The inflammable and explosive nature of methane, although useful in this instance, presents a constant threat in the sewers themselves (Fig. 12.20).

Fig. 12.19 *Digested sludge can be sold as a fertilizer*

Fig. 12.20 *Below Oxford Street, Manchester, a sewerman carries out a safety check for methane gas. From your physics lessons you may recognize the equipment he is using – a Davy safety lamp*

The production of methane in a digester and its use as a fuel are possible in other situations. For example, an area of a few hectares can support enough livestock to provide waste organic material for a digester. The treated sludge can be returned to the land as a fertilizer and the methane used as a fuel. (Approximately 3 m^3 of pig waste provides enough energy to boil five kettles of water.) Methane can also be compressed and stored in cylinders. In this form it may, with a simple modification to the carburettor, be used to drive a conventional petrol engine. These are yet more examples of how man can make use of waste in a recycling process. By such processes we prevent pollution and help to conserve the earth's natural energy resources.

Answers to questions in the text

1 In the mouth. Jaws and teeth chew (masticate) the food.

2 Folds produce a large surface area within a small space. There are more glands and greater secretion of gastic juice than would be produced by a flat surface in the same space.

3 The appendix becomes infected and inflamed. Thus it is extremely painful. Usually an inflamed appendix is removed rapidly so that there is no chance of it bursting and releasing poisonous material into the abdomen, causing infection of other organs (peritonitis).

4 The blood; tissue fluids.

5 Fish need dissolved oxygen in the water to live. Salmon are particularly large fish and need a high concentration of dissolved oxygen.

THINGS TO DO

Your teacher may be able to arrange for your class to visit a sewage-works.

1 Draw an outline of your home in cross-section. Indicate:
(a) the position of the lavatory, sinks and bath
(b) the waste-pipes from the sinks and sewage pipes from the lavatory
(c) the pipes that carry rain-water from the gutters
On a second diagram which shows the ground-plan of your home indicate:
(d) where the rain-water enters the drains
(e) the path of the drains and their junction with the main sewer
(f) the position of any manhole covers and inspection chambers

2 Draw a diagram of a lavatory pan. Explain the need for the S-bend on the waste-pipes of lavatories, sinks and baths.

3 Find out about the construction and action of a septic tank. Why do septic tanks have to be installed on some properties? How are they maintained by the local authority?

4 Take a cube of dry bread and chew it in your mouth for some time. What do you begin to taste? To confirm your theory spit the bread out into a container. Add water and stir. Allow the particles of bread to settle, then pour off the liquid. Using a 'Clinitest' tablet (page 128) test the solution for sugar. You might suggest that sugar was in the bread even before it was chewed. If you make this suggestion you have realized the need for a biological control. Therefore an unchewed piece of bread should be tested in the same way. Any difference between the two results must then have been brought about by the action of the enzyme in saliva.

5 Take a cup of pasteurized milk, pour it into a basin and add to it a teaspoonful of rennet. Place the basin in a saucepan that contains water and heat it carefully until it is lukewarm. Stir and leave the milk to cool. The rennet coagulates the protein in the milk, therefore a solid white precipitate (the curd) is formed. The curd is suspended in a watery solution, the whey. As a control another sample of milk should be treated in the same way but no rennet should be added (Examples 4 and 5 illustrate the action of digestive enzymes. Other tests need special solutions and should be carried out under the guidance of your teacher.)

6 Ask your butcher to supply you with some tripe. This is part of the stomach of a cow or sheep. You can examine the internal surface. Cooked in milk with onions it makes a delicious meal! The cleaned guts of pigs are used as sausage skins.

THINGS TO WORK OUT

In Questions 1–7 write down the *one* answer which you think is correct.

1 Roughage is needed in the
- **A** colon
- **B** oesophagus
- **C** appendix
- **D** duodenum
- **E** stomach

2 A part of the digestive system that does not produce a digestive juice is the
- **A** stomach
- **B** ileum
- **C** rectum
- **D** pancreas
- **E** cavity of the mouth

3 After proteins have been digested they are absorbed into the bloodstream as
- **A** glucose
- **B** amino-acids
- **C** fatty acids
- **D** glycerol
- **E** salts

4 The function of bile juice is
- **A** to convert complex sugars to simple sugars
- **B** to coagulate milk
- **C** to sterilize the food
- **D** to emulsify fats
- **E** to break down proteins to peptones

5 The chemicals that are responsible for the chemical processing of food within the digestive system are called
- **A** acids
- **B** alkalis
- **C** enzymes
- **D** mineral salts
- **E** neutralizers

6 Urea is a waste product of
- **A** respiration
- **B** protein metabolism
- **C** fat metabolism
- **D** digestion
- **E** anaerobic respiration by bacteria

7 'Bio-gas' contains a high percentage of
- **A** oxygen
- **B** carbon monoxide
- **C** methane
- **D** sulphur dioxide
- **E** nitrogen

8 Describe how a piece of lean meat is processed in
(a) the mouth
(b) the stomach
(c) the small intestine

9 (a) Where in the body are proteins broken down into amino-acids?
(b) Where in the body are amino-acids built up into proteins?
(c) Give *two* reasons why proteins have to be broken down by the body.
(d) What type of chemical is an enzyme?
(e) What is the action of an enzyme? Give *two* conditions that may limit its action.

10 Test-tube A contains starch solution and saliva.
Test-tube B contains starch solution.
Test-tube C contains distilled water and saliva.
All three test-tubes are heated in a water-bath at 30 °C for 10 minutes. When tested, reducing sugar is found to be present in A but not in B or C.
(a) Explain this result in A.
(b) Explain why B and C were also used in the experiment.
(c) Where would such a reaction take place in the digestive system?

11 (a) Describe the movement of food along the oesophagus.
(b) The material within the small intestine contains a lot of water. Give *three* places where this enters the system.
(c) Where and how is water taken out of the digestive system?
(d) If a lot of water remains in the faeces what name is given to this condition? What is it a symptom of?
(e) What other condition is related to the amount of water in the faeces?

12 (a) Distinguish between the action of aerobic and anaerobic bacteria.
(b) Draw a simple diagram and describe briefly the various processes that take place in the treatment of sewage.

13 Describe in detail how the starch in a cube of bread eventually becomes stored as glycogen in the liver. Under what conditions would glycogen be released from the liver?

14 Describe how the cells of our body obtain the food and oxygen that they need for respiration. Describe the process of respiration itself.

15 (a) What is meant by homoeostasis?
(b) How does the liver help to maintain the concentration and composition of the blood?
(c) What part does the liver play in nitrogen metabolism?
(d) What materials bypass the liver and how is this achieved?
(e) Describe two other functions of the liver.

13 Excretion

Fig. 13.1 *Refuse collected from our homes is either dumped at the Local Authority tip and buried, or burned in a large incinerator*

Fig. 13.2 *Uncollected rubbish remains on the street during a strike (1979)*

TABLE 13.1 **THE ORGANS THROUGH WHICH MATERIALS ARE DISCHARGED FROM THE BODY**

ORGAN		Materials discharged: Excretory products	Materials discharged: Other material
Lungs	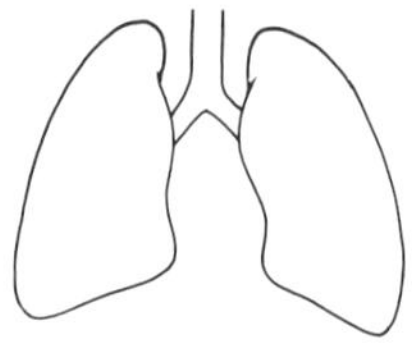	Carbon dioxide	Water vapour
Skin	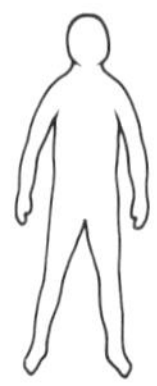	A small amount of urea	Water Some salts
Kidneys		Urea	Water Salts of sodium, potassium and calcium Drugs and poisons
Large intestine	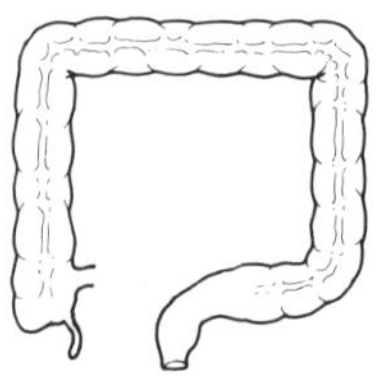	Products from destroyed red cells Dead cells Secretions from digestion	Undigested food (e.g. cellulose and surplus protein) Intestinal bacteria Some salts Drugs and poisons

13.1 **Waste and the organs that dispose of it**

The crane in Fig. 13.1 is disposing of communal waste at a local incinerator. Both as individuals and as a community we create huge quantities of waste. If all this waste is allowed to accumulate in our midst it can create considerable problems. Some people have experienced how awkward life becomes when collection of refuse ceases (Fig. 13.2). We must dispose of our waste correctly if it is not to pollute or poison our environment and affect our well-being. Thus household refuse is collected, transported and disposed of at local-authority tips. It may even be taken to a refuse plant where some items that have another use are reclaimed (for example, metals which can be recycled).

The body is the same as a community in that it is always producing waste. If allowed to accumulate within the body this waste would become poisonous (toxic) and threaten the living processes. We have seen in the previous chapter how some waste is discharged from the body through the large intestine of the digestive system. Like household refuse the body's waste is a mixture. It contains material which has not been digested or absorbed, together with water, dead cells, mucus and 'left-overs' such as the colouring substances of bile, produced when red blood cells are destroyed in the liver. It also contains some materials that the body has eaten to excess. Anyone who has taken iron-pills will observe that their faeces become blacker, because of the high content of iron salts.

It is not only solids that have to be disposed of. The lungs are concerned with the removal of carbon dioxide, a waste product of tissue respiration. Waste substances that are produced by the chemical activity of body cells are often referred to as *excretory products*. This distinguishes them from materials such as cellulose that are not the result of the body's activities but merely pass through unchanged. Carbon dioxide and urea are excretory products. Water is not an excretory product. It may be lost from the body as part of another process such as sweating (page 188) or when it forms a solution with an excretory product (urea in urine).

Question 1

(a) In what organ is urea produced?
(b) Is urea a breakdown product of (i) fats, (ii) amino-acids, or (iii) glucose?

Table 13.1 shows the four organs that excrete substances from the body. It is the *urinary system* that deals with the excretion of urea.

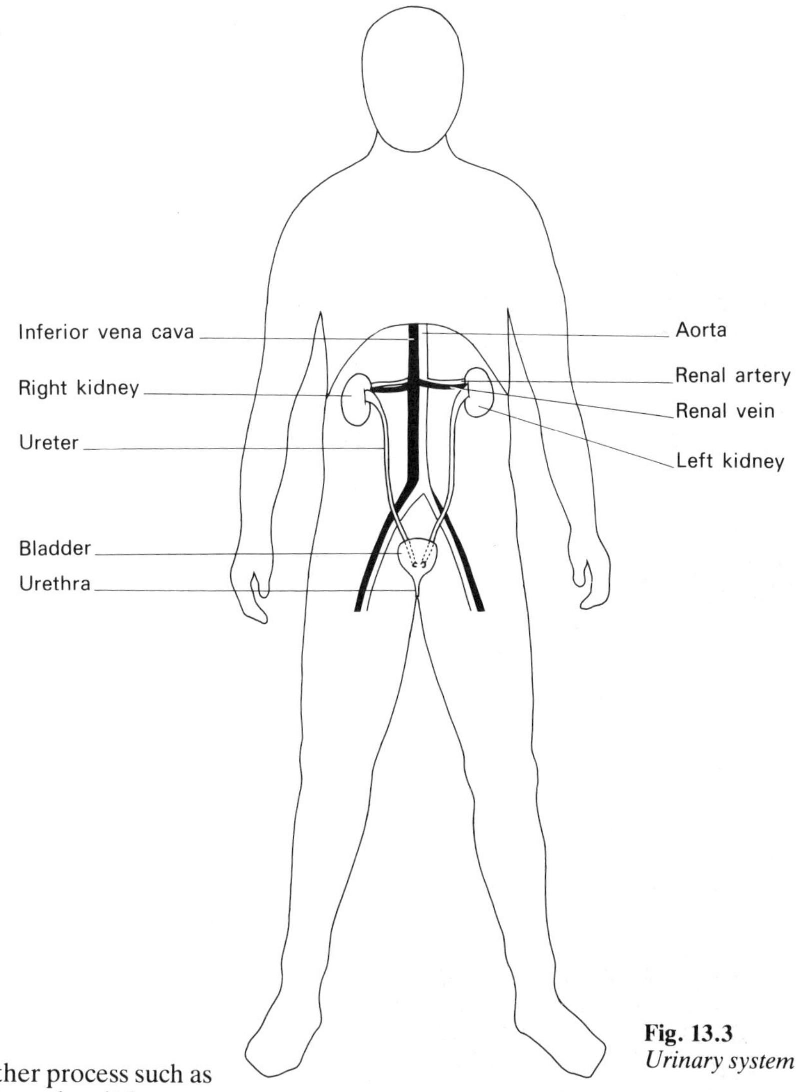

Fig. 13.3
Urinary system

13.2 **The urinary system**

Fig. 13.3 shows the urinary system in the body. It is not visible in a conventional X-ray picture (Fig. 13.4(a)); only bones are clearly seen by this method. However, by using a

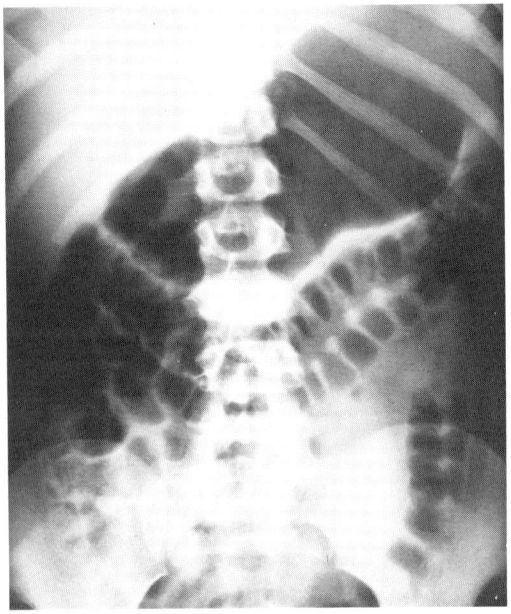

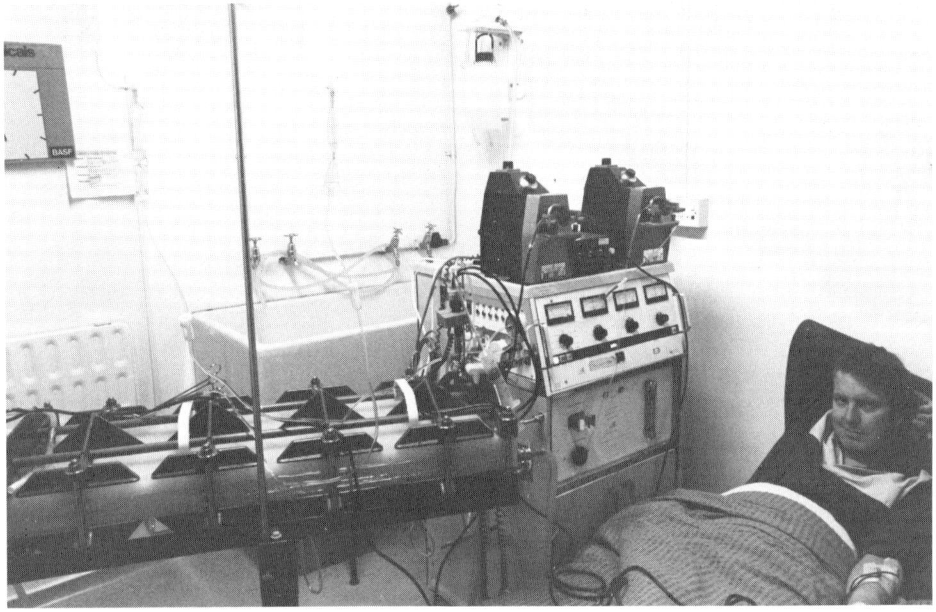

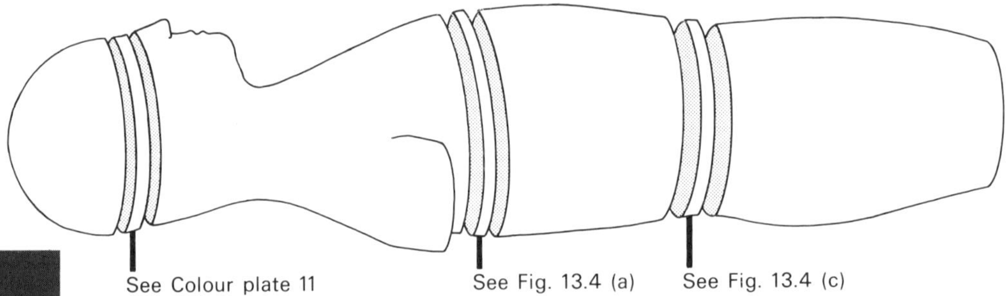

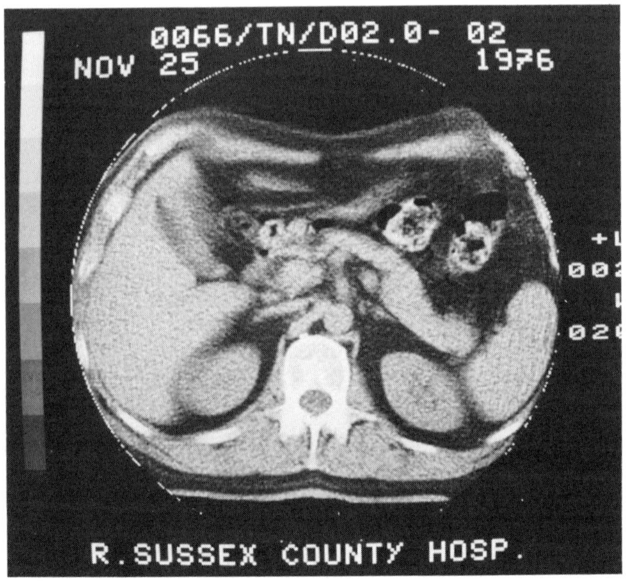

Fig. 13.4 *(a) Conventional X-ray of the abdominal region. (b) Computed tomography uses highly sensitive detectors to record the intensity of a narrow beam of X-rays after they have passed through a person's body. The X-ray tube and detectors revolve around the person 'scanning' the body over 180°. Thousands of measurements are fed into a computer, which in seconds converts this information into a television-type picture of a cross-section through the body. (c) Cross-section through the abdomen produced on a body scanner*

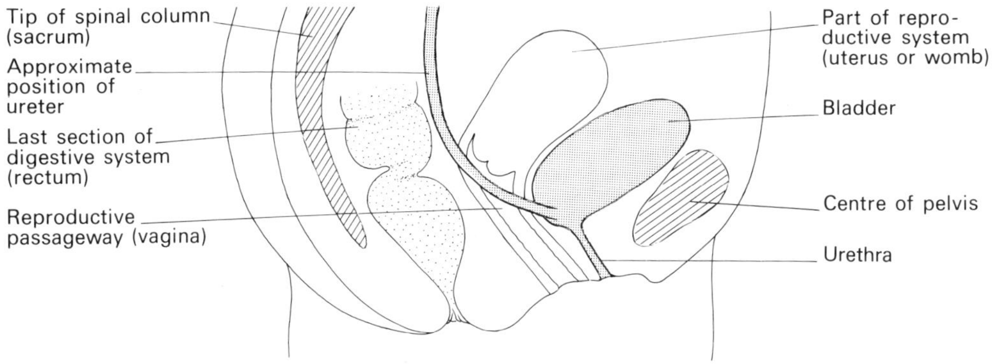

Fig. 13.5 (Left) *Position of the bladder in a woman*

technique known as computed tomography (Fig 13.4(b) and (c)) the kidneys and other soft tissues become visible. The kidneys are about 10 cm long, 5 cm wide and 3 cm thick. The are dark red in colour and are positioned on the back wall of the abdominal cavity, behind the peritoneum. Their concave surfaces face the backbone. At the point where this concave surface curves sharply inwards (the place where you would sit at a kidney-shaped dressing-table) there are three tubes. This point on the kidney is the *hilum* and two of the three tubes are blood-vessels. The *renal artery* brings blood which is rich in oxygen to the kidney. This blood also contains the excretory product urea. The renal vein carries blood away from the kidney. The third tube is the *ureter*, which travels downwards along the back of the abdominal cavity and connects the kidney to another organ, the *bladder*. There are three openings in the wall of the bladder. The two at the top to the left and right are the openings of the ureters. The third opening, at the base of the bladder, leads to another tube, the *urethra*. The bladder lies behind the pelvic bones. Sometimes its top surface comes above the pelvic bones because it swells upwards when liquid enters it (like a hot-air balloon anchored to the ground and swelling upwards and outwards as it inflates). In the female the reproductive passageway known as the *vagina* lies between the rectum and the bladder (Fig. 13.5). As this is the passageway that a baby travels along during birth you can see that at this time the rectum and bladder tend to get rather squashed! A female has three openings in the 'floor' of her abdominal cavity. At the front is the opening of the urethra, in the middle the opening of the vagina and towards the back the opening of the anus (Fig. 13.6).

The urethra in the male is about five times longer than that in the female, and it is contained within the organ known as the *penis* (Fig. 13.7). It serves a double purpose: transporting urine from the bladder and at certain times sperms (pages 2 and 251) from the reproductive organs (page 242).

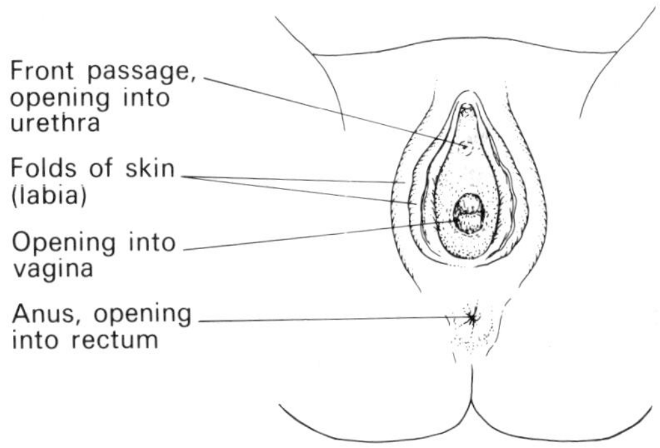

Fig. 13.6 *Floor of the abdominal cavity in a woman*

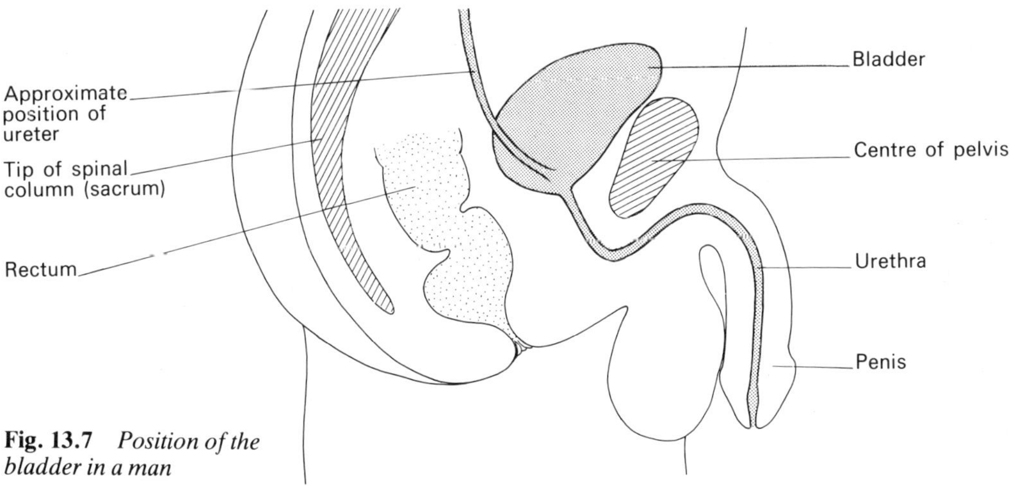

Fig. 13.7 *Position of the bladder in a man*

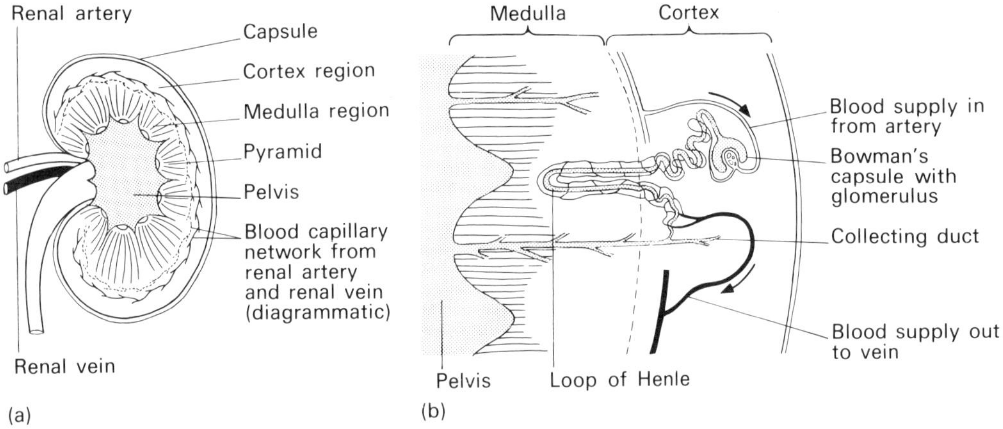

Fig. 13.8 *(a) Diagrammatic cross-section of kidney. (b) Enlargement of medulla and cortex region. (c) Function of the nephron*

13.3 **Inside the kidney**

Fig. 13.8(a) shows a kidney in cross-section. You can see that it has a thin protective outer layer, the *capsule*. Within this are two distinct regions, the *cortex* and the darker *medulla*. The medulla does not have a straight inner edge. It is arranged to produce a number of 'hills' or *pyramids* which protrude into an open area, the *pelvis*. The pelvis is in fact the wide, funnel-shaped end of the ureter.

The renal artery and vein branch out to serve each pyramid and form numerous capillaries. Fig. 13.8(b) shows some of these capillaries within the cortex region: they have become intertwined, like a tangled ball of string, to form a knot of capillaries, a *glomerulus*. Each glomerulus is surrounded on nearly all sides by the end of a very fine tubule. This end of the tubule is formed like a cup and is known as the *Bowman's capsule*. From the Bowman's capsule the tubule travels downwards towards the medulla region. On the way it forms numerous coils like a rope laid carelessly on the floor. Within the medulla region, the tubule becomes straighter and forms only one loop (*loop of Henle*) before travelling back into the cortex layer. Once more within the cortex it forms coils, then finally it enters a duct, together with many other tubules. This collecting duct has an opening at the tip of the pyramid. The length of the tubule from the Bowman's capsule to its opening into the duct is known as a *nephron*, and there are approximately 1 million of these within each kidney. Each nephron is a most ingenious mechanism. It 'cleans' the blood and at the same time makes sure that no substance which is needed by the body is thrown away. The large number of nephrons in each kidney forms the large surface area that is needed for this complex process to take place.

13.4 **The purpose of each nephron**

Blood reaches the glomerulus through an *arteriole* (a very small artery). This section of the arteriole is wider than that which leaves the glomerulus. Thus the pressure of the blood within the glomerulus is increased. Under such pressure small molecules are squeezed out of the capillaries and through

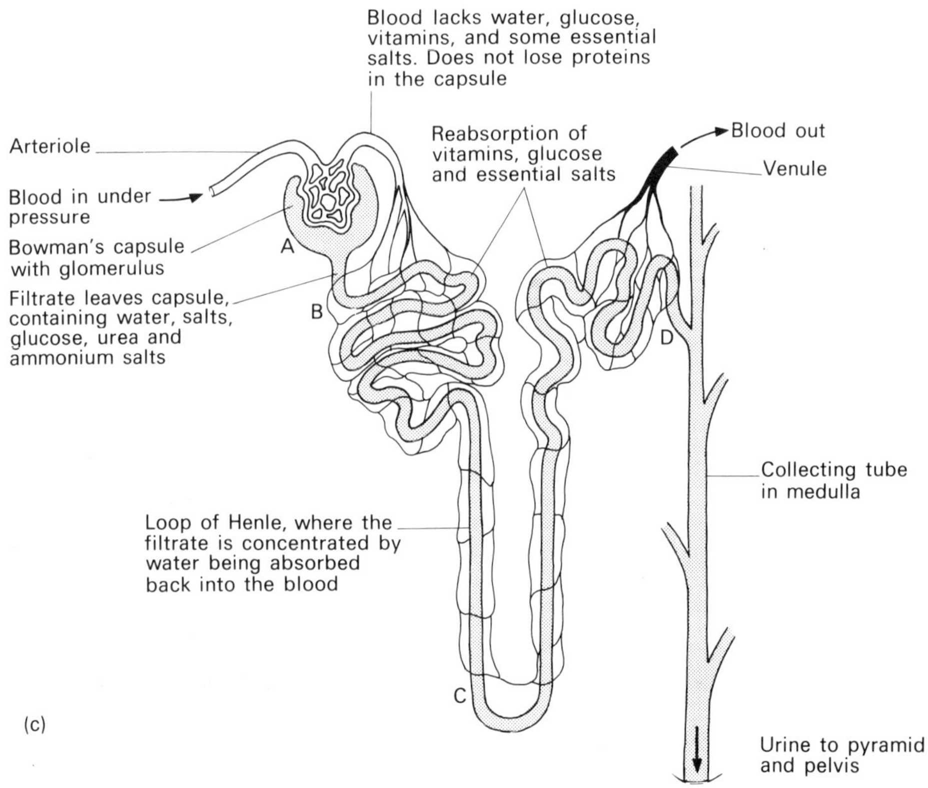

the thin walls of the Bowman's capsule (A in Fig 13.8(c)). The *filtrate* that collects in the Bowman's capsule by this process closely resembles blood plasma (page 69), but without its proteins. The molecules of the proteins are too large to filter across.

Question 2 You may have heard the expression 'Don't throw the baby out with the bath water'. How might this expression be applied to the kidneys?

As the filtrate passes along the sections of the nephron labelled B, C and D in Fig. 13.8(c) a second process takes place. The cells within the coiled sections of the nephron select and reabsorb (take back) the substances that are needed by the body, such as glucose, various salts, vitamins and water. Water is reabsorbed in great quantities in the loop of Henle. These materials are reabsorbed by the blood flowing in the capillaries that surround the coiled section of the nephron. The cells of the tubules can even secrete various substances into the tubule which have been unable to enter by the previous process of filtration.

We can now see that the function of the kidneys is not merely the excretion of urea. They also regulate the amounts of salts and water within the blood, maintaining its concentration and composition. This particular function of our kidneys is essential for homoeostasis.

When the liquid reaches the end of the nephron it is fairly concentrated. Its composition as it enters the collecting duct is as follows:

96 per cent water;
2 per cent dissolved salts, for example the chlorides and sulphates of calcium, sodium and potassium;
2 per cent urea.

This liquid is known as *urine*. Meanwhile the blood that leaves the kidney in the renal vein is free of the toxic urea. If the kidney fails to remove urea it accumulates and is toxic to the body's tissues.

13.5 Kidney failure and the artificial kidney

Kidney tissues, like any other tissues, can become inflamed, a condition known as *nephritis*. If a kidney becomes severely damaged by continuous inflammation or through infection, it may stop functioning. A person may survive with only one kidney working, but when both fail the accumulation of poisons, excess salts and imbalance of the tissue fluids cause great strains on other organs. A severe illness known generally as *uraemia* develops. *Chronic renal failure* causes death if it is not treated by a form of *dialysis* or by a *transplant*.

Dialysis is the term used to describe the process of cleansing a person's blood artificially. It is described in Fig. 13.9 and, as you can see, the exchange of molecules takes place by diffusion and through the selective action of a semi-permeable membrane. The figures shows that dialysis can remove only a certain amount of the waste that accumulates in the blood. It is not as efficient as the body's normal kidney action. Normal kidneys cleanse one-fifth of the body's blood every minute of the day and night. Dialysis is used on average three times a week, each time for a period of ten hours. Therefore the body of

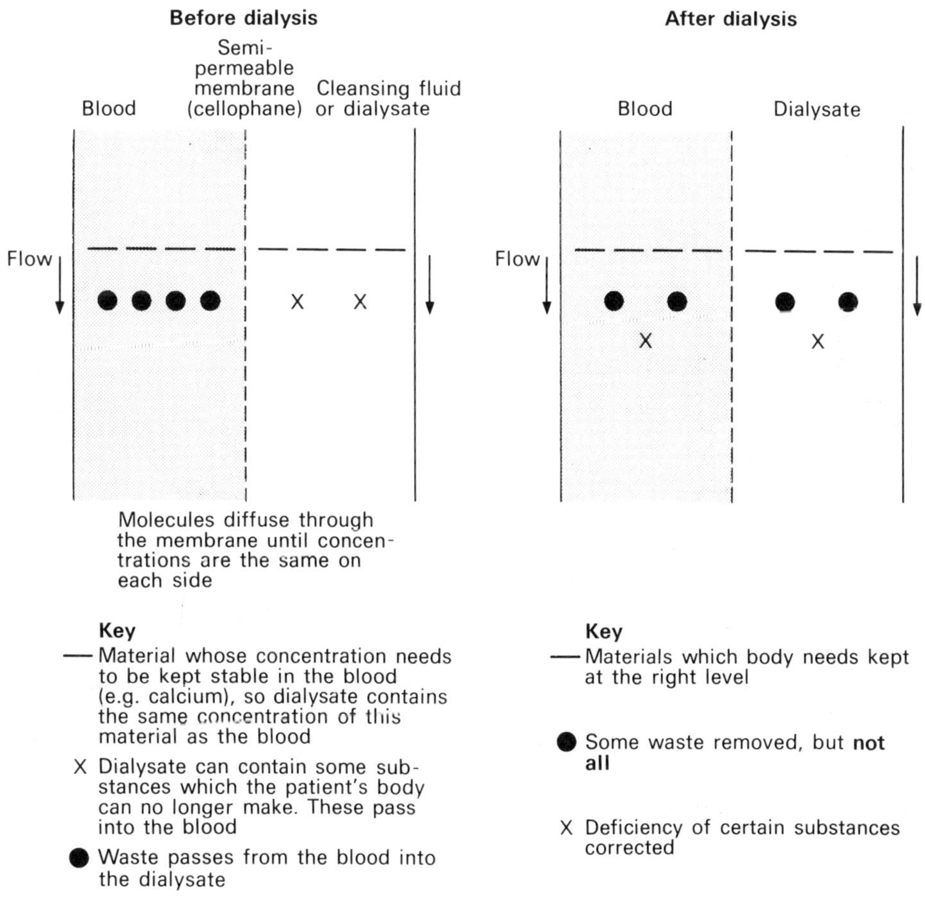

Fig. 13.9 *The principle of dialysis*

a person who relies upon dialysis adjusts to the accumulation of waste within the bloodstream, as long as it does not exceed certain limits. To make sure that these limits are not exceeded a person on dialysis has to observe a strict diet (page 00). This diet particularly limits the intake of fluid, salt and potassium. An excess of potassium in the blood can damage the heart.

Fig. 13.10 shows one type of *artificial kidney* that is used in dialysis. It looks like a sandwich with two separate layers of cellophane stretched tightly across parallel boards. The cellophane acts as the semipermeable membrane. The blood and dialysate flow on opposite sides of the semipermeable membrane.

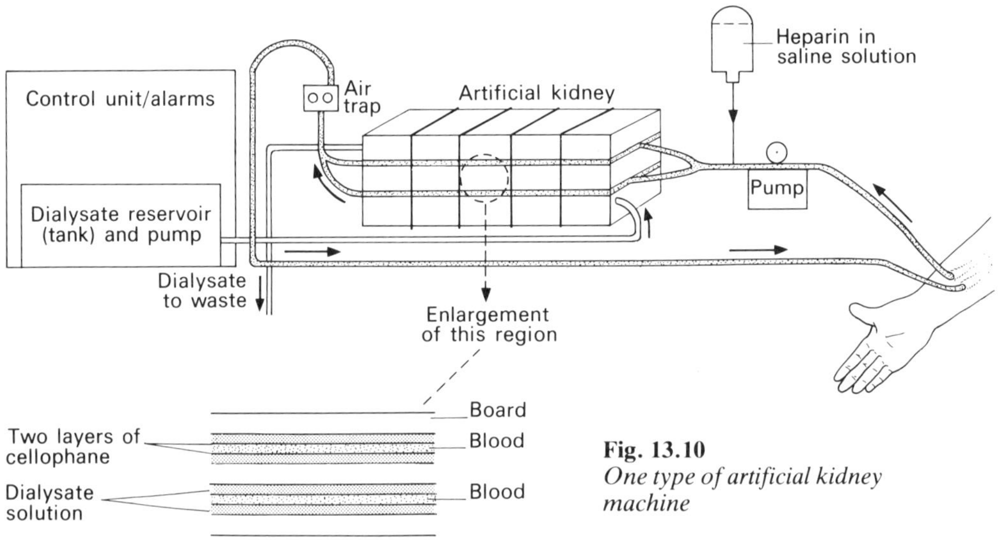

Fig. 13.10
One type of artificial kidney machine

Question 3 What does the cellophane represent in the normal kidney?

You can see in Fig. 13.10 how the person has to introduce two needles through the skin into a particular vein in the arm to make the connection between the body's blood system and the artificial kidney. Arteries are deep below the skin so it would be very difficult to insert a needle into them. Instead an artery and a vein are linked together near the wrist. This join is called a *fistula* and it allows arterial blood to flow into the vein, so that a source of arterial blood is brought near to the skin surface. Also the vein becomes swollen, making it visible and easier to find for needling.

A pump is connected into the circuit of patient and artificial kidney, to assist the pumping action of the heart. Another pump pushes the dialysate solution from its tank through the artificial kidney. It then flows from the kidney and down the drain. A certain amount of heparin is introduced into the blood to prevent it coagulating.

Between dialyses a person can generally live a normal active life. Things may not run smoothly all the time but each person will receive all the medical help and general support that he needs from the staff of the kidney unit of the hospital to which he is attached. His family can and will give their support, particularly when the kidney machine is installed in the home. There is no doubt that undergoing dialysis does demand a certain routine in a person's life and great personal determination and perseverance when faced with physical discomfort.

13.6 **A transplant**

A transplant is another answer to chronic renal failure. The healthy transplanted kidney is positioned within the groin region of the abdominal cavity. It is connected to the bladder, and to the blood system through the iliac vein and artery. When it is successful it allows the recipient a more normal way of life than does dialysis. The problem with any transplant is *rejection*, i.e. that the recipient's body will regard the implanted organ as foreign tissue. The body's defence system, the white blood cells and antibodies, attack the transplant organ, interfere with its functions and eventually destroy it. Obviously there will be less chance of this if the donor's tissues and blood group match those of the recipient. Also drugs are administered to suppress the body's defence systems. However, this makes the recipient vulnerable to other infections, so isolation is essential at this time. A person who has suffered chronic renal failure will undergo extensive tests to find out all the information needed by the doctors. When a donor kidney becomes available this information will be used to decide who should receive it. This network of information is both national and international. In some cases a kidney donated and removed within an hour of death may be transported thousands of miles to a suitable recipient. The more kidneys that are donated, the greater the chances of a transplant for a person with chronic renal failure (page 75).

13.7 **The bladder**

Although approximately 170 litres of water filter each day into the Bowman's capsule only a hundredth of this volume is finally released as urine. This is squeezed along by muscular contraction (peristalsis) in the walls of the ureters. It is therefore squirted into the bladder every few minutes. A *sphincter muscle* at the point where the urethra connects with the bladder is contracted and stops the urine going any further. The bladder is like a muscular bag: it stretches as more and more urine collects inside it. Altogether it can hold an average 0.6 litres without being overstretched. However it is usual to feel a distinct urge to release the urine before this volume is reached. During urination the sphincter muscle relaxes and the bladder contracts, forcing the urine out.

Generally infants under two years of age urinate automatically, i.e. the sphincter muscle relaxes as soon as the bladder muscles are stretched beyond a certain point. Hence the need for nappies! As we saw earlier, a heating system is usually set to operate automatically. However, we can always take over and override the system by operating the switch ourselves. Similarly the body in some instances can override an automatic system. Children gradually learn control over the sphincter muscle but it would be dangerous to the kidneys if we ignored the urge for a very long time. There is a point at which the body will not allow us to control the situation any longer!

13.8 **Changes in the volume and composition of urine**

You may have noticed that you pass a greater volume of urine on some days than others. If you consider your activities during those hours you will probably see why. Some examples are given in Fig. 13.11.

After a lot of activity in hot weather or after not drinking for several hours we usually feel thirsty. Although thirst cannot be fully explained it is a feeling that occurs when the blood has a low concentration of water and a high concentration of salts (page 69). When water is drunk it quickly enters the system and dilutes the blood. The kidneys can then restore the natural balance.

We can live without food for up to twenty-five days, but without water for only ten days. In the desert this time would be reduced to about two days.

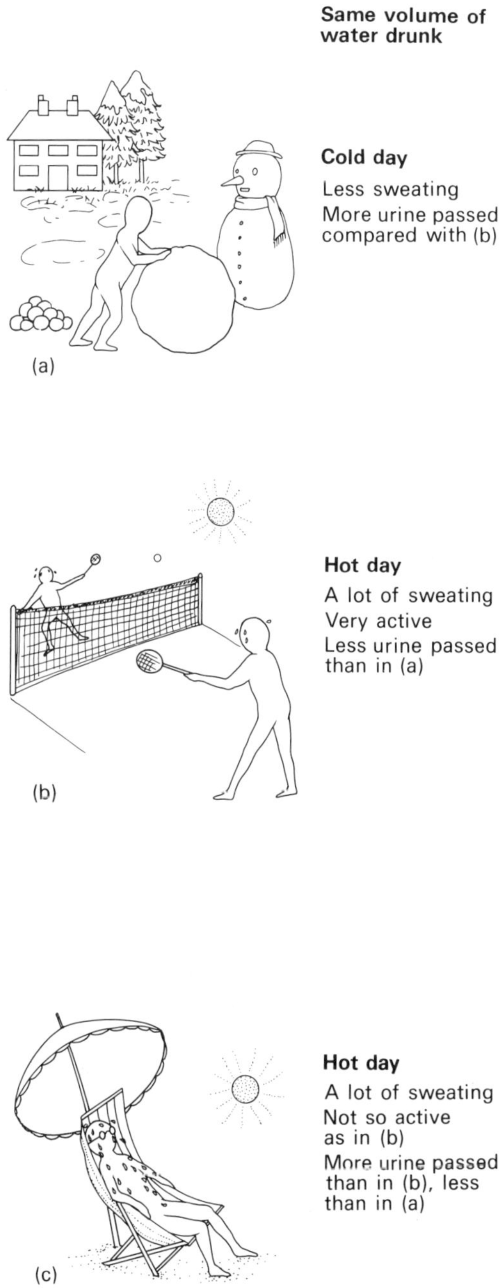

Fig. 13.11 *Changes in volume of urine*

Question 5 Why would a person feel thirstier if they drank sea-water?

The composition of urine is a good indication of the general health of the body. To look at, urine is yellow and clear as it leaves the body at body temperature. If allowed to cool it may become cloudy as deposits of salts settle out as solid particles. It has a distinct odour which becomes very strong if the urine has been kept in the bladder for a long time. This is because the urea in the urine has been broken down to form ammonia. Substances that should not appear in urine are glucose, blood and protein molecules like albumin. Its pH value should be around 6. Any tests carried out on urine will indicate its pH value and the presence of any abnormal substances. In Chapter 2 we saw that urine samples are taken from expectant mothers. Their kidneys are dealing with the excretory products from the foetus as well as from themselves. It is important to check that they are coping satisfactorily. An increase in blood pressure can strain the kidneys so that some materials are not removed from the blood and other vital ones are. Thus proteins may be detected in the urine and there may be an increase in weight due to water accumulating in the tissues. In pregnancy this condition, known as *toxaemia* or *pre-eclampsia*, must be avoided. If it is detected the expectant mother will be expected to rest and to follow a low-salt diet.

Glucose in the urine indicates the condition of *diabetes* (page 181). The glucose level in the blood is so high that the tubules cannot reabsorb all of it from the kidney filtrate.

The fine control that the kidneys have over the composition of the blood is in its turn controlled by the endocrine system, which we will be looking at in Chapter 14.

Answers to questions in the text

1 (a) The liver, (b) amino-acids (by deamination).

2 The kidneys at this point have removed waste products from the blood. They have also removed many precious substances that the body needs.

3 The cellular walls of the nephron and blood capillaries.

4 The salt from the sea-water would enter the blood, making it highly concentrated. Thus the urge would be to drink more in order to try to dilute the blood.

THINGS TO DO

1 Draw a map of your area and indicate any places where rubbish may be tipped. Find out if your local authority reclaims any materials, such as metal and rags.

2 Make a record of the collection of materials for recycling which goes on in your area. The most common is the collection of waste paper by local organizations in order to raise funds. You may like to take part in such a scheme.

3 Design a poster which you think will convince others that the reclamation and recycling of our resources is vital.

4 Obtain a sheep's kidney from your butcher. Cut it in half lengthwise and examine its internal structure.

5 A person can become dehydrated through drinking sea-water because more water is drawn from the body cells and excreted by the kidneys. Find out about the desalination equipment that is carried as part of a survival kit at sea. How does a desalination plant work?

THINGS TO WORK OUT

In Questions 1–5 write down the *one* answer which you think is correct.

1 Before it is expelled from the body urine collects in the
- **A** bladder
- **B** ureter
- **C** urethra
- **D** pelvis
- **E** Bowman's capsule

2 Urine does *not* normally contain
- **A** water
- **B** proteins
- **C** urea
- **D** calcium sulphate
- **E** sodium chloride

3 Of the following articles the one that is likely to remain intact at a refuse site after many years is
- **A** a wooden packing-case
- **B** a collection of newspapers
- **C** a plastic bottle
- **D** a piece of corrugated iron
- **E** an old coat

4 Which *one* of the following is an excretory product?
- **A** carbon dioxide
- **B** water
- **C** cellulose
- **D** dead cells
- **E** mucus

5 Urea is a substance which is excreted in urine and it is produced in the liver from the breakdown of surplus
- **A** proteins
- **B** fats
- **C** sugars
- **D** starches
- **E** mineral salts

6 Assuming that a person keeps to the same diet, what changes will take place in the volume and concentration of urine produced each day during a hot summer? Name and explain *one* other factor which will arise under these conditions.

7 How is dry refuse disposed of? State *two* dangers of leaving refuse uncollected.

8 (a) Explain why proteins and glucose are not found in a healthy person's urine.
(b) What disease may cause the presence in the urine of (i) protein, (ii) glucose.

9 (a) Draw and label a simple diagram to show the structure of a single nephron in the kidney.
(b) Using a diagram explain how the kidneys cleanse the blood and at the same time keep its composition within very fine limits.
(c) Describe *two* consequences of kidney failure.

14 Hormones and the Endocrine System

14.1 Messengers

It is Friday afternoon and Mr Jones the music master has just been told by the headmaster that the school band has been invited to play at the hospital fête on Saturday afternoon. Mr Jones must inform all the players of the arrangement and tell them what action they should take. But the players are scattered all over the school, taking different lessons! There is one way Mr Jones can communicate with them. He can send another pupil round the school to look for each player and to deliver a message. The message tells them what Mr Jones wants them to do. Sending messages in this way is one method by which we can communicate with one another. Communicating with one another enables us to co-ordinate our activities.

14.2 The chemical messengers of the body – hormones

It was impossible for Mr Jones to have a direct physical link with every pupil in the school. Similarly, it is impossible for each cell in the body to have a direct communication link with every other cell. Instead the body uses chemical messengers. These can travel through the channels – the blood-vessels – that are to be found in every part of the body. The chemical messengers are called *hormones*.

Hormones are produced by specialized *glands* or groups of cells which are found among the tissues of another organ, for example the kidney. Fig. 14.1 shows the position of some of these specialized glands. The hormones are secreted directly into the blood as it flows through the capillaries within the glands. The glands do not pass their hormones into the bloodstream by means of tubes or *ducts*, which is why these glands are often called *ductless* or *endocrine glands*. Glands which have ducts are referred to as exocrine glands.

Question 1 In Fig. 14.1 you should be able to spot an organ which acts as both an exocrine and an endocrine gland. Which is it?

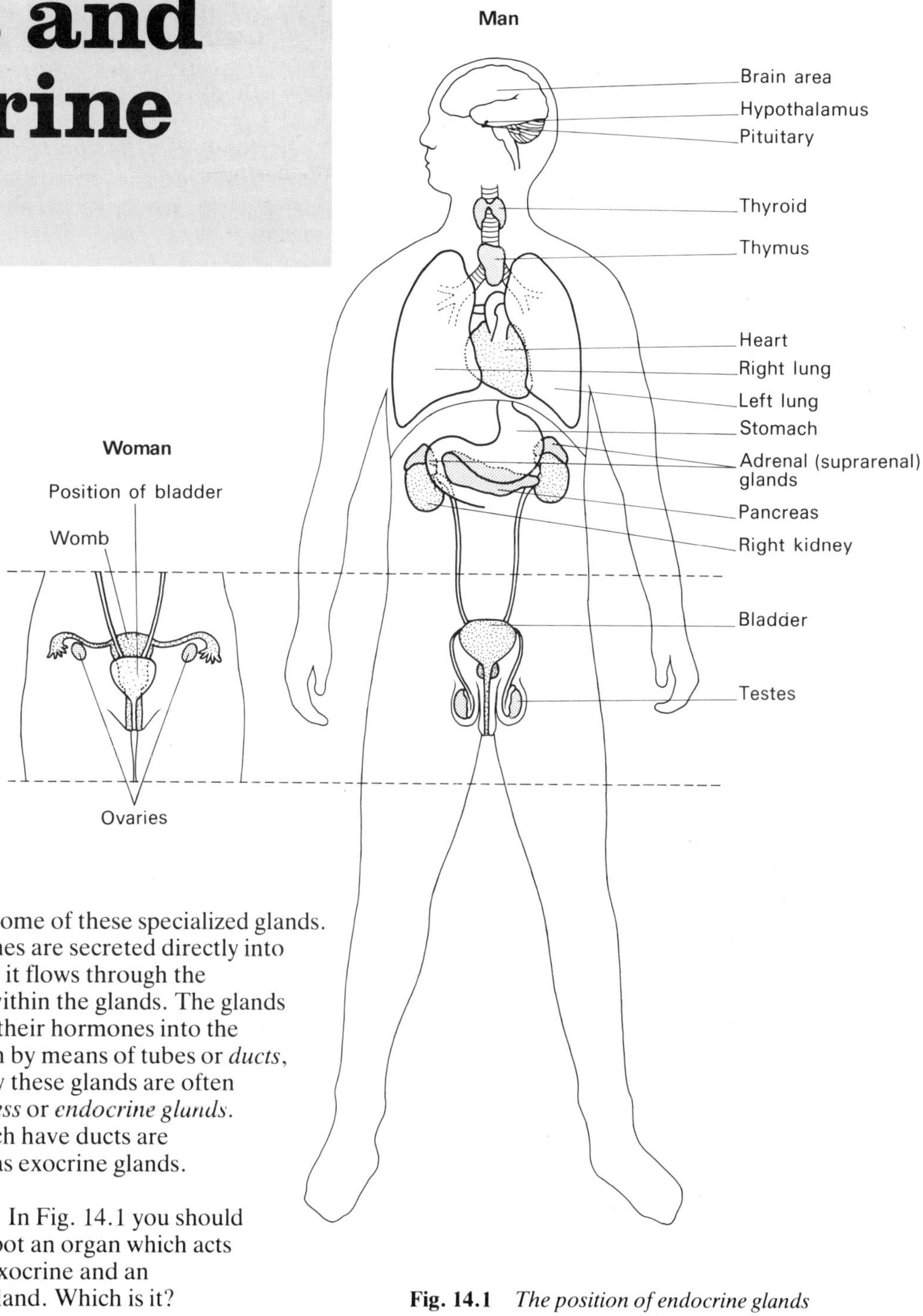

Fig. 14.1 *The position of endocrine glands*

The message from Mr Jones has a specific destination; anyone who was not a member of the school band would ignore it. Similarly, in the human body the only cells that will act on instructions received from a hormone will be those for which the message was intended.

Hormones play a key role in the process of keeping the internal environment of the body constant (homoeostasis). For instance, we have already discussed in Chapter 13 how the kidneys regulate the concentration of water in the blood. This osmoregulation is carried out according to instructions received from the *pituitary gland*. This gland, about the size of a pea, is positioned on a stalk at the base of the brain (Fig. 14.2). It produces *antidiuretic hormone* (ADH). When this hormone reaches the kidneys it causes the tubule cells to reabsorb and return water to the bloodstream. In this way ADH reduces the volume of urine that is produced by the body. If the pituitary gland detects that the blood in the body contains too much water it will not release so much ADH. As a result the kidneys will extract a greater amount of water and the volume of urine will increase.

Some substances that we take into our bloodstream, such as excessive alcohol and caffeine (found in coffee), can lower the production of ADH from the pituitary. This is why large quantities of these drinks do not seem to quench our thirst. Instead they cause our bodies to release a greater volume of water as urine. Similarly, too much alcohol can dehydrate the body rather than give it water. This often leads to what is commonly called a 'hangover': a headache, dry mouth, and a need for another drink, which should be water!

14.3 Hormones – undersecretion and over-secretion

Table 14.1 gives some of the other hormones that are produced by the pituitary. It also gives some of the hormones released by the other endocrine glands and describes their action.

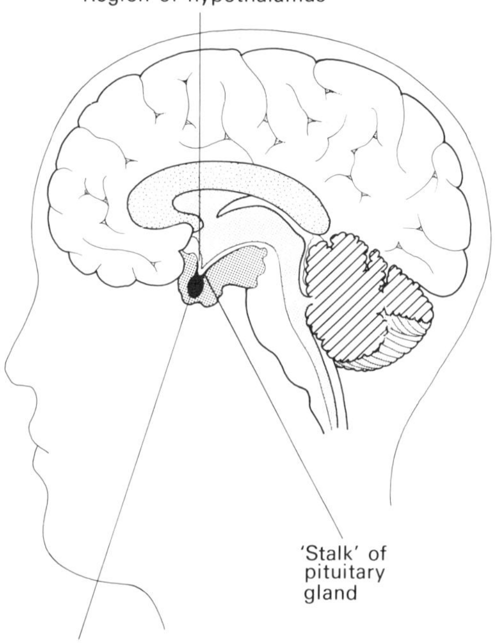

Fig. 14.2 *Slice through the brain to show the position of the pituitary gland*

TABLE 14.1 **ENDOCRINE GLANDS AND THEIR ACTION ON THE BODY**

Gland	Structure	Hormone	Action	Effects of over-secretion (hypersecretion)	Effects of under-secretion (hyposecretion)
Pituitary	Size of a pea, situated on a stalk at the base of the brain	Hormones that influence other glands	Influence growth, thyroid gland, adrenal gland, sex organs	Too much growth hormone: (a) in childhood – 'over-growth' or *gigantism* (b) In an adult – *acro-megaly*. Regions of the body such as jaw, nose, hands and feet continue growing	Too little growth hormone in childhood – *midget* or *dwarf*. Person develops as an adult in every way other than size
		Prolactin	Causes mother's breast milk to be produced after the birth of a baby		
		Antidiuretic hormone (ADH)	Influences urine production		
		Oxytocin	Causes muscles of the womb to contract during birth (page 254). Causes breast to give milk when the baby suckles		

TABLE 14.1 **continued**

Gland	Structure	Hormone	Action	Effects of over-secretion (hypersecretion)	Effects of under-secretion (hyposecretion)
Thyroid	Two lobes either side of trachea, with a connecting piece in front	Thyroxine	Controls body's metabolic rate. Influences physical and mental development. (Iodine is needed to produce thyroxine)	(a) *Hyperthyroidism* or *exophthalamic goitre* — thyroid enlarged, eyes bulge, heartbeat fast, restlessness and excessive sweating. Treatment is by surgery or drugs (b) *Simple goitre* – arises through iodine deficiency	*Hypothyroidism*: (a) in childhood – poor physical and mental development (b) in middle age – *myxoedema*. Tissues become puffy and the person becomes slow and lethargic. Treatment: give thyroxine
Thymus	Mass of pink-coloured tissue. Large when we are young, but gradually decreases in size		Hormones which in some way affect the body's ability to make lymphocytes in the lymph glands (page 92). Influences immunity to disease		
Adrenals (suprarenals)	Yellowish objects about 2–3 cm in length	Adrenaline	Brings the body to 'action stations' in times of stress, fright, etc. (a) Heartbeat increased (b) Blood capillaries in some tissues are constricted (e.g. skin), so more blood is available to muscles, heart and lungs (c) Breathing rate speeds up (d) Pupils of eyes dilated (e) Liver releases glucose (f) Tissue respiration increased	*Cushing's syndrome* – obesity, hypertension and diabetes	*Addison's disease* – excess loss of salt and water from the body. Low blood pressure. Skin becomes 'patchy', with darker brown areas. Treatment: give hormone
		Steroids (e.g. cortisone)	Influence level of blood-sugar. Assist body's defence system of the blood		
		Others	Influence sexual development		

Table 14.1 continued over . . .

<table>
<tr><th colspan="6">TABLE 14.1 ENDOCRINE GLANDS AND THEIR ACTION ON THE BODY continued</th></tr>
<tr><th>Gland</th><th>Structure</th><th>Hormone</th><th>Action</th><th>Effects of over-secretion (hypersecretion)</th><th>Effects of under-secretion (hyposecretion)</th></tr>
<tr><td>Pancreas</td><td>Cells within the pancreas known as the islets of Langerhans</td><td>Insulin</td><td>Influences storage of glucose in the liver as glycogen. Also influences how efficiently the tissues use glucose in the blood (blood-sugar)</td><td></td><td>Sugar diabetes (diabetes mellitus, page 181)</td></tr>
<tr><td>Testes
(male)</td><td>Oval-shaped bodies found outside the body in bags of flesh – scrotal sacs. Adult length about 4 cm (page 242)</td><td>Androgens (e.g. testosterone)</td><td>Influences production of reproductive cells (page 246). Influences development of secondary sexual characteristics (page 244, e.g. growth of beard)</td><td colspan="2" rowspan="3">In both sexes there is a secretion of male hormones (androgens) and female hormones (oestrogens). The balance between these secretions influences the body's secondary sexual characteristics. If this balance is upset the sexual characteristics will be affected: over-secretion, or an administered dose, of androgens in a woman can produce more masculine characteristics such as well-developed muscles</td></tr>
<tr><td rowspan="2">Ovaries
(female)</td><td rowspan="2">Found just inside the top line of the pelvic girdle. Oval-shaped, firm bodies. Adult length 2.5–4 cm</td><td>Oestrogens</td><td>Influences female's monthly reproductive cycle (page 244). Influences development of secondary sexual characteristics (page 244, e.g. development of breasts)</td></tr>
<tr><td>Progesterone</td><td>Prepares lining of womb (page 245). Controls the release of female reproductive cells</td></tr>
</table>

These hormones represent only a few of those that are known and whose action within the body can be identified. A few of them can even be manufactured in the laboratory. However, there are many that remain unknown and we can only guess at their existence. This can be seen through the side-effects that occur when an area of the body becomes critically damaged. An example of this is seen in a person suffering from chronic renal failure, who has to rely on dialysis (page 173). The normal kidney produces a hormone which stimulates the production of red blood cells, so a patient with kidney failure can easily become anaemic. Other side-effects which have been observed in dialysis patients seem also to be linked with a lack (deficiency) of hormones from the kidney.

Within a healthy body the secretion of hormones is finely balanced to maintain homoeostasis. It becomes obvious how fine this balance is when either an *oversecretion* (hypersecretion) or an *undersecretion* (hyposecretion) of a particular hormone occurs. Some of the conditions that arise are listed in Table 14.1. As more becomes known about hormones so the treatment of such conditions becomes more effective. If symptoms can be spotted in the very early stages of the condition then treatment has a better chance of halting its advance. For instance, it is known that a baby with undersecretion of the thyroid (hypothyroid) is generally 'floppy': the skin is dry, rough and wrinkled; the tongue seems too large for the mouth; growth and progress are very slow.

Question 2 What services are always available to a mother so that she can check on her baby's healthy physical development? (Clue: see Chapter 2.)

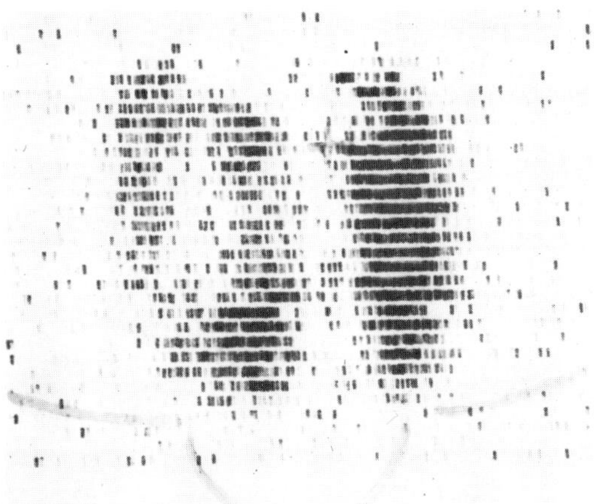

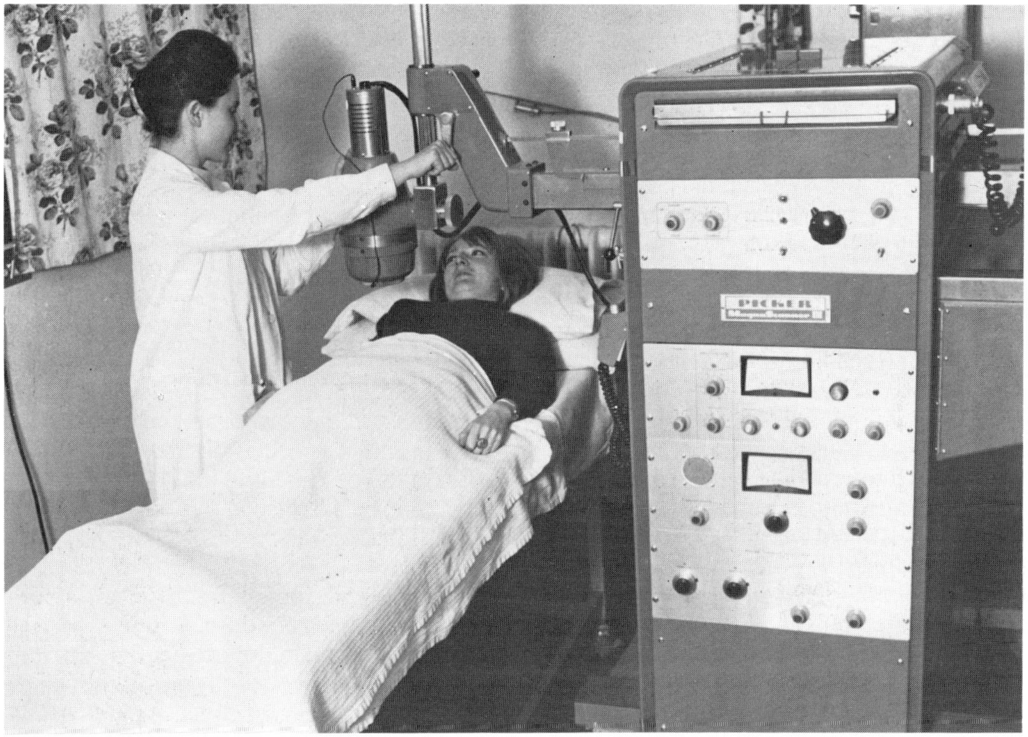

Fig. 14.3 (Above) *Scanner used to detect radioactive elements with body tissues.* (Left) *Scan of a defective thyroid gland*

It is known that if the body lacks iodine it is likely that the thyroid gland will become enlarged (*simple goitre*). This condition is more common in certain areas of the world such as Switzerland, northern Italy and parts of India and America. In these areas there are fewer iodine salts in the soil and in the water-supply. This deficiency can be corrected by adding potassium iodine to the water-supply, and it is in fact added to some forms of table salt. It is also possible to measure how much iodine the thyroid gland is absorbing from a person's food. The person is asked to eat food containing a measured amount of radioactive iodine. Then by using a scanning instrument which detects radioactivity, the uptake of iodine can be traced (Fig. 14.3), and any malfunction of the gland will become obvious. Similar tracer techniques can be used to check on the normal or abnormal passage of substances as they pass through the body.

14.4 Diabetes

When we eat, the carbohydrate content of our meal is broken down in the digestive system (page 157). The result of this is glucose, which is absorbed into the bloodstream, causing the blood-sugar level to rise. In response to this, the special cells that act as an endocrine gland within the pancreas secrete the hormone *insulin* into the blood. Insulin causes the body cells to use glucose and the liver to store any that is left over. When there is no insulin from the pancreas (or too little) the cells do not use the glucose and the liver fails to store it. Therefore the blood-sugar level rises. Eventually symptoms of diabetes can be observed. These are excessive thirst, loss of energy, loss of weight, an increased incidence of infection and the passing of large quantities of urine. Tests on the urine will reveal that it contains sugar. However, if this disorder continues undetected, the high blood-sugar level will cause a state of unconsciousness slowly to develop. This is termed a *diabetic coma* and before 1921, 50–60 per cent of people with diabetes died in such a coma. In 1921 Frederick Banting and Charles Best, working in Canada, isolated insulin. Since then daily injections of insulin used with a restricted diet have given diabetics the means of controlling their condition. The insulin used in the injections is extracted from the pancreas of cattle and pigs.

Once insulin has been injected into the body (Fig. 14.4) it causes the blood-sugar level to be lowered. However, the body has no control over the action of the insulin, so the blood-sugar level gradually gets lower and lower. Therefore it is essential that diabetics have very regular meals so that their blood-sugar level never becomes too low. Similarly, strenuous exercise increases the activity of insulin. This does not mean that a diabetic cannot take exercise; it just means that a diabetic may need to eat a carbohydrate snack just before the exercise. So diabetics usually carry extra carbohydrate around with them in the form of sugar, sweets or chocolate. They may also make use of this supply if they begin to feel dizzy, sweaty or shaky, which are symptoms of a low level of blood-sugar. If these symptoms are not treated quickly by the diabetic, or by someone present who can give assistance there could be a rapid loss of consciousness.

Diabetes is rare in schoolchildren. Out of the 600 000 diagnosed diabetics in Britain only 30 000 are under the age of sixteen. Diabetes may develop at any stage during a person's lifetime. The reasons why this happen are unknown but obesity, age and heredity seem to have some influence. With medical advice and an understanding of the nature of diabetes, diabetics can learn to balance their condition so that they may lead a full and active life.

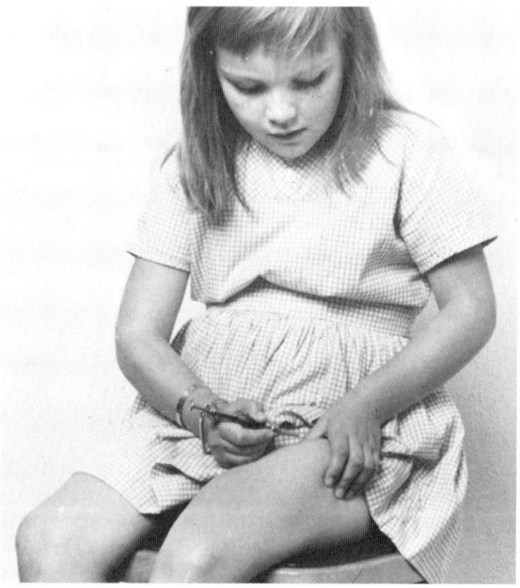

Fig. 14.4 *A diabetic child giving herself an injection of insulin*

14.5 **How we can use hormones**

Hormones need not be given only to treat problems of undersecretion: some are very useful in the treatment of other diseases. Corticosteroids, which are chemicals manufactured from *cortisone*, are very useful in the treatment of *rheumatoid arthritis*.

The most widespread use of any group of hormones is in what is commonly called 'the pill'. This is taken under medical supervision by a woman who wishes to control the monthly cycle of her reproductive system. By doing this fertilization (page 251), or conception after intercourse, is prevented. There are many types of contraceptive pills; most contain the sex hormones *oestrogen* and *progesterone* (in slightly different quantities). We shall be discussing their action further in Chapter 19.

14.6 **Other organs and tissues that secrete hormones**

The placenta, the organ which links the embryo to the mother's womb (page 17), produces progesterone and an oestrogen. Progesterone is important in preventing the release of further female productive cells. It also helps the womb to develop in order to meet the demands of the embryo (page 17).

In Chapter 3 we mentioned the effects of histamine and its release in the tissues as an allergic response, or as a response to injury. Histamine is a hormone which causes the blood capillaries to become wider, i.e. *dilate*, in that particular area. As a result more blood reaches the area, causing swelling and inflammation, and the body's defences are strengthened.

Other hormones that are released in the tissues or within the cells control the action of enzymes. Hormones also play a very important role in the other co-ordinating system of the body, the nervous system (page 222).

14.7 **The endocrine system and nervous system as partners**

The endocrine system and nervous system work as partners whose business is the co-ordination of the body's activities. These activites, both internal and external, are aimed at survival. The secretion of hormones is not under the control of the will – it happens automatically (page 232). In most cases the action of this system is much slower than that of the nervous system. In some instances the action of the endocrine system brings about long-term changes such as growth or sexual development (page 244).

Hormones do not accumulate within the body. They are broken down and excreted along with other excretory products. Therefore their effects will continue only as long as they are being secreted.

If you look again at Table 14.1 you will see that several hormones produced by the pituitary gland stimulate other endocrine glands to release their own hormones.

Question 3 Can you name two glands that secrete their hormones under the influence of the pituitary?

We can consider the pituitary gland as being the 'captain of the team', directing the players (other members of the 'endocrine team') into action on the field. However, on the sidelines is the manager watching the team's overall performance, deciding what form of play is or is not succeeding and

Fig. 14.5 *The 'endocrine team'*

sending on replacements when they are needed (Fig. 14.5). The manager of most of the body's hormone system is the *hypothalamus* (turn back to Fig. 14.2). The hypothalamus is a small portion of the brain, and it functions as part of the nervous system and as a gland. It contains special nerve cells which are able to secrete hormones. These hormones travel along the fibres of the special nerve cells and enter the pituitary gland. The hormones can either be stored by the pituitary or released. ADH and oxytocin are two of these hormones. The hypothalamus also produces hormones called *releasing factors*. These are secreted into the bloodstream and when they reach the pituitary they instruct it to release a particular hormone. Through the hypothalamus, the brain can communicate with the endocrine system and exercise control over the body's activities.

The hypothalamus has 'a finger in every pie'! There is very little that takes place in the body that is not in some way influenced by the hypothalamus. This applies to both physical and emotional matters (page 232). For instance, it helps to keep the body's temperature constant. This is one of the most critical factors in the body's mechanisms and we shall discuss it in the next chapter.

Answers to questions in the text

1 The pancreas. It acts as exocrine gland when it secretes pancreatic juice into the digestive system. It acts as an endocrine gland when it secretes a hormone called insulin directly into the bloodstream.

2 The baby is examined by a doctor after birth. A mother can visit her local child health centre or her doctor for advice. She will be asked to visit the health centre at certain times so that a health visitor or doctor can check on the baby's development.

3 Any two of the following: thyroid gland; adrenal glands; testis; ovary.

THINGS TO DO

1 Collect as much information on hormones as possible from newspaper articles and advertisements. Do this for other organisms as well as for ourselves, for example, plant hormones. Try to indicate the use of plant hormones in horticulture.

2 Carry out a survey in a chemist's shop of the specialist foods which are available to diabetics.

3 Hide from a friend, and when they are least expecting it, jump out behind them. Make a record of their body's reactions, for example, the size of the pupils of their eyes, their breathing rate and pulse rate. Do not try this experiment near a busy road or where startling them might cause an accident.

4 Imagine you are a script-writer and producer of a play (a) on stage, (b) on radio. Write part of the script and at the same time note down the instructions you will give to the actors, sound-effects team and lighting-team on what they should do to produce a sensation of fear in the audience.

5 Ask your butcher for the throat sweetbread of a young calf. This is part of the animal's thymus gland. Any good cookery book will tell you how to cook this particular form of offal. For instance, when it is skinned and fried in butter the result is delicious. Sometimes the pancreas is also referred to as sweetbread or stomach bread.

6 'The thrills of the fair'. Describe how you think certain amusements at a fairground arouse our feelings.

THINGS TO WORK OUT

In Questions 1–5 write down the *one* answer which you think is correct.

1 *One* of the following is *not* part of the endocrine system:
- **A** the pancreas
- **B** the pituitary gland
- **C** the ovaries
- **D** the thyroid gland
- **E** the salivary glands

2 The pituitary gland produces a hormone known as
- **A** insulin
- **B** adrenaline
- **C** oestogen
- **D** ADH (antidiuretic hormone)
- **E** cortisone

3 Oversecretion of the hormone from the thyroid gland cases a condition known as
- **A** diabetes
- **B** hypothyroidism
- **C** exophthalmic goitre
- **D** gigantism
- **E** Addison's disease

4 The hypothalamus is
- **A** a small gland situated over the heart
- **B** a small portion of the brain
- **C** a gland situated in the neck region
- **D** a portion of tissue within an ovary
- **E** a portion of tissue within the pancreas

5 The hormone which speeds up the action of the heart during times of fear or danger is secreted by the
- **A** thyroid gland
- **B** adrenal glands
- **C** islets of Langerhans
- **D** pituitary gland
- **E** kidneys

6 (a) Distinguish between an endocrine and an exocrine gland.
(b) Redraw the chart below and supply the missing information.

Gland	Hormone	Activity
	Thyroxine	
Pancreas		Regulates blood-sugar levels
	Androgens	
	Oestrogen	Influences woman's monthly cycle
		Activates the body during danger

7 (a) How is the body's basal metabolic rate affected by the activity of the thyroid gland?
(b) How does a deficiency of iodine in a person's diet affect the function of the thyroid gland?

8 (a) Draw a simple outline of the human body and pin-point the position of the pituitary gland, thyroid gland, adrenal glands, pancreas, hypothalamus and ovaries. Name *one* hormone produced by each gland.
(b) What is histamine? Describe *one* of its functions.
(c) What is meant by a diabetic coma? Describe *two* ways in which diabetics control their condition.

15 Temperature Regulation

15.1 **Body temperature**

When we iron clothes it is essential that the iron is at the right temperature. If it is too hot we put a hole in that nylon shirt. If it is too cold we cannot remove the creases. The temperature of the iron is regulated by the thermostat, which is controlled by a dial on the outside of the iron. Once this is set correctly for a particular fabric, the thermostat within the iron will automatically switch the current on and off. This keeps the temperature of the iron within the correct range for that fabric. In Chapter 12 we saw that enzymes will work only within a certain temperature range. The body's own thermostat, the hypothalamus, has to keep the body at the correct temperature for the natural activity of the cells. This temperature, known as *body temperature*, is 36.9 °C (98.4 °F).

When we are in good health the body is capable of keeping this temperature constant, despite temperature changes in the external environment. We are said to be *homoiothermic* (warm-blooded). This gives us a considerable advantage over creatures whose body temperature changes with that of the surroundings. *Poikilothermic* (cold-blooded) creatures can survive only in certain environments. If the temperature drops below normal they are forced into non-active states. Man, however, is able to survive and remain active in most environments.

In other science lessons you have probably already discovered that heat is a form of energy and that we get a flow of heat when one area is hotter than another. A hot object has a higher temperature than a cold object.

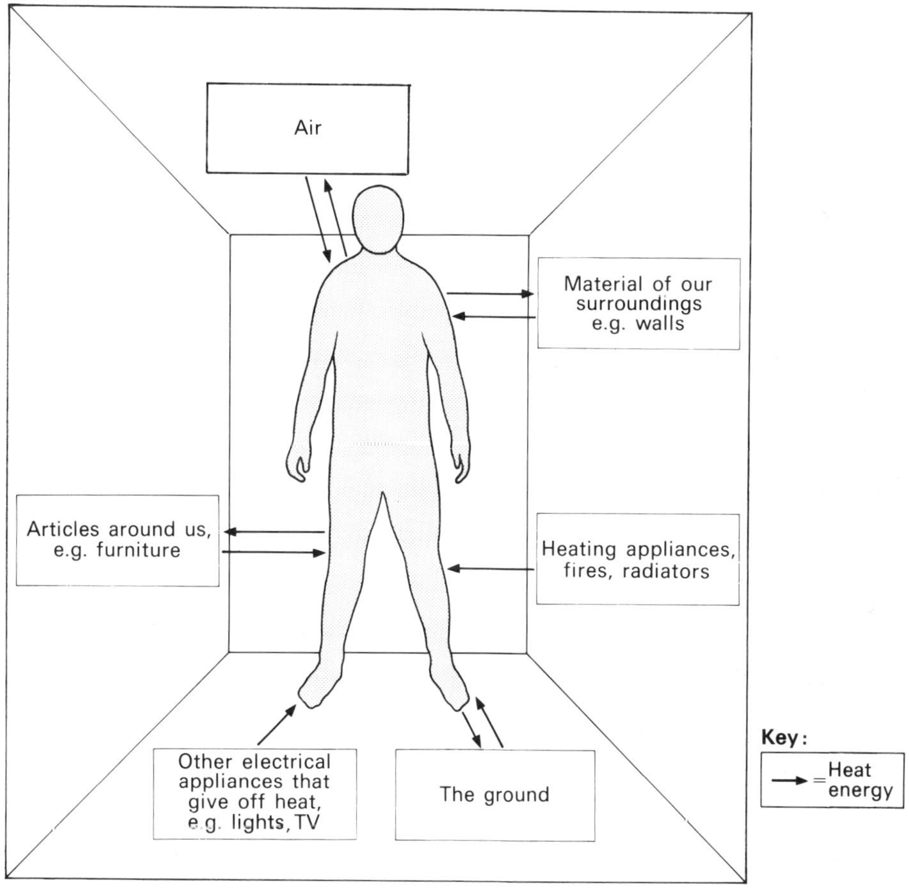

Fig. 15.1 *There is a constant exchange of heat between our bodies and our surroundings*

A fire, being hotter than the air in the room, will cause heat to travel into the room in an effort to equalize the temperatures. Therefore if there are a number of objects in a room, each at a different temperature, there is a continuous flow of heat between them (Fig. 15.1). Such heat exchange will stop only when everything is at the same temperature.

15.2 The body's heat gains and heat losses

In Chapters 4 and 9 we saw how energy is released in the cell from the 'instant fuel' ATP. By means of tissue respiration, glucose is broken down to obtain the energy needed to rebuild the cell's supply of ATP (Fig. 15.2). If there is sufficient oxygen the glucose molecules will be oxidized completely, forming water and carbon dioxide, and releasing a quantity of energy. Less than half of this energy is used to rebuild molecules of ATP. The remaining energy is released mostly in the form of heat. Complete oxidation of glucose does not always take place but even so a large amount of heat is released.

Question 1 What is the chemical powerhouse in the body that releases a lot of heat?

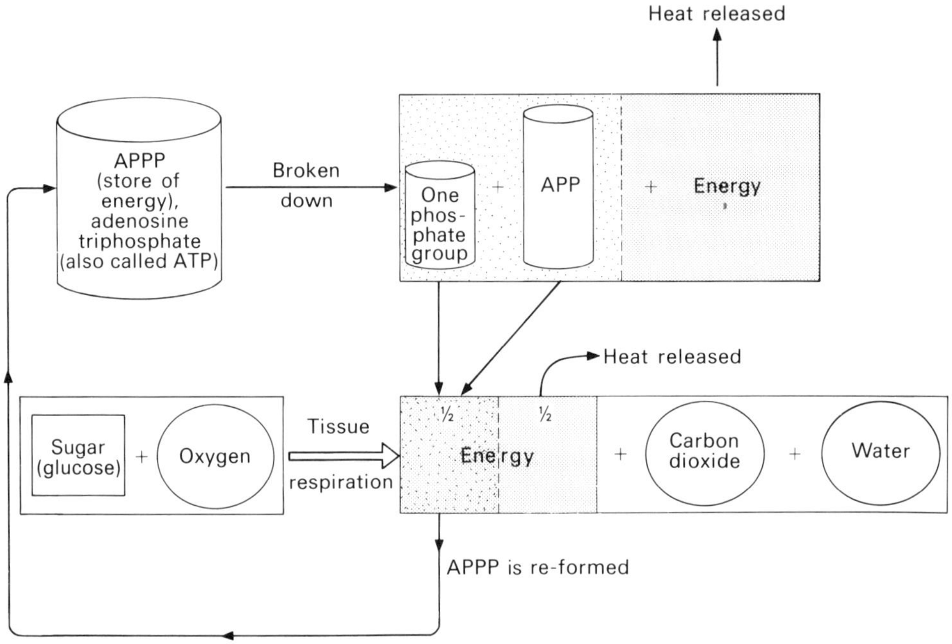

Fig. 15.2 *The release of energy from the chemical activities within a cell*

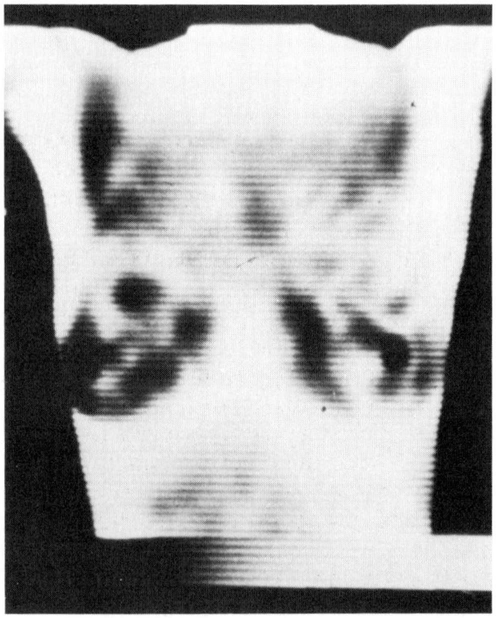

Fig. 15.3 *A heat picture or thermogram of the upper part of a woman's body. This is a photograph of the television-type screen of a special scanner which converts radiant heat (infra-red rays) into a visible picture. The light areas of the body are warm and the dark areas are cool.*

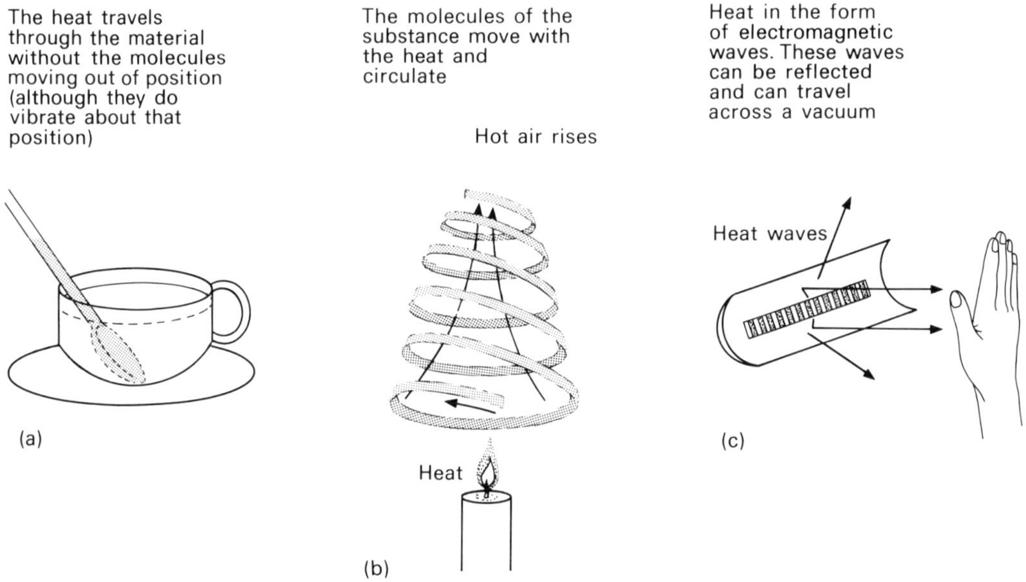

Fig. 15.4 *Three methods of heat travel*

It is estimated that a man at his basal metabolic rate will release 7140 joules of heat energy a day. Fig. 15.3 and Colour plate 8 are photographs taken with a special camera. They show very clearly that heat is given off from the human body.

Heat can travel away from our bodies by three methods: *conduction, convection* and *radiation* (Fig. 15.4). We are aware of conduction taking place between our bodies if we huddle together for warmth.

The body also loses heat by a fourth method. As you know, water can exist in three states: ice, which is a solid; water, which is a liquid; and water vapour, which is a gas. Puddles left after rain gradually dry up as the water (liquid) changes its state to water vapour (gas). This change of state, known as *evaporation*, is brought about by heat from the surroundings being used to change the molecular structure of the water (Fig. 15.5).

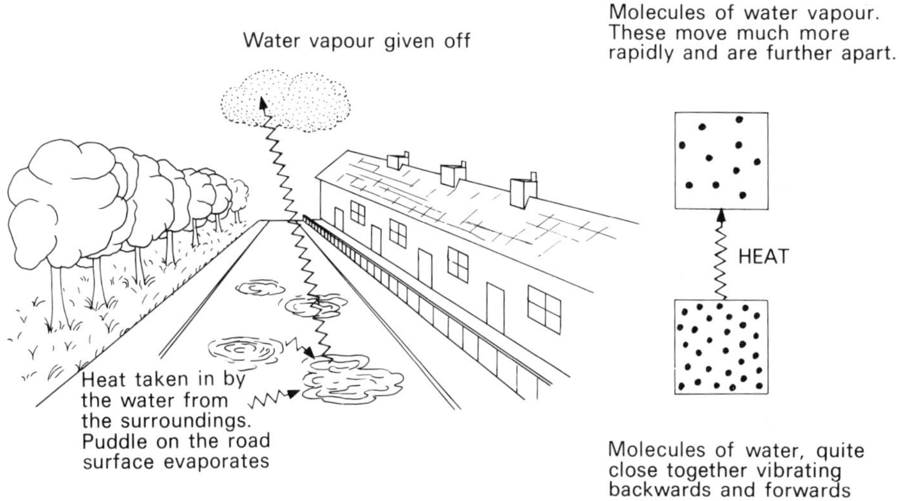

Fig. 15.5 *Evaporation*

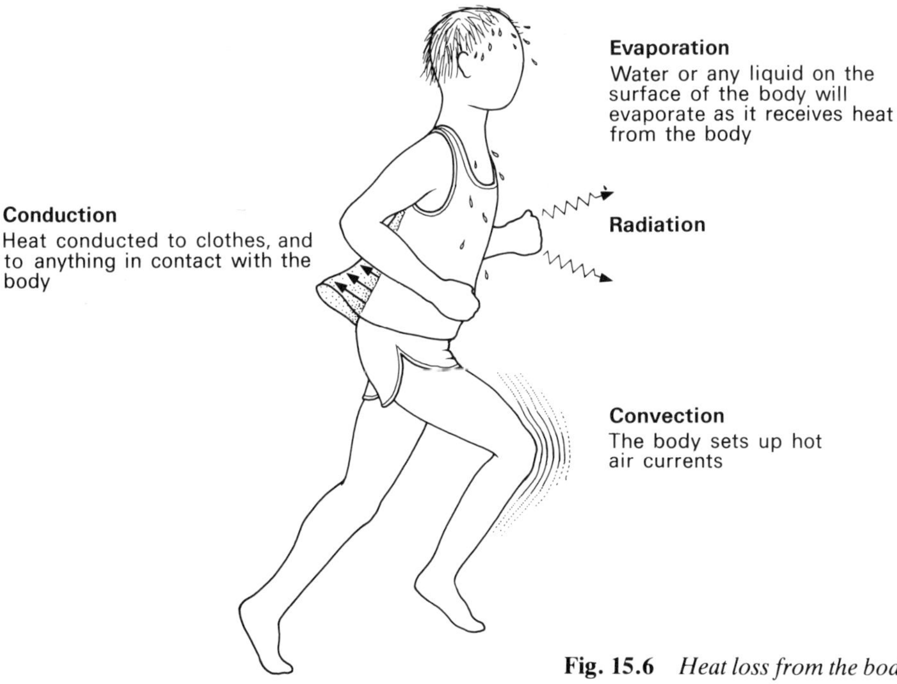

Fig. 15.6 *Heat loss from the body*

However, because heat is removed from the surroundings, the temperature falls (see page 196). You only have to wet your hair and leave it to dry in the air to feel how the evaporated water removes heat from your head and cools it down.

Thus, as we see in Fig. 15.6, our bodies are constantly losing heat by conduction, convection, radiation and evaporation. However, the body does not only lose heat: other circumstances may cause it to gain heat. For instance, when the body is very active it produces more heat than when it is still. Heat can also be absorbed from the surroundings. The body has to monitor all of these heat losses and gains, and adjust the mechanisms within the body which keep its internal temperature constant. The hypothalamus is the body's thermostat. It monitors the temperature of the blood, then automatically sends out instructions to bring about the necessary change in heat loss or gain. This monitoring and adjustment goes on every minute of every day. Most heat is lost or gained at the surface of the body, i.e. the *skin*. It is mostly within the structure of the skin (Fig. 15.7) that the adjustment takes place.

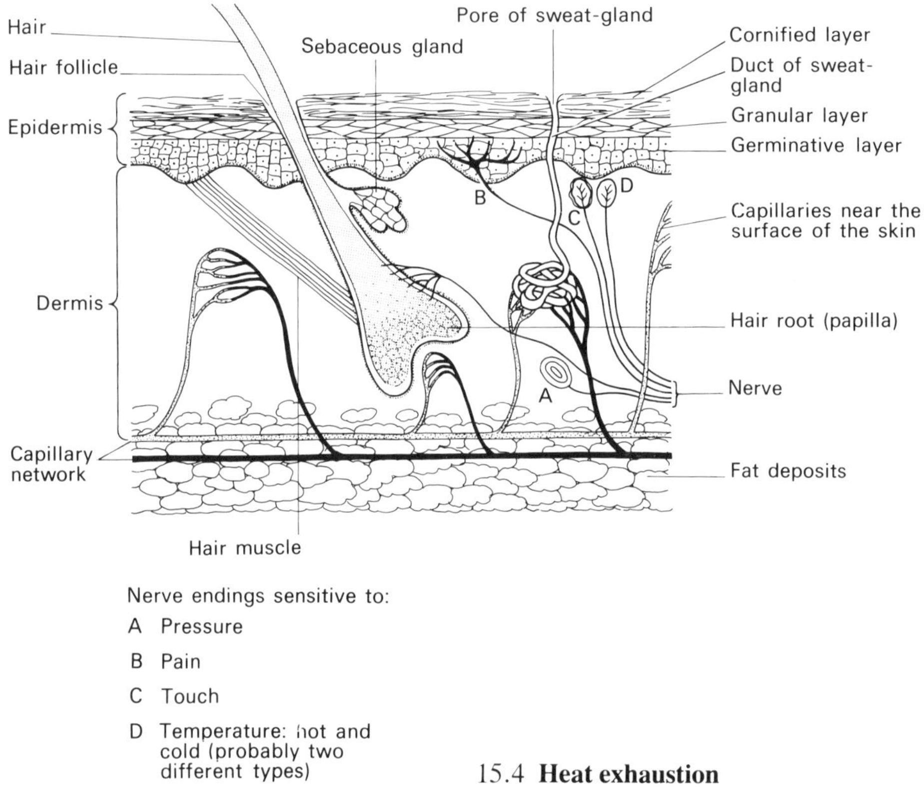

Fig. 15.7 *Internal structure of the skin*

15.3 What adjustments take place when we overheat?

The woman in Fig. 15.8(a) has decided to sunbathe for the afternoon. As she lies in the hot sun, she is gaining more heat than she is losing. Her blood temperature begins to rise. It takes a rise of only a fraction of a degree to alert the hypothalamus, which instructs the capillaries within the skin (Fig. 15.7) to dilate (page 60). This allows a greater volume of blood to reach the surface of the skin, so that more heat is lost from it.

If the temperature of the blood continues to rise (Fig. 15.8(b)) the hypothalamus stimulates the *sweat-glands* into vigorous activity. Fluid is filtered out of the blood and into the sweat-glands. It then oozes out to cover the surface of the skin as *sweat*. Like wet hair drying in the air the sweat evaporates, removing heat from the body, and the body cools. Sweat consists mostly of water, plus a few salts (particularly sodium chloride) and a little urea.

We are all aware of the fact that we sweat (perspire) when the weather is hot or when we become very active physically. However sweating can continue only as long as the body can replace the water it loses (Fig. 15.8(c)). When we sweat we feel thirsty.

Question 2 How does the quantity of water in the blood change when we sweat?

If we do not drink but carry on sweating, the body's cells become very dehydrated. Under these circumstances the body automatically 'shuts down' the sweating process. This starts a dangerous chain of reactions. With no sweat to cool the body down its internal temperature becomes higher; enzymes work faster, producing more energy and more heat. Under these conditions of *heat stress* or *heat-stroke*, the person collapses (Fig. 15.8(d)). If no water is given, or if no attempt is made to cool the body down, survival becomes impossible. A similar condition in hot climates can occur owing to loss of sodium chloride. Those not used to the heat in India or Africa have to take extra salt.

15.4 Heat exhaustion

Heat exhaustion, when we feel dizzy and sick after a long spell in the sun, is a mild form of heat-stroke. It can be avoided in hot weather, or when we go on holiday abroad, if we take enough rest, drink extra water and take more salt than usual on our food. It is also advisable to remember that 25 per cent of our heat loss through evaporation and sweating takes place from the head, particularly the forehead and neck. So if we wish to wear something to shade our heads it should have a fairly open weave (e.g. a straw hat). This will ventilate the head so that evaporation can continue.

In some countries, not only is it hot but it is also humid. This means that the air is *saturated* with water and like a wet sponge cannot absorb any more. This makes sweating difficult and we feel wet and clammy. Air-conditioning systems may be installed to make living conditions more comfortable and to prevent heat exhaustion.

Question 3

(a) Why are guardsmen wearing bear-skins more likely to faint on a very hot day?

(b) How will an electric fan help to keep people cool?

Treatment for people suffering from heat exhaustion or heat-stroke is straightforward. We should attempt to cool their bodies down by fanning or by constantly bathing the skin with water. When they become conscious they should rest, drink plenty of fluids and take a little salt.

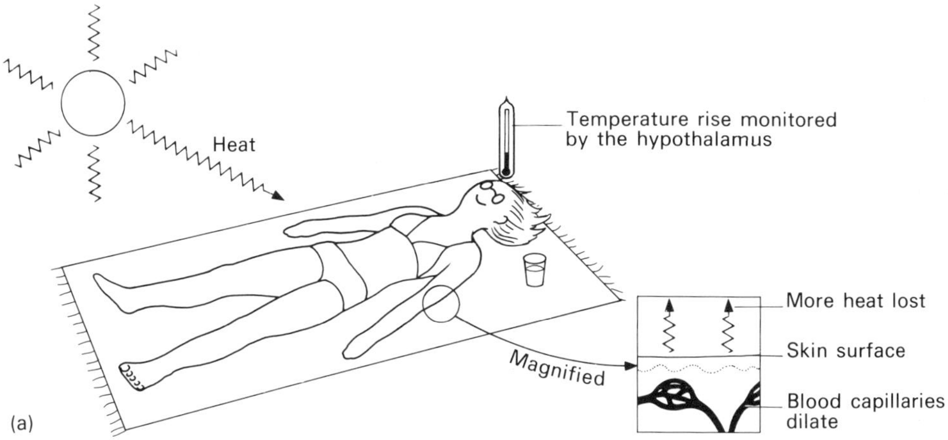

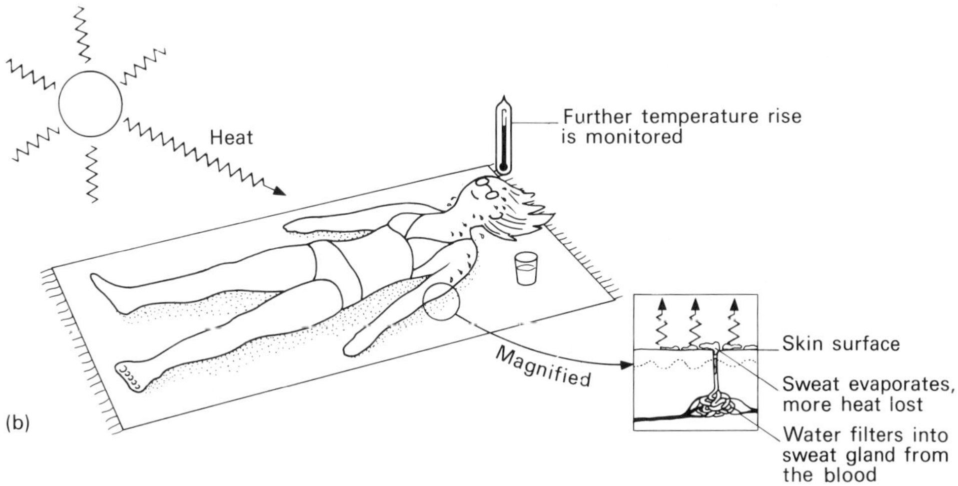

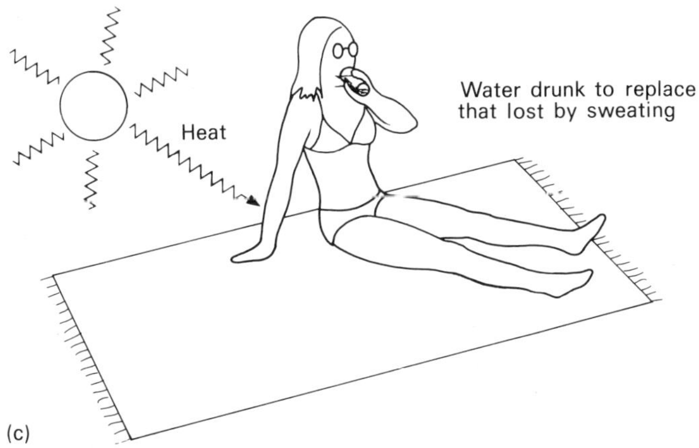

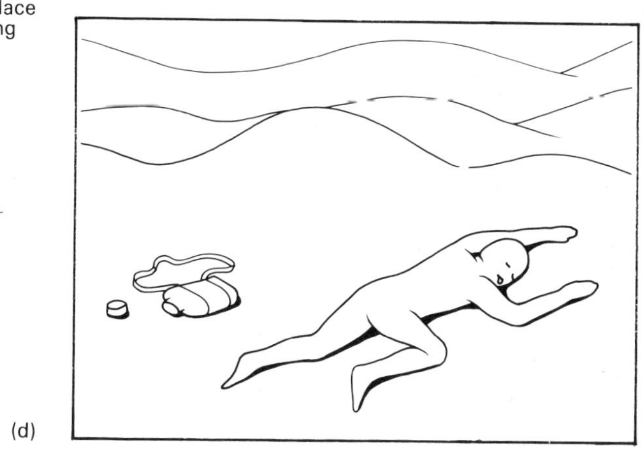

Fig. 15.8 *How the body adjusts when we overheat*

15.5 **What adjustments are made when the body loses too much heat?**

If you have ever stood at the side of a swimming-bath shivering and with your teeth chattering you will know that water easily takes heat from the body. As it does so the hypothalamus detects the fall in blood temperature. Therefore it instructs the liver to work harder to produce more heat, and the capillaries in the skin to constrict (Fig. 15.9). The decrease in blood supply to the skin makes us look pale and blue. If the blood temperature continues to fall, muscles are instructed to contract regularly and generate heat by their activity. This produces spasms or shivering which we cannot control. The contraction of the minute muscles of the hairs in the skin produces 'goose-pimples'. This does not help the body to generate heat or prevent heat loss. If we were more hairy mammals it would be useful for raising hairs on the surface of the skin, which would then trap a layer of air to act as an insulator. So this response, of use to our ape-like ancestors but not to us, is an example of a *vestigial* mechanism.

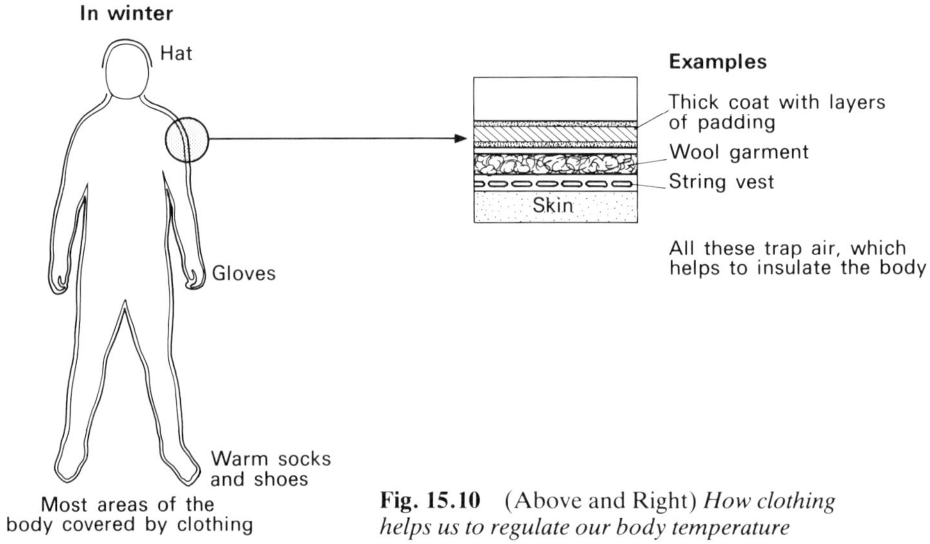

Fig. 15.10 (Above and Right) *How clothing helps us to regulate our body temperature*

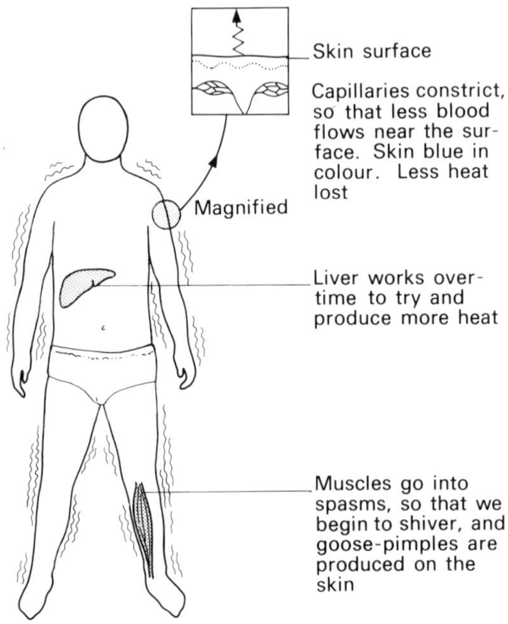

Fig. 15.9 *How the body adjusts when we have lost too much heat*

If all of these actions fail and the person cannot help himself, as in the case of a man who falls overboard into an icy sea or is lost on a cold mountain at night, then a condition called *hypothermia* develops. At a blood temperature of 30 °C the person goes into a coma and the muscles become stiff. Survival is usually impossible below a blood temperature of 27 °C, although some races can survive harsh conditions better than others. For example, an Eskimo is better adjusted than we are to surviving in a very cold climate.

15.6 **How can we help to keep our temperature constant?**

Our bodies cope with keeping our temperature constant automatically. However, the brain also interprets signals from the hypothalamus. We have a highly developed brain (see Chapter 18), so we are capable of understanding when our body is under stress. We are able to think about what action we can take to protect ourselves. This may be a simple voluntary action, such as taking off a coat or a sweater when the sun comes out. It may be a more complicated action, which involves building a shelter to protect ourselves from *exposure*, the general term for a state of hypothermia.

Clothing ourselves is not merely social convention, or a way of improving our appearance. It is also an important way of helping us to control our body temperature. Some fabrics and types of clothing prevent heat loss: they are *insulators*. Other fabrics and types of clothing allow us to lose heat through them (Fig. 15.10). On page 196 there are some tests you can carry out on different types of fabric to help you see how they can assist temperature regulation. Whenever you take part in some special activity, for example a hike or walk on the moors, you will require the right clothing: survival may depend upon it. It is also important to remember that wet clothing loses a lot of moisture by evaporation. Therefore clothing made wet by sweat or rain, for example, cools the body very rapidly.

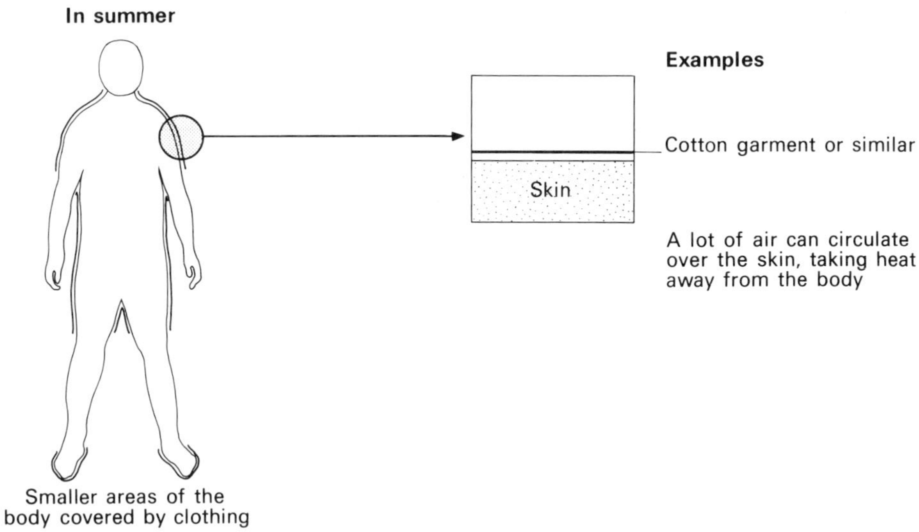

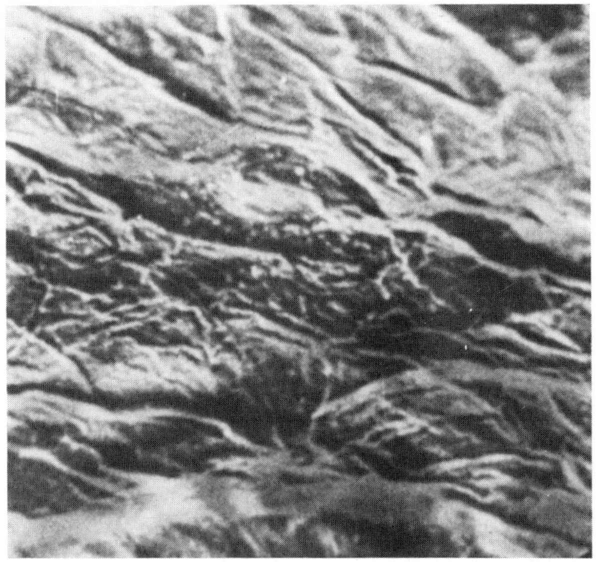

The elderly and very young can easily suffer from hypothermia. The temperature-regulating system of a premature baby is not ready to function, so it has to be kept in a controlled temperature. The elderly, who may be unable to move around easily, or who may have insufficient fuel, need help from other people to see that they are kept warm during the winter.

Question 4 Why do you think long-distance swimmers put a thick layer of grease all over their bodies?

15.7 The skin layers

We have already seen how the skin is vital to the temperature regulation of the body. An adult's skin weighs approximately 3 kg and covers an area of about 1.5 m^2. It is a remarkable covering material (Fig. 15.11). It is waterproof and it protects the body against invasion by other organisms. However, unlike other waterproof coverings – plastic, for instance – we do not get wet with condensation inside it. The skin allows moisture to escape through it at all times. During the day we can lose about half a litre of fluid through the skin. Physical exercise will greatly increase this loss by sweating.

As we know to our cost, our clothes gradually wear thin! Our skin never does and you can see why if you look at Figs 15.7 and 15.12. The upper layer of the skin, the *epidermis*, is in fact made up of three layers. The lowest of these three layers, known as the *germinative layer*, consists of cells which are always rapidly dividing. They are nourished by tissue fluid. The new cells gradually push the old ones upwards towards the surface of the skin. During this process the internal structure of the cells changes. At this point the cells form the *granular layer*. The cells lose their nuclei and a hard protein known as *keratin* forms in their protoplasm. (Your finger-nails consist mostly of keratin, so you can see how hard it is.) By the time they reach the top surface, the cells are flat, full of keratin and dead. They form a protective, waterproof layer, known as the *cornified layer*, which is constantly replaced as the

Fig. 15.11 *The surface of human skin as seen using an electron microscope. The folds and fissures can harbour dirt and bacteria*

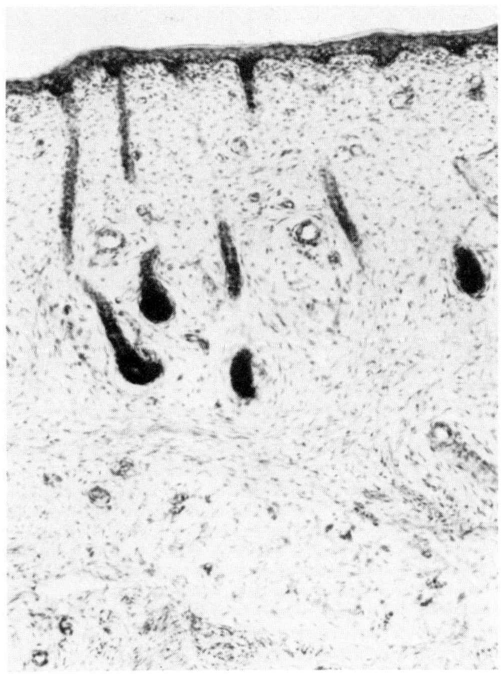

Fig. 15.12 *A vertical section through skin*

uppermost surface rubs off. This can be seen most clearly when flakes of dead cells are rubbed from the body as we towel ourselves dry after a bath. Dead cells are even shed from the scalp, a condition called *dandruff*.

This top, outermost layer of skin forms patterns which are very individual.

Question 5 How can these patterns be useful to a detective? Fig. 15.13 might help.

Beneath the three layers of the epidermis is the *dermis*. This is made up mostly of fibrous and elastic tissue, which make it an ideal 'stretch cover' capable of flexing with every movement.

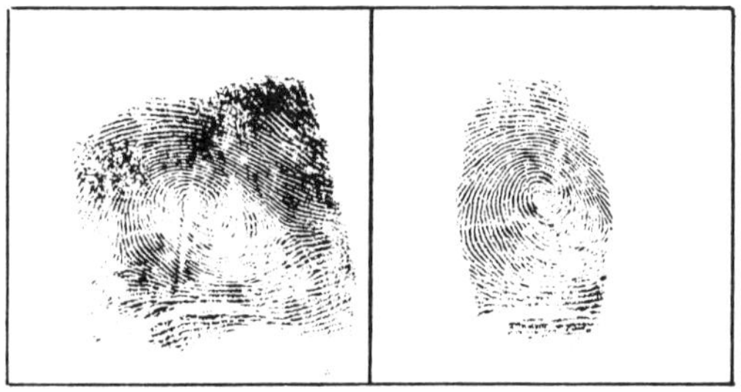

"Twinned loop" patterns — Identical

"Twinned loop" patterns — Not identical

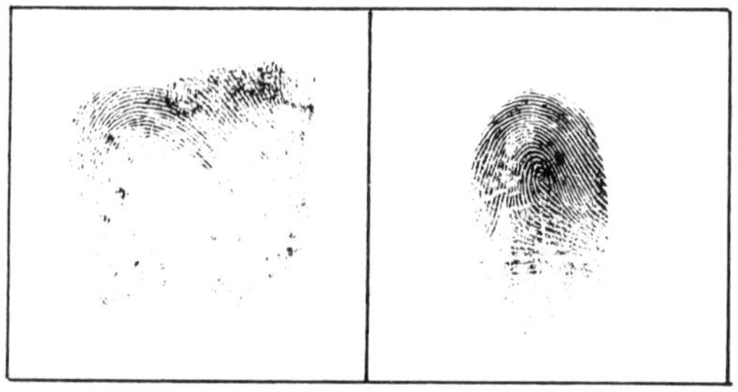

"Loop" patterns, converging type — Identical

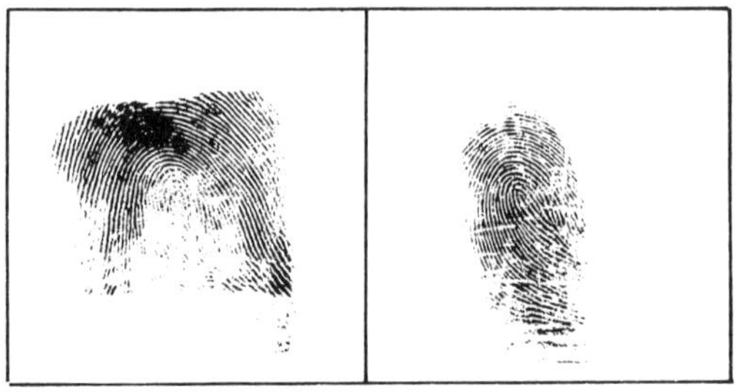

"Loop" patterns, converging type — Not identical

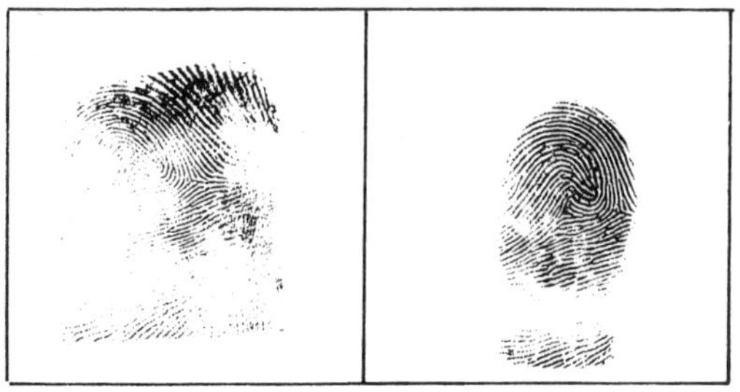

"Lateral pocket" patterns — Identical

Fig. 15.13 *Fingerprints*

15.8 **Skin glands**

We have already discussed the function of sweat-glands. They act like wells sunk into the deepest layers of soil to obtain water. Fluid passes from the blood capillaries into the coiled end of the sweat-gland and then rises to the surface of the skin through the duct or *pore*.

Whether you have a dry skin or a greasy skin depends upon the action of the second group of glands called *sebaceous* glands. There are large numbers of these glands in the skin covering the face and hairy regions of the body such as the scalp. If you run your fingertips across your forehead you can easily feel the greasy secretion of these glands, which helps to lubricate the skin. Too much *sebum* or too little can create problems.

Many teenagers develop greasy skins. The excess grease clogs the pores, allowing bacteria and dirt to collect easily within them (Colour plate 9). As a result, a person with a greasy skin has a tendency to develop *acne*. *Blackheads*, which are pores blocked with grease and dirt, become visible. The sebaceous gland itself may become infected so that it swells and becomes inflamed, forming a hard spot or pimple. Squeezing or picking the offending spot only increases the chances of introducing more dirt with further infection and possible scarring of the skin. As yet there is no permanent cure for this condition, but removal of grease and dirt by frequent washing with soap and water can help. Although many teenagers are upset by this problem, they may be reassured by the fact that many people have suffered from it and the vast majority appear simply to have grown out of it! However, if the condition is severe and scarring a real possibility, a doctor may prescribe a course of treatment.

It does not take long for bacteria to set to work on the skin's secretions. The activity of the bacteria produces the characteristic smell of sweat, often referred to as B.O. Taking a regular bath or shower will prevent this, as well as protecting the body against invasion by parasites (page 108). Washing of the skin should be sufficient to prevent excess grease, dirt and bacteria. However, it should not be done so much that the natural balance of the secretion is upset. Once again it is up to the individual to make the decision on his own health and well-being.

As we grow older our skins tend to become drier. Moisturizing creams can only lubricate and put moisture on to the surface of the skin, they cannot penetrate any further.

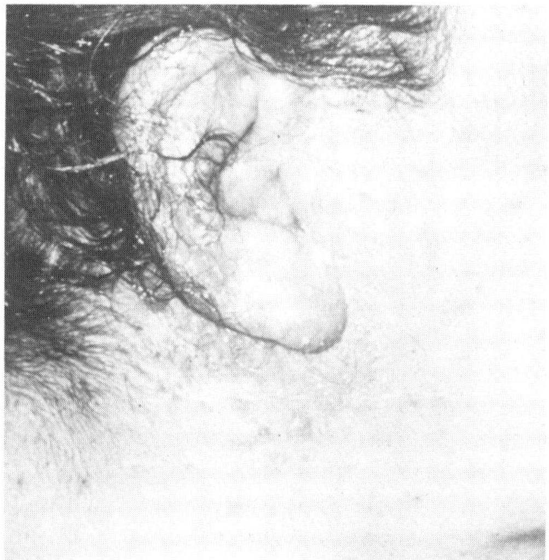

15.9 Damage to the skin

Dermatitis is the word used to describe inflammation of the skin. It can have many causes, for example when the skin becomes irritated by constant contact with a strong chemical (Fig. 15.14). Hairdressers can suffer from such a condition, but the wearing of gloves when using chemicals can help to prevent it. *Eczema* is another form of skin inflammation. It is often related to other conditions like asthma and hay fever. It has been suggested recently that an increase in this condition in babies and young infants is due to the tendency of mothers to bottle-feed their babies.

Sunlight can be beneficial to the body because it converts a substance (dehydrocholesterol) found in the skin into vitamin D. However this benefit applies only when there is reasonable exposure to sunlight. The ultraviolent rays in sunlight (and artificial sun-ray lamps) can, in excess, burn and blister the skin to quite a depth. A sun-tan is in fact the skin's way of protecting itself. Pigment cells, present in the lowest layer of the epidermis, produce the pigment *melanin*. This enters the surrounding cells to produce the characteristic tan, which filters the sunlight and protects the lower layers of the skin.

The activity of the pigment cells determines the tone of the skin. For example, in an African these cells are very active, producing ebony-coloured skin. In a European the cells are not so active, giving a paler skin. It is an individual's genetic make-up that determines the extent to which these cells become active.

Burns may be caused by hot objects, hot liquids (scalds), chemicals or electricity. Any burns which cover an area larger than the palm of your hand need professional treatment. If the casualty is a child it is wise to seek a doctor's advice. The aim of any treatment is to stop further damage, to stop the entry of germs and to replace lost fluid tissues (Fig. 15.15). Burns can scar for life if they affect the lower layers of the skin. Like all forms of accident it is best to try to prevent them by using thought and common sense. You might like to discuss this with reference to Table 3.2, which gives figures for accidents in the home.

Fig. 15.14 (Left) *A skin reaction has occurred because of certain chemicals in a particular hair dye*

Fig. 15.15 (Below) *Treatment of burns or scalds*

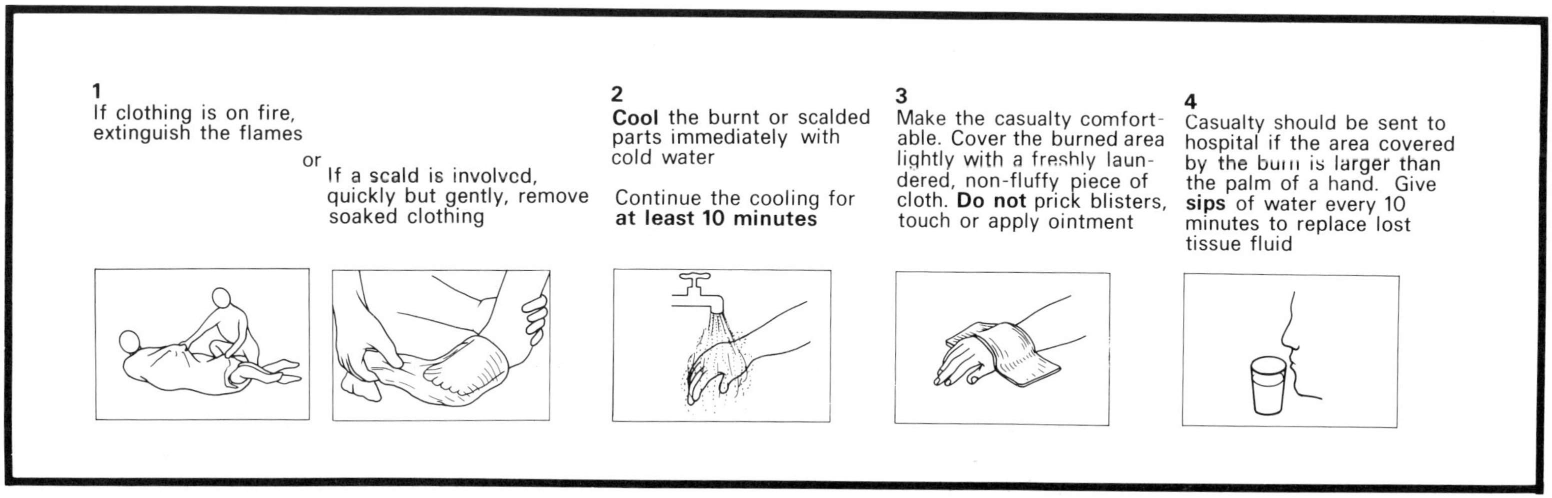

Fig. 15.16 *House of horrors*

Question 6 Fig. 15.16 shows a home in which there are several chances of a person becoming seriously burned. Can you spot the danger zones?

15.10 **Hair**

Hair grows continuously on most areas of the body, more in some places than others. You only have to pluck out a hair from your scalp or eyebrow to see its *root* (Fig. 15.7). A new hair grows from the base of the *hair follicle*, which is formed by a depression in the germinative layer. At the root of the hair the cells are alive. Gradually as more new cells are added, these older cells are pushed upwards, develop keratin and die. Within the root the cells also produce pigments which colour the hair. The pigments that are produced are determined by our genes (page 10). Hair goes grey when these cells no longer produce pigments and become transparent. This usually happens as we grow older.

Another condition that may occur with age is baldness. A hair on the head usually grows for about three to four years before it falls out and a new one begins to grow. If a new one fails to grow, baldness develops. This condition, like the colour of our hair, is decided by our genes.

Some people may be troubled by a lack of hair, others by too much. We have seen on page 244 that hormones cause hairs to grow on certain regions of the body as we become sexually mature. Boys have to start shaving if

Fig. 15.17 *Washing our hair removes dirt and grease*

they want to remove facial hair. Girls may apply the same technique to hair on the legs and under the arms, or they may use creams or have it removed professionally. Hair does not grow more strongly because it has been shaved or plucked: it may just feel coarser because its fine end has been 'blunted'. Plucking, if not done in the right direction, can alter the shape of the hair follicle.

Older women may find that they develop certain hairs on the face which they want removed. Electrolysis, done by a fully trained professional, involves putting a needle down through the hair follicle and into the root. A small electric current is then passed through the root to kill it. It can take several treatments before that one hair is completely prevented from growing again.

To most people the state, condition and styling of the hair on their scalp is important to their appearance. As shown in Fig. 15.7 the sebaceous glands secrete sebum into the hair follicle. This lubricates the hair, making it supple. However, this grease accumulates and easily collects dirt. Most people find it necessary to wash their hair twice a week to keep it looking at its best (Fig. 15.17).

TABLE 15.1 **DISTRIBUTION IN THE SKIN OF SENSITIVE NERVE ENDINGS**

Sensation on the skin	Label in Fig. 15.7	Especially sensitive regions, where nerve endings are close together	Less sensitive regions, where nerve endings are far apart
Pain	B	Most numerous type and generally widespread in every region	
Touch	C	Tongue Lips Tip of nose Tips of fingers External regions of sex organs	Middle of the back Thigh Forearm
Cold	D	Front of upper arm Elbow Back of hands	Middle of the back Thigh
Heat	D	Front of upper arm Elbow Back of hands	Middle of the back Thigh
Pressure	A	Found deep in the fatty layers of the skin and inside the body around the organs	Generally widespread in most regions

15.11 The skin as a 'detector'

We have seen how the hypothalamus can detect and monitor changes in blood temperature. Can the body detect similar changes in temperature outside the body? Think of going swimming. You walk to the side of the swimming-bath and probably your first thought is whether the water is warm or cold. To find out you probably dip a toe or finger into the water first. Immediately, the brain receives the information from the finger or toe that it needs to decide the temperature of the water. However, as soon as you dive into the water, the same information is transmitted to the brain from all over the body surface. Organs which detect changes in the external environment are called *sense-organs*. The skin is a sense-organ, and it is said to detect or be *sensitive* to hot, cold, pain, pressure and contact (touch). The skin's sensitivity depends upon receptors located in the dermis of the skin. These 'receivers' are *nerve endings* (Fig. 15.7). On page 196 there are some ideas on how you can map these receptors of the skin in different areas of the body. You will probably find that their distribution can vary as is indicated in Table 15.1. One type of receptor may be concentrated in a particular area because it is most useful there.

Similar nerve endings are distributed internally throughout the body, transmitting information to the brain on the well-being and activity of the organs (page 222). Therefore the body is constantly monitoring its internal and external environment. If the body acts either internally or externally on the information it receives we say that it has *responded*. Any internal or external change that produces a response is termed a *stimulus*. The ability of the body to receive stimuli and to respond to them is essential to its survival.

Answers to questions in the text

1 The liver.

2 There is less water in the blood, so the blood has a higher concentration of salts.

3 (a) Bearskins cover much of the forehead and neck, and prevent heat loss. So the guardsman's body overheats and he suffers from heat exhaustion. This is in addition to the problem of circulation in the lower limbs when we stand still for long periods (page 64).
(b) A fan helps to circulate the air. Moving air speeds up the evaporation of sweat.

4 The grease acts as an insulator and prevents heat loss.

5 On the tips of the fingers these patterns form what are called *fingerprints*. We leave these patterns on things we touch. Each individual has a different set of fingerprints, so we can each be identified by them.

6 (a) Handle of saucepan over the side of the cooker within reach of the toddler.
(b) Girl in night-dress near to an open fire.
(c) Person smoking in bed.
(d) Electric fire beneath bathroom towels.
(e) Boy playing with matches.

THINGS TO DO

1 (a) Use an office ink-pad to record your fingerprints.
(b) Run water over your skin and observe its waterproof properties.
(c) Rub your forehead with a tissue and observe the grease.
(d) Wear rubber gloves for half an hour. What collects in the gloves?

2 Examine the palm of your hand with a strong magnifying glass. Repeat the observation after strenuous exercise. Using newspaper advertisements or articles illustrate how tennis players maintain a good grip or prevent sweating interfering with their performance.

3 Carry out and record the following both before and after strenuous exercise:
(a) mouth temperature
(b) surface temperature of the skin
(c) skin colour
(d) appearance of the blood-vessels in the skin
(e) evidence of sweat

4 Examine a clinical thermometer and find out how it should be used.

5 To observe the cooling effect of an evaporating liquid, clamp a thermometer in a vertical position. Soak a ball of cotton wool in water, then tie it round the bulb of the thermometer. Set up a second thermometer without the cotton wool, to act as a control. Observe the temperature of the thermometer as the water evaporates. You can speed up the evaporation by blowing air across the cotton wool.

6 Make sure your hand is warm. Lick part of the back of your hand, then blow across the whole surface. Observe how the area which is wet feels cooler than the rest.

7 Obtain a medium-sized can with a push-on lid (a tin which contained golden syrup would be ideal). Drill a hole through the centre of the lid so that it will hold a centigrade thermometer. Tie a piece of fur round the can. Fill the can with warm water and replace the lid and thermometer. Record the temperature of the water every minute and plot the results on a graph. Repeat the experiment using different materials to cover the can. You could even try smearing a thick layer of grease on to the can. The graphs should indicate which type of material is most suitable for clothing during winter months.

8 Use illustrations, drawings and samples of material on a chart to show how the very young and elderly should be clothed in winter and in summer. Also indicate any other precautions which should be taken to prevent heat stress and hypothermia.

9 Imagine you are taking a group of children on an all-night hike. Write out (a) a list of equipment and clothing and (b) a survival guide, e.g. what anyone should do if they are lost and separated from the rest of the party.

10 Blindfold someone. On the floor place a number of materials which have different surfaces. Ask the person to identify the material by feeling their surfaces with their fingertips. Repeat the experiment with the person using their nose, elbow, toes, etc. Compare their accuracy.

11 Blindfold another person. Place a knitting-needle in hot water. Use the hot tip of the needle to find out at what points on their body the person is sensitive to heat. On a diagram mark the location of the 'heat-spots'. Compare the concentration of heat spots in different regions of the body. Repeat the experiment but this time use a knitting-needle which has been placed in ice and water. In this case locate the 'cold spots'.

12 Push two pins through a piece of stiff card so that their points are 5 cm apart. Blindfold another person and then gently tap various regions of the body and record whether one or two points are felt. Move the pins closer together and repeat the procedure. Go on doing this until the pins are only a fraction of a millimetre apart. The regions most sensitive to touch will be able to detect two points even when the pins are very close together.

13 A cube of side 1 cm occupies a volume of 1 cm^3 and has a surface area of 6 cm^2. A cube of side 2 cm occupies a volume of 8 cm^3 and has a surface area of 24 cm^2. Explain the importance of this fact when we consider the heat loss of a baby.

THINGS TO WORK OUT

In Questions 1–3 write down the *one* answer which you think is correct.

1 The pigment produced by the pigment cells of the skin is known as
- **A** sebum
- **B** melanin
- **C** keratin
- **D** histamine
- **E** pepsin

2 The average body temperature of a healthy adult is
- **A** 38.5 °C
- **B** 35.9 °C
- **C** 36.9 °C
- **D** 37.5 °C
- **E** 36 °C

3 A burn should be treated by
- **A** smearing on antiseptic cream
- **B** smearing on vaseline
- **C** covering with cotton wool
- **D** cooling in cold running water
- **E** bursting the blisters that develop

4 (a) Name *four* methods by which the body loses heat.
(b) Name *one* organ in the body that generates a large amount of heat.

5 (a) Explain how the evaporation of sweat from the surface of the skin causes cooling.
(b) Explain how sweat is produced by the body.
(c) Give *one* region of the body which sweats a lot.
(d) What causes body odour?

6 (a) When a person is very cold the tone of their skin becomes much paler (even blue in fair-skinned people). What causes this to happen and how does it help the body to control its temperature?
(b) Describe *two* other changes that can take place when our bodies are very cold. State whether these other changes assist temperature control.
(c) What is hypothermia? Describe *one* situation in which a person may run the risk of hypothermia, then state how it may be avoided.

7 (a) What is the skin sensitive to? With each answer state the *most* sensitive region on the body for that particular sensation.
(b) Is the skin a reliable guide to temperature? Describe an experiment that will support your answer.

8 (a) How does the humidity of the atmosphere affect sweating?
(b) Why do fully waterproof cagoules have small holes under the arms?
(c) What fabrics and types of clothing are most suitable for (i) hot sunny days, (ii) cold winter days? Explain your choice.

9 What is the function of each of the following?
(a) the hypothalamus in controlling body temperature
(b) sebum
(c) the germinative layer of the skin
(d) the cornified layer of the skin
(e) nerve endings in the skin

10 (a) Draw a diagram, with labels, to show the structure of skin. Include on the diagram a sweat-gland, a hair, blood-vessels, nerves and a sebaceous gland.
(b) Describe the protective function of the skin.
(c) Describe *one* benefit and *one* harmful effect of sunlight on the skin. What is the function of melanin?
(d) Give *three* reasons why the skin should be kept clean.

16 The Senses – Taste, Smell and Hearing

16.1 The tongue

Fig. 16.1 is a picture of a baby six months old. Most babies at this age want to put everything in their mouths, and this means that the mother must be both watchful and understanding. She needs to be understanding because, as we have seen in the previous chapter, the lips and tongue are particularly sensitive to touch and the baby is using this sense to find out about the object in contact with his mouth.

When the baby is given food (Fig. 16.2), his tongue will give him more information. It will detect whether the food is hot or cold and what sort of texture it has. But as you can see in Figure 16.2 the baby is detecting something else which may determine whether he swallows the food or spits it out. The other thing he is detecting is the *taste* of the food.

The tongue, in addition to nerve endings which are sensitive to hot, cold, pain and touch, has taste receptors. These receptors are *sensory cells*. They all look alike: each has a sensory hair which projects from the top surface of the cell. Each cell is a detector of only one kind of taste: whether the food is *salty*, *bitter*, *sweet* or *sour*. This detection can be carried out by the sensory cells only when the food is dissolved as a solution.

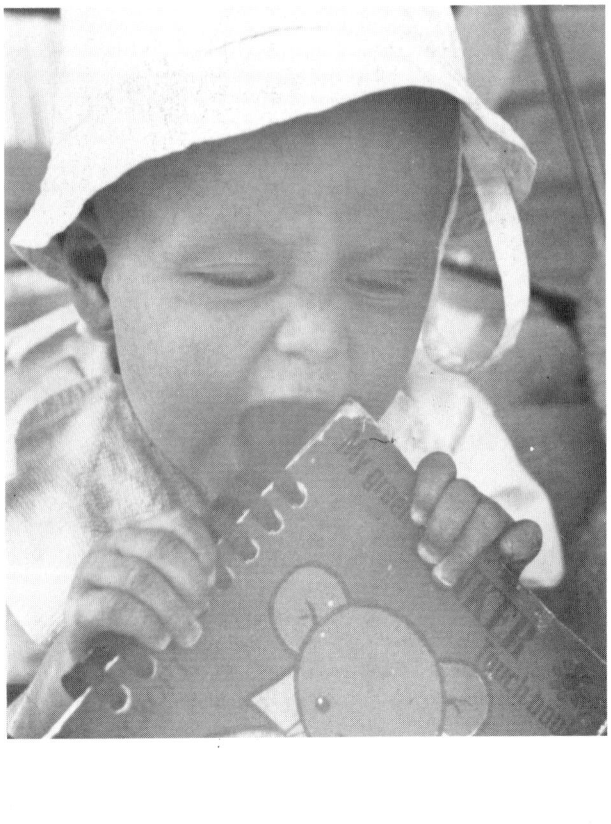

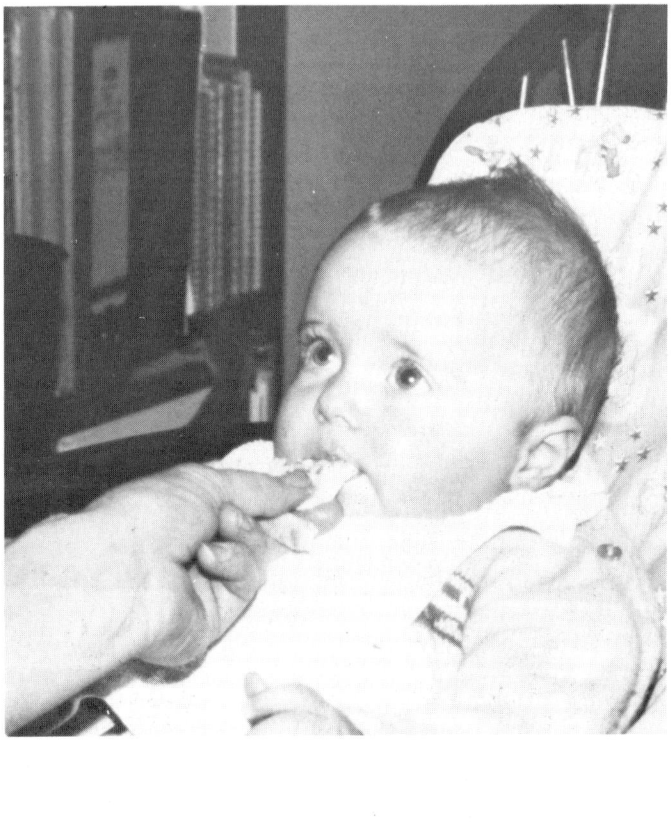

Fig. 16.1 (Far left) *Our tongues and lips are very sensitive to touch. A baby uses this touch sensation in his exploration of the things around him*

Fig. 16.2 (Left) *A baby's tongue detects the taste of food, as well as its temperature and texture*

Question 1 What is secreted into the mouth to help dissolve the food? (Clue: see Chapter 12.)

The sensory cells are grouped together with supporting cells to form a structure known as a *taste-bud* (Fig. 16.3). Altogether there are about 9000 taste-buds on the tongue and in other areas of the mouth, such as the soft palate.

The tongue is a very muscular organ with an upper surface that is not flat, but raised in a series of small bumps (papillae). These are of three different shapes, as seen in Fig. 16.4. Taste-buds are arranged along the sides of types (b) and (c).

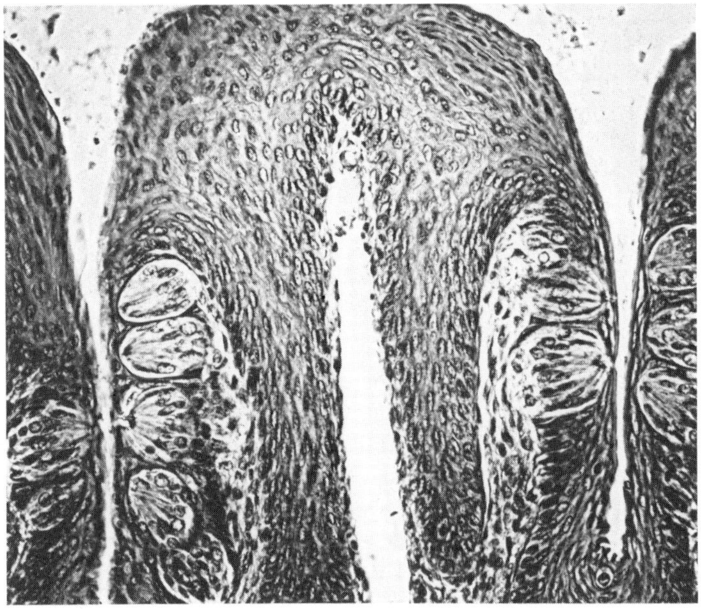

Fig. 16.3 *Taste-buds in the tissue of the tongue*

16.2 **Taste**

When a substance is dissolved in the mouth, the chemical drips into the channels of the tongue. The solution then enters the pores of the taste-buds, where it comes into contact with the sensory cells. The taste-buds that detect the same taste are grouped together on certain areas of the tongue. These areas are shown in Fig. 16.5 (see page 207). Each taste-bud will send its information on to the brain. We can then take action on this information, for example we can add more sugar to our food, or decide that the milk is too sour to drink.

We all know that our food has more to it than just being salty, sour, bitter or sweet. What about all those different-flavoured ice creams? If the tongue is not tasting the flavours how are we detecting them? A clue to the answers to these questions lies in the way our appetite for food fails when we are suffering from a heavy cold. We start to feel ready for food when we smell it cooking, or when it is being served. When the nose is blocked we cannot smell the food, and when we eat, it appears to have lost its flavour. So we do not taste flavours, we smell them. Smell and taste work together to allow us to enjoy all the sensations of food.

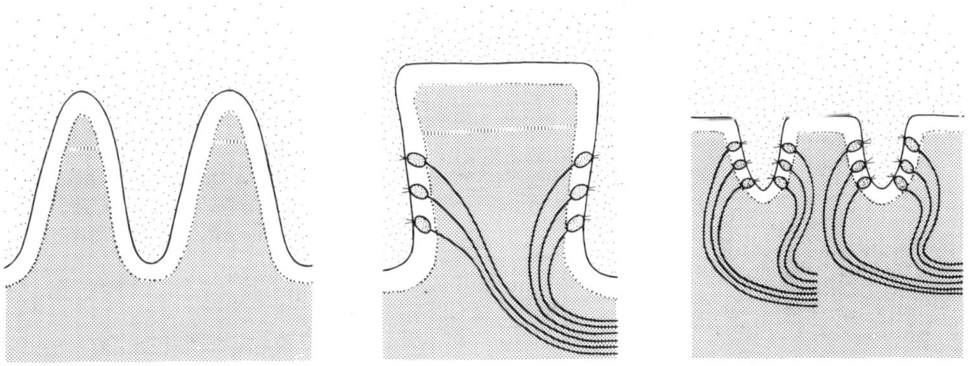

(a) (b) (c)

Fig. 16.4 *Papillae on the tongue*

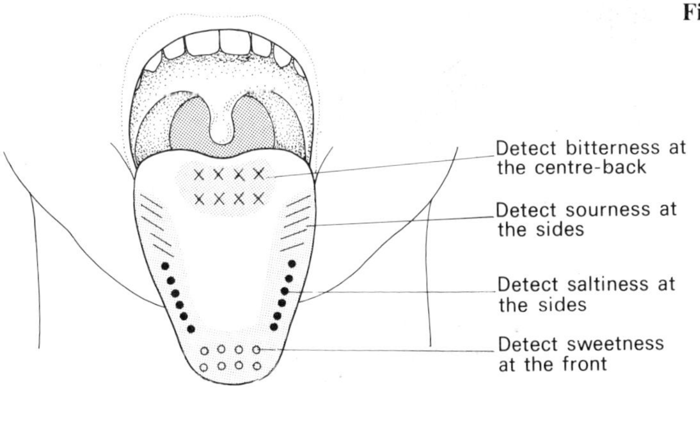

Fig. 16.5 *The taste regions of the tongue*

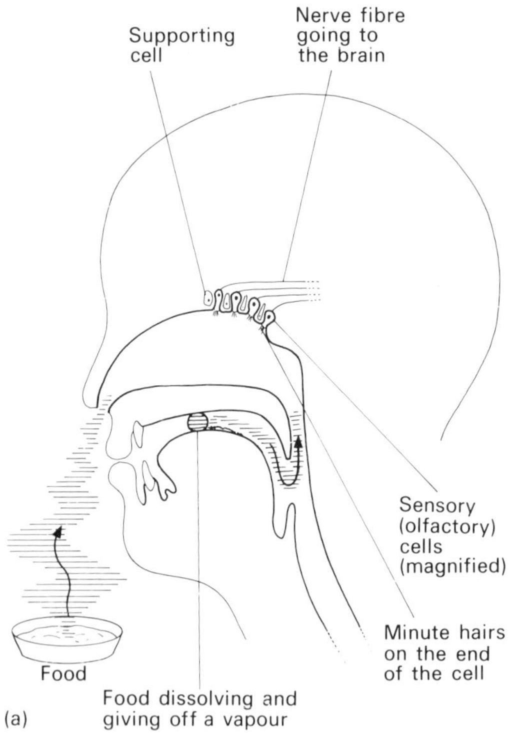

16.3 The nose and smell

When a chocolate-flavoured ice cream dissolves in the mouth, the tongue detects its sweetness. Meanwhile the chocolate flavouring chemical is vaporizing, i.e. giving off a gas (Fig. 16.6(a)). This gas travels along to the back of the mouth, then rises into the nasal passages. On the upper surfaces of the top nasal passage there is a small area covered by cells which are receptors for *odours* (Fig. 16.6(b). These cells have minute hairs projecting from them. They are exposed to the air and gases inside the nasal passage. The membranes in the nasal passages also secrete mucus (page 45). Gases and very small particles, if they are soluble, can dissolve in this mucus and thereby stimulate the sensory hairs. By this method we can detect or smell thousands of different flavours and odours.

The human nose has a keen sense of smell. Someone wearing perfume only has to enter a room for us to detect it, even at a distance. It is estimated that the smelling or *olfactory* region of the nose can detect one molecule of an odour even when it is mixed with 5000 molecules of air.

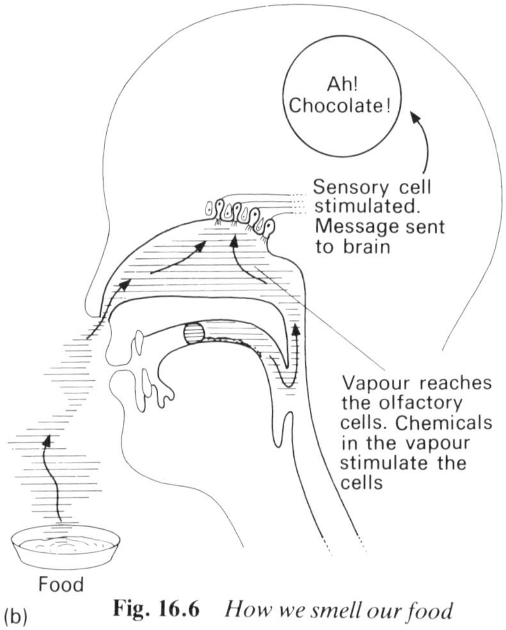

Fig. 16.6 *How we smell our food*

Some people have a keener sense of smell than others. The man in Fig. 16.7 has what might be termed a 'trained nose'. His ability to detect and recognize odours has developed through years of experience. When you want to recognize a smell, you may have noticed that you sniff. Sniffing draws air and gases into the nose much more sharply than during normal breathing. It also directs them upwards in the nasal passages and on to the olfactory area. This gives a higher concentration of molecules to 'smell at'.

It is important that we use our sense of smell wisely. It is a sense which can be used to alert us to danger, for instance we can recognize dangerous fumes and take avoiding action. However, sniffing any fumes that are given off at room temperature from volatile chemicals is extremely dangerous. They affect the body in a similar way to drugs and introduce such poisonous substances into the body's systems that they can cause serious disorders and even death.

Question 2 As you enter a room you detect that someone has put some highly scented flowers in a vase on the table. You sit down to read the paper for quarter of an hour. When another person comes into the room and comments on the nice smell from the flowers, you realize that you have become unaware of them. What does this suggest to you about your sense of smell?

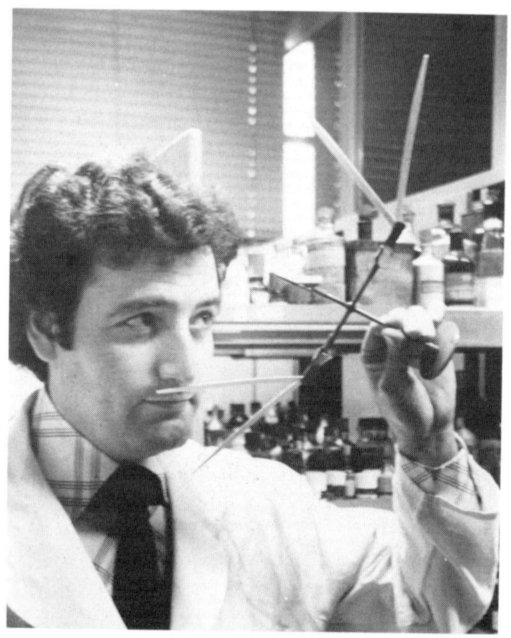

Fig. 16.7 *This man has what might be called an 'educated nose'. He is employed in a soap factory to detect the right scent to use in the soaps. He can identify at least 700 different 'scent notes'.*

16.4 **The ear**

The ear helps our bodies in two ways: it provides us with a sense of hearing (Fig. 16.8) and with a sense of balance.

Banging on a drum makes the drum skin move up and down. This backward and forward movement is called a *vibration*. Air molecules surround the drum skin, and these are made to vibrate. The movement of the air particles sets up a series of pressure waves which travel through the air as a *sound-wave*. Another membrane will be made to vibrate if this wave hits it. Membranes that receive vibrations are found in most *sound receivers*, for example a *microphone* or an *ear*. The ear is the body's receptor of sound-waves.

One ear is clearly visible on the middle man in Fig. 16.8. The flap of skin attached to the side of the head is called the *pinna*. It consists of skin, cartilage and some muscular fibres. Some people can waggle their ears quite well. In animals the movement of the pinna makes an effective sound collector but in man there is little use for it. If we wish to distinguish a faint sound we usually cup our hand around our ear (Fig. 16.9). Years ago, people who were hard of hearing used a similar idea, the ear-trumpet. Both the hand and the trumpet help to focus the sound into a tube known as the *ear canal*.

Fig. 16.9 *This sign is to be displayed where a deaf person may seek assistance*

Fig. 16.8 *A local water authority's night sounding team. These men have very sensitive hearing. They sound the road surface at night to detect and pinpoint hidden leaks.*

Fig. 16.10 shows a diagrammatic section of the ear. You will see that it is divided into three areas: the *outer*, *middle* and *inner ear*.

The outer ear

This area is made up of the pinna and the ear canal. The ear canal is about 2.5 cm long and is lined with a delicate membrane. This membrane is a continuation of the skin of the pinna. It is protected by a secretion of wax, which is produced by its own glands. These glands are closely related to the sebaceous glands of the skin (page 188). The amount of wax that is secreted varies from person to person. Generally there is no need to remove the wax, for it flows out of the ear by itself, bringing with it any dust or small particles. However, some people have a tendency to produce a lot of wax, which forms a solid mass in the ear canal and can cause temporary deafness. This has to be removed by a doctor. He uses a special technique known as *syringing*: water is syringed under pressure into the ear to force the wax out. The removal of anything from the ear needs specialist treatment. The reason for this is very clear: stretched across the end of the ear canal is a very thin membrane, the *ear-drum*. Any object which is pushed down into the ear could eventually end up against this tight membrane, damaging it.

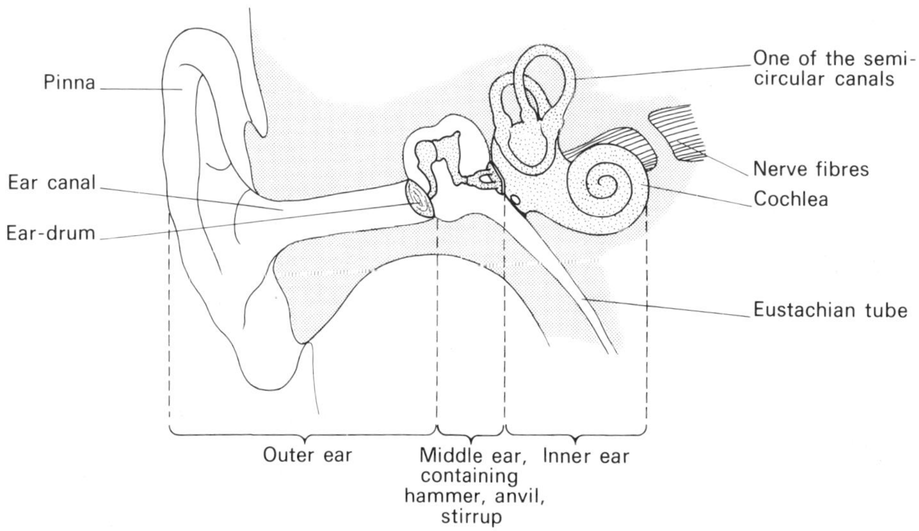

Fig. 16.10 *The ear: internal structure*

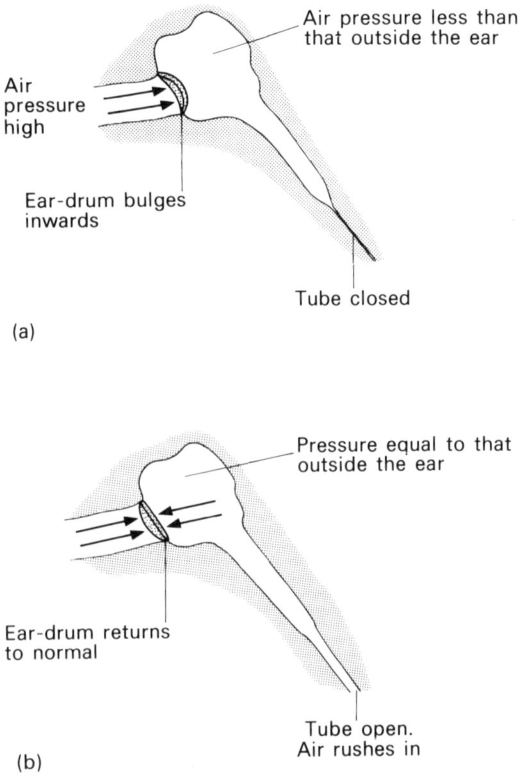

Fig. 16.11 *Action of the Eustachian tube if the air pressure outside the ear increases*

The middle ear

If you have ever been through a long tunnel rapidly on a train, or flown in an aircraft, you may have experienced some discomfort in your ears or temporary deafness. This disappears when the ears 'pop'. To encourage this to happen you may have been told to keep swallowing. Fig. 16.10 shows clearly how the ear-drum separates the outer and middle ear. The air in the external ear canal exerts a pressure on the ear-drum. Such a pressure could burst the ear-drum if there were not an equal and opposite pressure on the other side. The pressure of the air in the middle cavity is adjusted by means of a narrow tube, the *Eustachian tube*. This passes through the bones of the skull, connecting the middle ear with the back of the nasal cavity (page 45). Normally this end of the tube is closed, but the actions of yawning and swallowing cause it to open. This allows air to enter or leave the middle ear (Fig. 16.11).

Question 3 If the air presure becomes less outside the ear, what happens to the ear-drum? How is the ear-drum restored to its normal position?

In the opposite wall to the ear-drum are two smaller transparent membranes. These are called the *oval window* and *round window* (Fig. 16.12). The ear-drum is connected to the oval window by a line of three small bones, which are called the *hammer*, *anvil* and *stirrup* (Fig. 16.12). Synovial joints (page 32) exist between the bones, and ligaments support them within the middle-ear cavity. The hammer is connected to the tissue of the ear-drum. Therefore any vibration of the ear-drum is passed on to this first bone (Fig. 16.12). The joint between hammer and anvil is not very flexible, so these bones tend to move as one. Their movement is then transmitted through the more flexible synovial joint to the stirrup. This complex arrangement of tiny bones serves a purpose. Not only is the vibration carried across the middle-ear cavity, but it is also amplified. Thus a tiny movement of the ear-drum will cause the stirrup to create a sizeable movement of the oval window. This is necessary if the oval window is to set up vibrations in the inner ear which the sensory cells can detect.

The inner ear

If, like Alice in Wonderland, we were able to take a potion to make us small enough to pass through the oval window, we would find ourselves in a maze of tunnels and caverns. This is the *labyrinth* of the inner ear and lies within the bone of the skull. The other fact we would notice as we passed through the oval window would be that we were no longer surrounded by air, but were floating in a fluid (*perilymph*). The bony labyrinth contains

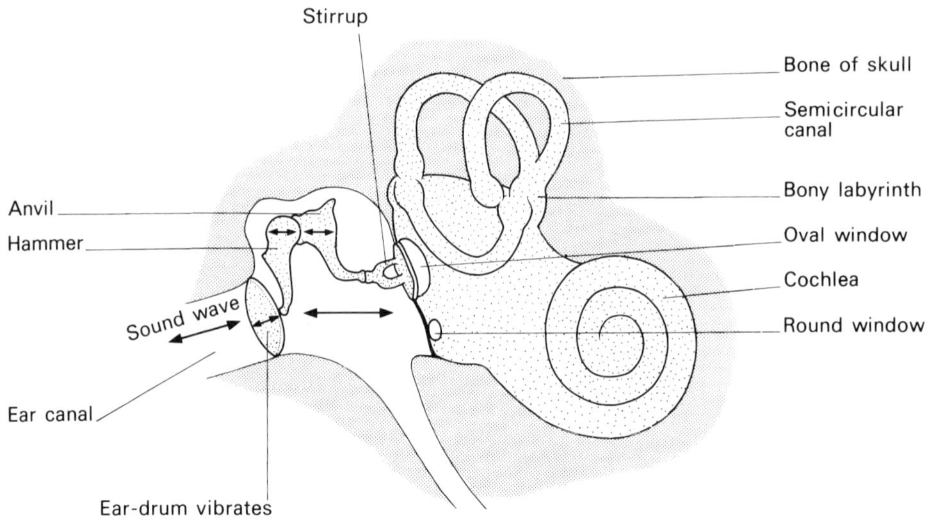

Fig. 16.12 *The middle and inner ear*

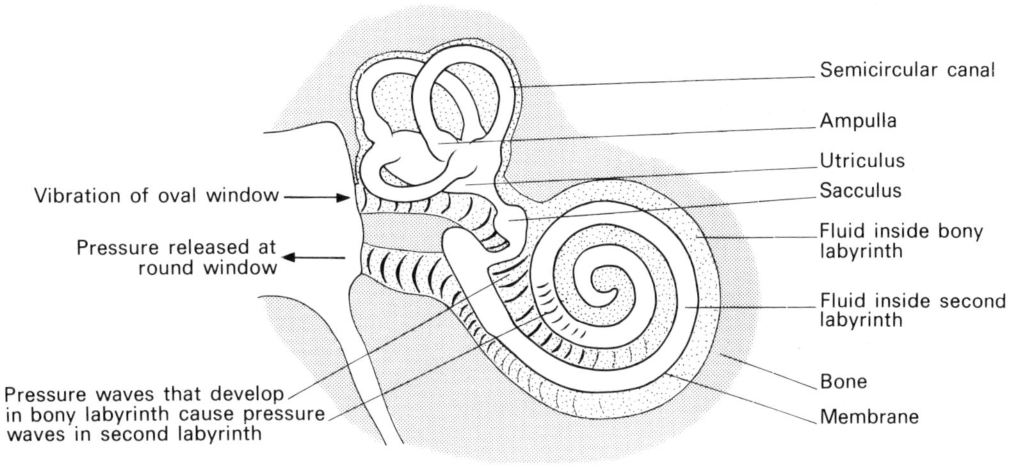

Fig. 16.13 (Left) *The inner ear in greater detail to show the second labyrinth (the bony labyrinth has been sliced open)*

not only a fluid but also another structure: a second labyrinth constructed not of bone, but from a membrane. Its shape is almost identical to that of the bony labyrinth, so it fits within it (Fig. 16.13) in the same way as an inner tube fits inside a bicycle tyre. The membranous labyrinth also contains a fluid (*endolymph*). The tunnels of the labyrinths have different functions. The tunnel which goes round and round in a spiral shape is the *cochlea*. This is responsible for our sense of hearing. The three tunnels positioned at 90° to one another are the *semicircular canals*, which are responsible for our sense of balance.

16.5 Our sense of hearing

The sound vibrations pass along the ear canal and set the ear-drum vibrating in sympathy with them. These vibrations are carried by the three bones across the middle ear to the oval window. The oval window vibrates and sets up pressure waves within the fluid in the bony labyrinth (Fig. 16.13). These pressure waves in turn affect the walls of the membranous labyrinth, moving them in and out and creating similar vibrations within their internal fluid. Inside this area of the cochlea there is a continuous row of sensitive hair-cells, which have the tips of their hairs fastened to a membrane (Fig. 16.14). Vibrations within the fluid cause these hairs to vibrate, and their sensitive cells pass the information on to the brain via the auditory nerve.

The number of vibrations that there are in every second is known as the *frequency* of a sound. (A bus service that provides two buses every hour is more frequent that one that provides one bus every hour.) The frequency of a sound determines its *pitch*. A high-pitched sound has a high frequency – more vibrations per second. A low-pitched sound has a low frequency – fewer vibrations per second. Different areas of the cochlea respond to notes of different frequency.

16.6 Frequencies and music – intensity and noise

Our ears can detect 1600 different frequencies ranging between 20 and 20 000 hertz (1 hertz or 1 Hz = 1 vibration or cycle per second). The sensitivity of ears varies from person to person and also changes with age. Usually, young children can detect higher frequencies than adults.

Few sounds are pure regular vibrations. Even vibrations from a musical instrument have slight distortions. This is due to the pressure of weaker vibrations (overtones). Overtones vary from instrument to instrument. Our ears detect these differences as the *quality* of a note, and use them to distinguish one instrument from another. Again, one individual may be more sensitive than another to small differences in quality. Those with most sensitivity are usually better at blending notes to produce music. We often say that they have an 'ear for music'.

When frequencies do not blend we get *noise*. Noise is sound that has no definite

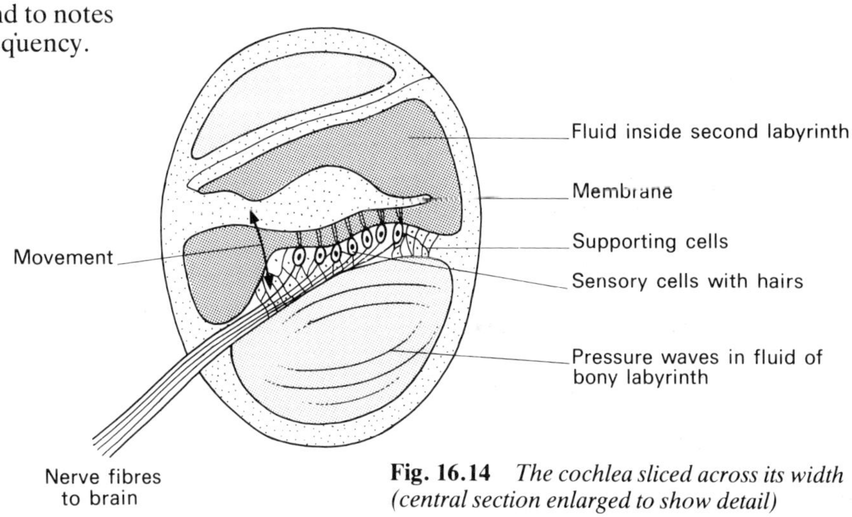

Fig. 16.14 *The cochlea sliced across its width (central section enlarged to show detail)*

Fig. 16.15 *The measurement of sound is important, as excess noise is considered as another form of pollution*

pitch and consists of irregular vibrations.

The third property of sound that the ear can detect is its *intensity* or loudness. Fig. 16.15 shows a man measuring the sound level of Concorde. Instruments measure sound levels in units called *decibels*. Table 16.1 lists some everyday sounds and their approximate sound levels expressed in decibels. It is essential to be able to measure sound in this way, for it can be as much a hazard to our environment and health as any other form of pollution. A high level of sound or noise around us not only irritates and annoys, it can also damage the ear.

16.7 **Deafness**

We only have to get our ears blocked with water during a swim to realize the problem that a deaf person faces all the time, i.e. the problem of hearing himself and others speak. When we use our vocal cords to utter a sound it is picked up by our ears, and our brain rapidly monitors the noise we are making.

TABLE 16.1 **EVERYDAY SOUNDS**

Sound source	**Sound level (decibels)**
Sound that is just audible (threshold of hearing)	0
Ordinary breathing	10
Whisper	20
Rustle of a newspaper	30
Noise in a town at night	40
Inside a moving train (window shut)	50
Ordinary conversation	60
Inside a moving train (window open)	70
Heavy traffic in city rush-hour	80
Pneumatic drill (near)	90
Underground train (near)	100
Thunder or take-off of jet (near)	110
Noise causing pain (threshold of feeling)	120

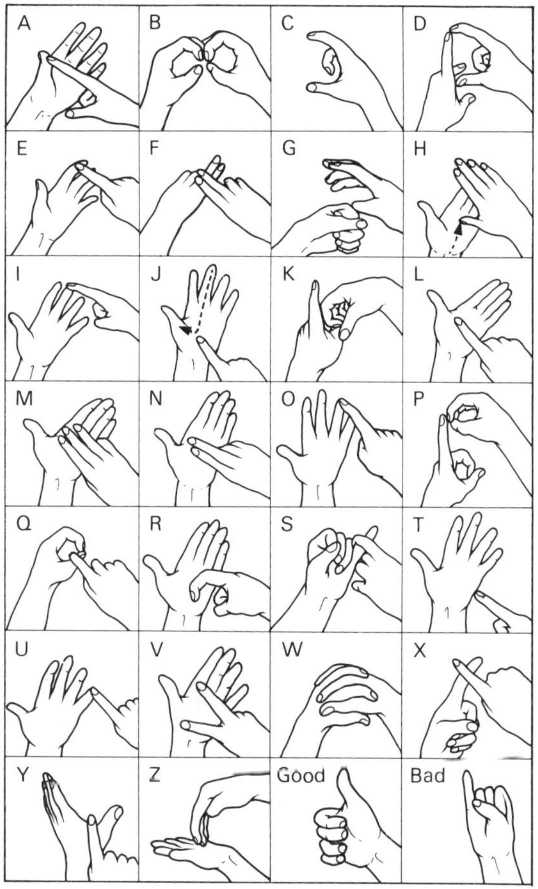

Fig. 16.16 *Sign language*

CHILDREN who have never heard a splash or the gurgle of water down a plughole and to whom the sight of a man holding up a starter's gun must seem a meaningless pantomime came to Manchester from all over the kingdom recently to train as competitive swimmers.

The best of them will be chosen for a deaf children's swimming squad to represent Britain in future European championships, including the deaf Olympics in 1981.

Getting them up to scratch involves extreme problems of communication.

No standard vocabulary of signals has yet been devised that can reach a squad once they are in the water. Each child has to be summoned individually and taught the necessary tricks.

'Once they get started in the water,' Mr Percival Deighton, organizer of the training session, told me, 'you have to virtually drop a bomb on them to stop them going. There is only one golden rule. Always have the light on your face so that they can see your lips.'

Fig. 16.17 *How the deaf are making a splash*

Because we can hear what we are saying we can adjust the noises that we make to form words, and so develop a language. A person who is born deaf (pre-lingual deafness) has no feedback of information from his ears to his brain on the nature of his voice, so he has to be taught other ways of 'knowing' the sounds that he is making. Through such specialist teaching those born deaf may learn some speech and language by which they can make themselves understood.

Some people may go deaf after they have learned speech and language. They too will have difficulty in registering their voice and may have to learn alternative methods of doing this. Like those born deaf they will be unable to hear other people and will have to use other methods such as *lip-reading* to interpret what is being said. Equipment such as hearing-aids may help a person who has a small degree of hearing (partial hearing). A hearing-aid amplifies, i.e. increases the volume, of the sound reaching the ear. It may be used by the hard of hearing: those people whose hearing has deteriorated in later life. Deaf people can communicate through lip-reading and *sign language* (Fig. 16.16). However, other people should be aware of the concentration and effort that is needed and be prepared to take time to talk slowly, with their faces turned towards the deaf person. As some form of deafness affects one adult in every twenty it is a disorder that many of us are likely to experience. It is a disorder that we need to be aware of and to understand (Fig. 16.17).

Some forms of deafness can be relieved by surgery. This applies to a form of *middle-ear deafness*, when the three sound-conducting bones are not sufficiently free to move. Another form of deafness is *nerve deafness* caused by a disorder in the inner ear or brain.

Various things can affect the quality of hearing during our lifetimes. An infection of the middle ear, as well as producing painful earache, can cause temporary deafness. Antibiotics are an effective treatment for this form of infection, and usually hearing is gradually restored. However, there may be complications if the infection is so severe that an accumulation of pus bursts the ear-drum. Usually the ear-drum will heal normally and there will be no loss of hearing. But if it fails to heal, or there is thickening due to scar tissue, then hearing may be impaired. Injuries to the head, such as a heavy fall, may result in similar damage to the ear-drum, particularly if there is bleeding from the ear.

A loud noise, such as a gunshot, can create such a large pressure wave that the cochlea is damaged. The deafness is usually temporary and hearing returns. However, constant exposure to loud noise can produce permanent damage. This may become evident only in later years. The Health and Safety at Work Act 1974 now requires precautions to be taken at work to protect people from excessive noise which could lead to deafness. A factory inspector will, among many other checks, be able to see whether or not the regulations are being observed. For instance, the legal limit for noise is 90 decibels. Above this level protective ear appliances such as plugs, ear-muffs or helmets must be worn.

16.8 The ear and balance

In the swelling (*ampulla*) at each end of a semicircular canal, there are sensory cells with hair-like projections. Movements of the head create motion and pressure changes within the fluid of the semicircular canals (Fig. 16.18)). The sensitive hairs are 'tugged' in different directions and the cells are stimulated to send information to the brain. When we spin rapidly the movements within the fluid carry on for some time after we have stopped spinning. So although we are standing still, the brain is still receiving messages of movement from the ear. This produces a reeling and giddy sensation.

Fig. 16.18 *How our ears help our balance*

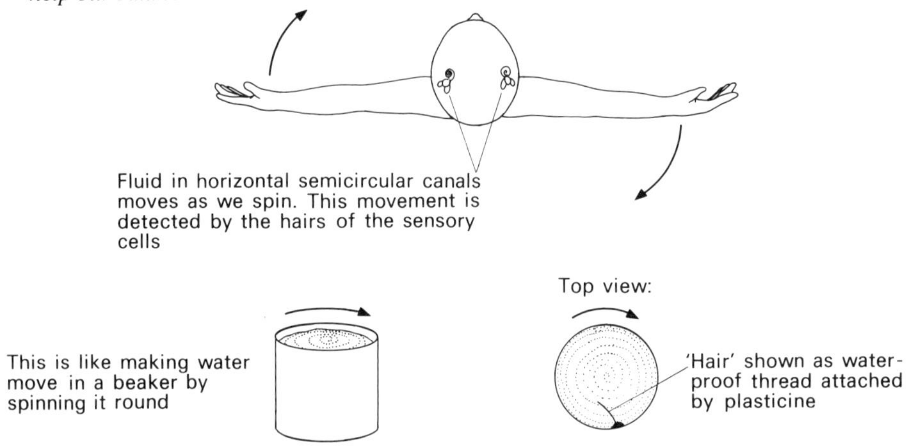

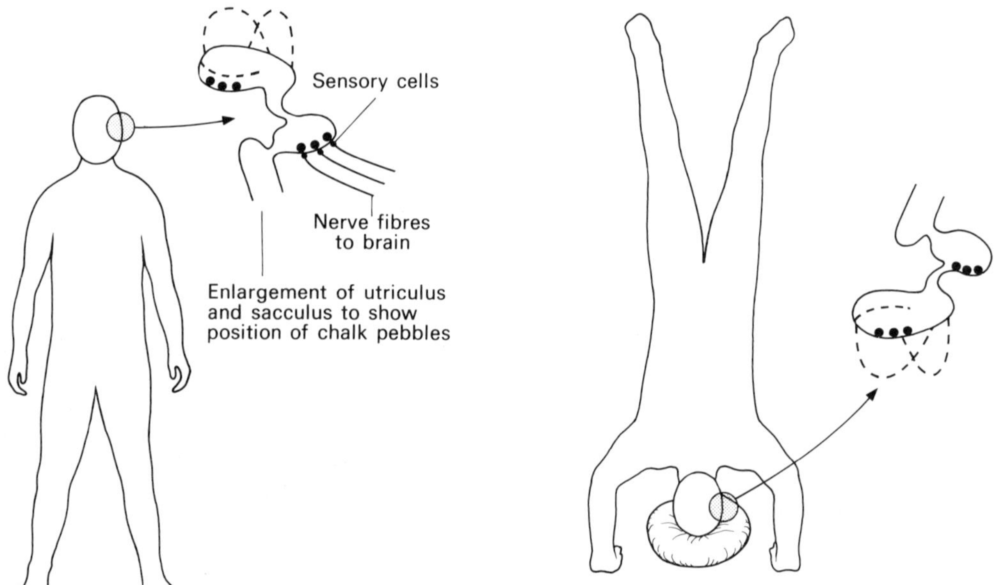

Fig. 16.19 *How we know whether we are standing on our feet or our heads*

Question 4 A ballerina when spinning moves her head in a series of jerks, rather than letting it spin with the rest of her body. How does this prevent a giddy sensation?

The *utriculus* and *sacculus*, the large sacs between the cochlea and the semicircular canals, assist us in keeping our posture. There are sensitive hair-cells which detect the position of tiny 'chalk pebbles'. These move under the action of gravity within the cavities (Fig. 16.19). This is how we can detect whether we are standing on our feet or our heads, or in which direction we are leaning.

The brain receives information on movement and balance from receptors within muscles of the limbs (page 222). However, if you try to stand with your eyes closed you will see how important our vision is to our sense of balance.

Answers to questions in the text

1 Mucus is secreted from the linings of the mouth and tongue. Saliva is secreted from the salivary glands.

2 Our sense of smell is easily tired or fatigued. We grow accustomed to a smell and become unaware of it after our first exposure to it.

3 The ear-drum will bulge into the ear canal because of the higher pressure in the middle ear. If air leaves the middle ear through the Eustachian tube the ear-drum will return to normal.

4 A sharp movement of the head will prevent continuous movement of the fluid within the semicircular canals. Therefore it will not be in motion at the end of the spin.

THINGS TO DO

1 Blindfold someone and either get them to hold their nose or clip it with an under-water-swimming clip.
(a) Obtain solutions of salt, sugar, vinegar and lemon. Place a drop of each solution on to different regions of the person's tongue. Ask them to identify the taste and record the results in the form of a diagram. The person should wash their mouth out with water after each test.
(b) Chop an apple, banana, potato, onion, orange and lemon into cubes. Give the person one cube at a time to eat and ask them to identify the taste. Record the results. Repeat the tests with the nose-clip removed. Compare the results. Repeat the test with food that has been liquidized. Observe whether the texture of the food helps in identification.
(c) hold a piece of highly scented soap under the person's nose. See if they can tell the exact moment at which you take the soap away.

2 Place a variety of substances (e.g. tea, coffee, mint, cloves) into shoe boxes with holes punched in the lids. Ask people to sniff at the holes and to identify the contents of the box. Observe whether the person's sense of smell deteriorates after a lot of testing.

3 Blindfold another person and sit them down on a chair in the centre of the room. Move to some point in the room and click your fingers or clap your hands. Get the blindfolded person to locate your position by pointing (they must not move their head). When they correctly locate the source of the sound, plot it on a plan of the room. If you move to different points in the room, you should be able to observe how our accuracy of locating the source of a sound can vary. Repeat the tests, this time with the person covering up one of their ears. Compare the results.

4 Stand next to a blindfolded person with a ticking watch in your hand. Move away slowly and ask the person to say when they can no longer hear the watch. Repeat the test with different people. Compare the results.

5 Carry out a series of activities in front of a blindfolded person, for instance unwrap a sweet. Ask them to identify what you are doing.

6 Construct a montage of illustrations to show how our senses of taste, smell and hearing (a) assist our survival, (b) give pleasure and enjoyment.

7 Carry out a survey of the provisions for the deaf in your area.

8 Draw a map of your area. Using a colour code map in the noise levels. Attach to the map any articles from local newspapers relating to the problems of noise.

9 'I can work better with music in the background.' This phrase often produces a family argument at homework time! Test it out. Write out fifty mental arithmetic questions and answers. Sit a person in a quiet room and ask them to answer questions which you select at random. Count how many questions are answered in two minutes. Repeat the experiment with a record playing in the background. Try the experiment with the sound at different levels. If you continue with the experiment, does the person's performance improve?

THINGS TO WORK OUT

In Questions 1–3 write down the *one* answer which you think is correct.

1 The quality of food that our tongues are *not* able to detect is its
- **A** bitterness
- **B** flavour
- **C** saltiness
- **D** sweetness
- **E** texture

2 The part of our ears that makes it possible for us to detect the frequency of a musical note is the
- **A** ear-drum
- **B** oval window
- **C** cochlea
- **D** ampulla or a semicircular canal
- **E** utriculus

3 The Eustachian tube is the channel which
- **A** connects the outer ear with the inner ear
- **B** extends from the pinna to the ear-drum
- **C** connects the middle ear with the nasal cavity
- **D** carries the auditory nerve
- **E** leaves the inner ear and ends in the skull

4 State where the following are positioned in relation to the outer, middle and inner ear:
(a) pinna
(b) anvil
(c) ear-drum
(d) round window
(e) organ of Corti

5 (a) Draw a simple outline of the human tongue. Pin-point on your diagram *four* regions and the tastes they are sensitive to.
(b) How does saliva help us to detect the taste of a food?

6 Describe how the ear-drum is protected against damage when a person enters a region of low air pressure. Give *one* example of when this might occur and what action can be taken.

7 (a) What is the function of the hammer, anvil and stirrup?
(b) Describe how the ear assists the balance of the human body.

8 A boy who is blindfolded is fed a jam-filled doughnut by another person. Describe how the boy detects the sugar on the doughnut and the fact that the jam is strawberry flavoured.

9 (a) Explain how the following may cause deafness. In each case state whether you think the deafness will be temporary or permanent:
(i) accumulation of wax in the ear canal;
(ii) a throat infection
(iii) injury to the auditory nerve
(iv) lack of movement between the bones in the middle ear
(v) an explosion
(b) What is the cochlea embedded in, and what does it contain?

10 Describe, using a diagram, how a sound received at the ear-drum causes a nerve impulse to travel along the auditory nerve to the brain.

17 Vision

17.1 The eye and how it is protected by the body

Each eye nestles in fatty tissue in the *socket* or *orbit* of the skull (page 31). These delicate organs are protected by a surrounding area of bone. Most of these bones are hollow, forming *sinuses* which are connected to the nasal cavity. If the connecting passageway becomes blocked the secretions that collect in the sinuses infect the sinus membrane, a condition known as *sinusitis*. The bony ridge over the top of each eye is covered by skin and hair, the *eyebrows* (Fig. 17.1). These are not merely decorative, but have a special function. We mentioned in Chapter 14 how the skin on the forehead contains many sweat-glands. Tennis players may even wear cotton sweat-bands round their heads to collect the sweat and prevent it entering their eyes. Under normal conditions the eyebrows are our natural sweat-bands!

Further protection is given to the eye by an upper and lower flap of skin, the *eyelids* (Fig. 17.1). These are lowered or raised across the eye by muscles. Throughout the day this movement takes place automatically about once every 2½ seconds. The eyelids shut if the eye sees an object approach too closely and they also close rapidly when anything comes into contact with the hairs that form the *eyelashes*.

The surface of the eye is normally kept clean and sterile (page 84) through constant washing by tears. Tears are a fluid containing a small amount of salt and an antibacterial enzyme. Thus tears have an antiseptic property (page 86), but most important they keep the surface of the eye moist and free from all fine particles. Tears are produced in the *tear-gland* (*lachrymal gland*), which is in the orbital cavity above the eye and below the eyebrow (Fig. 17.2).

Small tears are produced continuously and the movement of the eyelids helps them pass across the surface of the eye. We notice tears only when something irritates the eyes or when we are emotionally upset. Under these circumstances the secretion of tears is increased to cleanse the eye of the irritation or as an emotional response. This extra amount of tears is too great a flow for the drainage channels, the *tear-ducts* (*lachrymal ducts*) (Fig. 17.2). Some tears pass down the ducts and into the nasal cavity; the rest pour down the face. So when we cry we have to blow our noses to get rid of the tears and mop them from our faces.

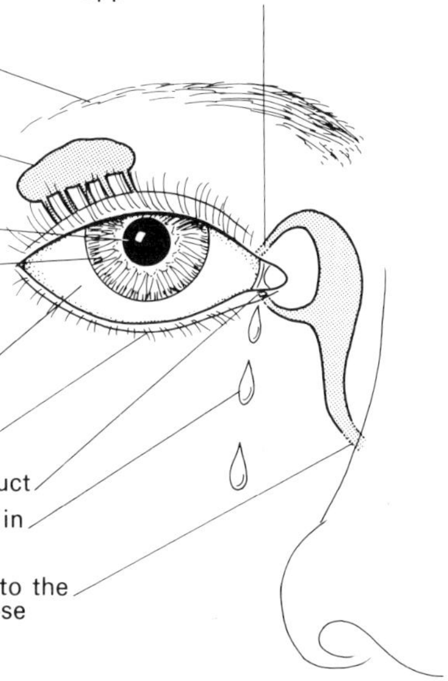

Fig. 17.2 *The tear-gland (lachrymal gland)*

When we close our eyelids to sleep, they are usually easily opened next morning. Along their edge are small sebaceous glands which secrete a fluid to prevent the eyelids sticking together. However, an infection of the eye may cause an increase in the secretion, which will stick the eyelids together. In such cases bathing the eye with some warm water which has been boiled will help to unstick the eyelids. The material, such as cotton wool, which has been used to wipe the eye should be thrown away because the infection will probably be contageous if it is bacterial in origin (page 80). A doctor may prescribe an eye ointment to cure the condition.

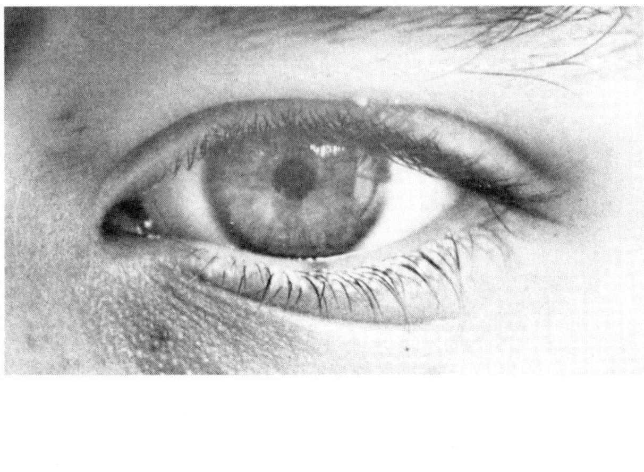

Fig. 17.1 *The eye*

17.2 **The eye and how we can protect it**

Where there is a danger of corrosive substances or flying particles entering the eye, toughened goggles should be worn. Safety regulations (Health and Safety at Work Act 1974) in workshops and laboratories demand this form of protection for anyone working there (Fig. 17.3). Unfortunately, a man who wears protective goggles at work may come home to his 'do-it-yourself' bench and expose his eyes to the same sort of danger without a second thought. Safety goggles can be bought for use in the home.

Question 1 What form of road transport might you use which demands the use of goggles?

Bathing the eye using an eye-bath can help to remove particles or solutions from the eye. Also, you can pour water gently over the eye or immerse the upper part of the head in water. As with all injuries the rule is not to do further harm or introduce infection. Therefore medical attention should be sought if a particle cannot be easily removed, if the eye becomes inflamed or if the injury involves a corrosive chemical.

17.3 **Eye development and eye muscles**

At birth the eyeball is about 14 mm long; the eye muscles can move the eye, but they do not work together, i.e. they are unco-ordinated. Gradual development of the eye and muscles takes place during early childhood. By about six months a baby is able to co-ordinate eye movements and concentrate its eyes on a particular object. The baby will also show the first signs of co-ordinating eye and limb movements so that objects can be picked up. By four years old, the eyeball has grown to seven-tenths of its full size of 24 mm; the child is able to recognize colours, judge depth and distance and see all objects, near or far. By about fourteen years eye growth is complete. Simple eye tests are carried out as part of the routine of checking a child's physical development both before and at school (page 20). It is essential to spot any problems as the eye develops as early as possible so that they may be corrected. For instance, some children have what is often called a lazy muscle which produces a *squint*.

Like all muscles of the body, eye muscles work in pairs. When the eye is looking forward all the eye muscles are relaxed. Fig. 17.4 and Table 17.1 show the external eye muscles and how they act together to turn the eye within the orbital cavity. Any lack of co-ordination by a particular muscle will cause a squint, and the person will be unable to direct both eyes towards the same object at the same time. Most squints can be treated successfully by wearing glasses or surgery to lengthen or shorten the defective muscles.

When we look at a book or any object close to us the muscles of the eyes have to work hard to turn the eyes inwards, i.e. *converge* them. If this is done for a very long time the muscles become tired and we become conscious of the strain. Very young children particularly should not do close work for very long periods of time.

Question 2 How can you rest the eyes during periods of close work?

17.4 **Binocular (stereoscopic) vision and field of vision**

To look at an object at any distance our eye muscles work automatically to turn each eye in the correct direction. However, the eyes are set apart by about 65 mm, so each will see the object from a slightly different angle (Fig. 17.5). The brain combines these two images so that we see the object in three dimensions (i.e. the three dimensions of

Fig. 17.3 *Protection for the eyes is essential in some situations*

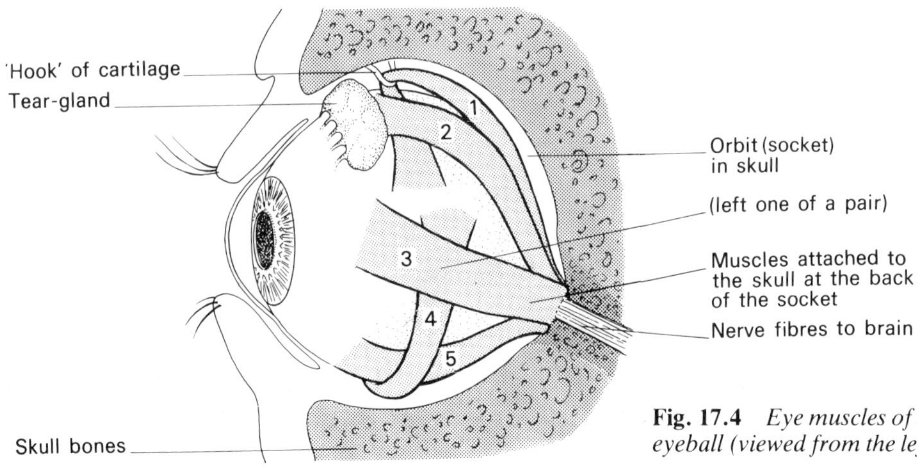

Fig. 17.4 *Eye muscles of the left eyeball (viewed from the left side)*

TABLE 17.1 HOW THE PAIRS OF MUSCLES MOVE THE EYE		
Movement of eye	**Pair of muscles involved in the movement (numbers as Fig. 17.4)**	**Action of each muscle (numbers as Fig. 17.4)**
Upwards	2 and 5	2 contracted 5 relaxed
Downwards	2 and 5	2 relaxed 5 contracted
Sideways, left	3 (left) and 3 (right)	3 (right) relaxed 3 (left) contracted
Sideways, right	3 (left) and 3 (right)	3 (right) contracted 3 (left) relaxed
Downwards and outwards	1 and 4	1 contracted 4 relaxed
Upwards and outwards	1 and 4	1 relaxed 4 contracted

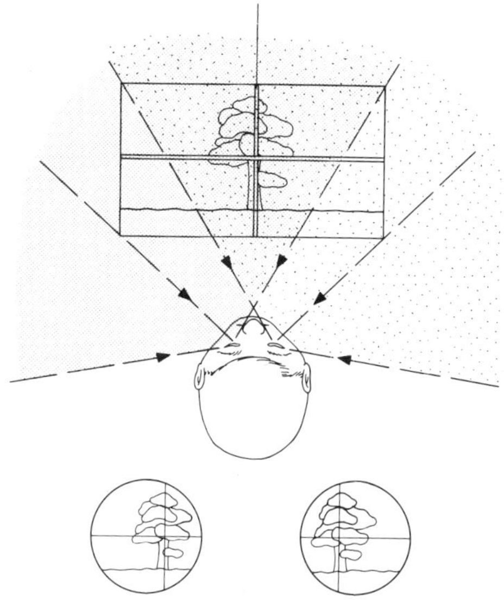

Fig. 17.5 *Stereoscopic vision*

length, breadth and depth). *Stereoscopic vision* helps us to judge distance and the exact position of objects in relation to each other and ourselves.

We 'see double' if the eyes are not working together. A physical defect may produce this effect, but it can also be brought on by tiredness, drugs or by an excess of alcohol. Naturally, when vision is hindered in this way a person is more likely to have an accident because it becomes much more difficult to judge distances.

Eyes are the body's receptors of light energy, which travels in the form of a wave. We receive light from objects which give out light, i.e. *light sources*. Materials which are not light sources are seen because they *reflect* light. Light cannot bend itself round corners, so if we keep our head still our eyes cannot receive light from behind us. If you stand looking forward with your arms outstretched

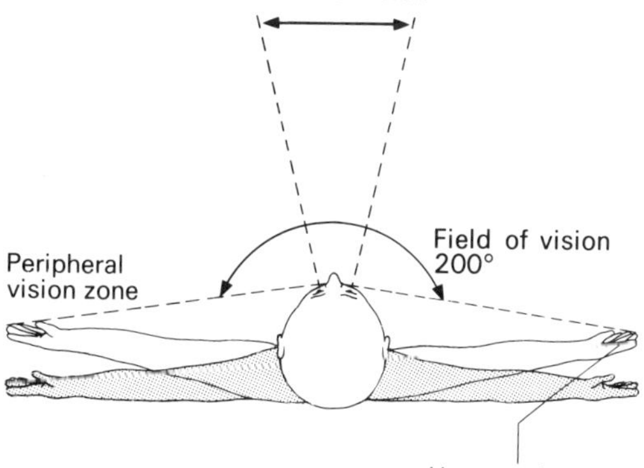

Fig. 17.6 *Field of vision*

as in Fig. 17.6 you will find that you can just see the movement of your fingers. Push your arms right back and the fingers disappear out of your *field of vision*. Light rays are received by the eyes if they fall within an area covered by an angle of 200° at the nose. However, our clearness of vision varies over this area. We can view objects that are straight ahead very clearly and accurately. Towards the edges (the periphery) of our field of vision, although we are aware of objects and their movement, we do not see them precisely. Nevertheless *peripheral* vision is important for, having spotted some movement out of the 'corner of our eye', we can turn quickly to see what is causing it. When we turn to look at a particular object in detail it is brought within a very narrow field of accurate vision. All light rays within this area fall upon the most sensitive area of the screen of the eye. Table 17.2, which compares the eye with the camera, describes the internal structure of the eye, as well as the particular function of the *yellow spot* (*fovea*) and *retina*.

TABLE 17.2 COMPARISON BETWEEN THE EYE AND THE CAMERA

STRUCTURE

EYE

The tough outer *sclerotic layer* forms a light-tight ball, except for a round transparent window at the front, called the *cornea*. The cornea is protected by a very thin, moist, transparent membrane, the *conjunctiva*. This contains blood-vessels which, when they become dilated, make the eye bloodshot. The cornea contains no blood-vessels. It protects the lens. The *choroid layer* contains many blood-vessels, providing nourishment for the retina and a black pigment to prevent reflection of light. The lens and cornea are nourished by the circulating *aqueous humour*. This transparent watery fluid also helps to regulate the eyeball's internal pressure. The thicker vitreous humour helps the eyeball to keep its shape

CAMERA

The body of the camera forms a tough, light-tight box. A coating of matt black paint prevents reflection of light inside the camera. The air pressure within the camera is equal to that outside the camera. The lens cap protects the surface of the lens

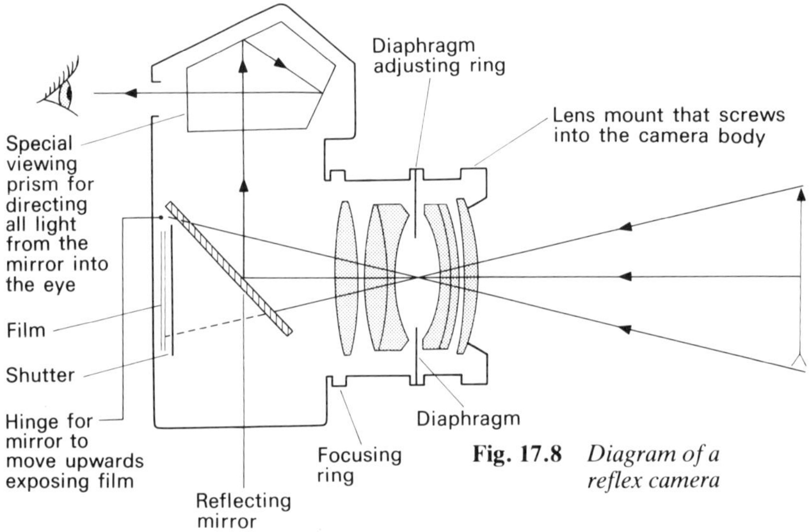

Fig. 17.8 *Diagram of a reflex camera*

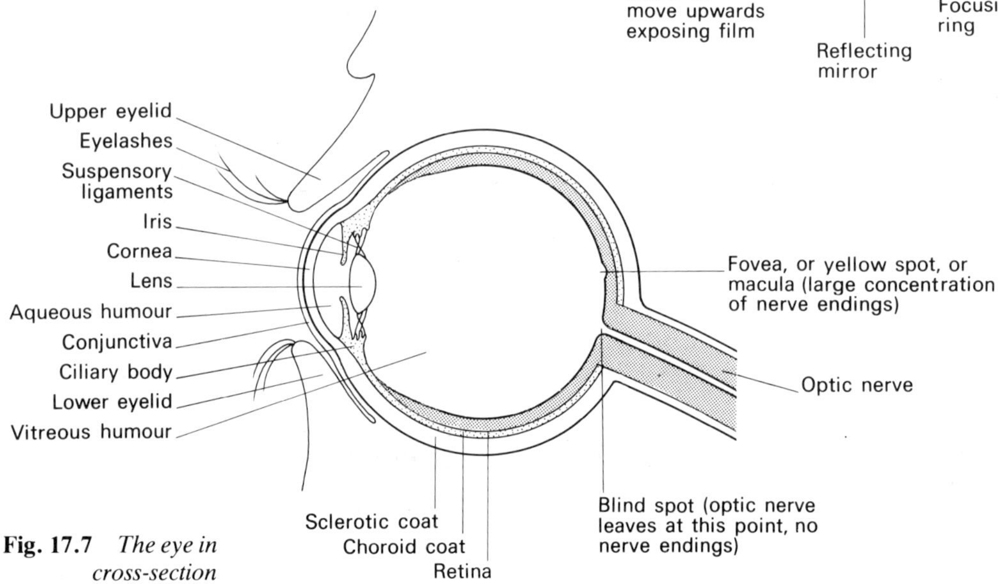

Fig. 17.7 *The eye in cross-section*

TABLE 17.2 **continued**

ENTRY OF LIGHT

EYE

The *eyelid* can cover the eye to prevent the entry of light. It is estimated that it can shut in 1/15 of a second

The *iris* is the coloured area of the eye, as viewed from the front. Its colour is inherited (page 13) and is seen when the pigment cells develop after birth. When these cells do not develop this area appears blue. The iris is a thick ring of muscular tissue. The hole in its centre is called the *pupil*. It is through the pupil that light enters the inner eye. The muscles of the iris, by contracting and relaxing, are able automatically to increase the size of the pupil in poor light (Fig. 17.9) and to decrease its size in bright light (Fig. 17.10). The iris is a muscular continuation of the choroid coat

Fig. 17.9 *Dilated pupil allows more light to enter the eye*

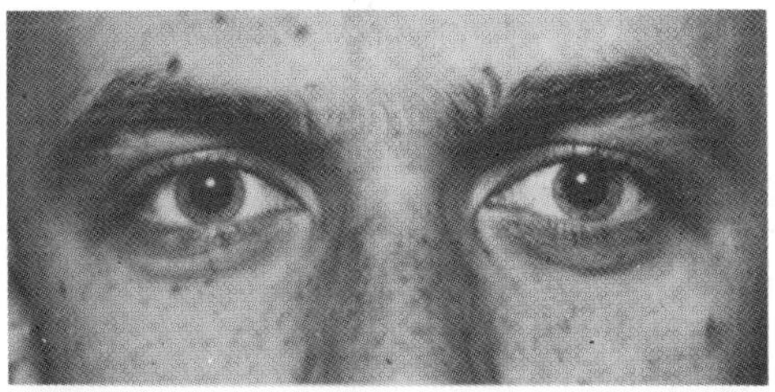

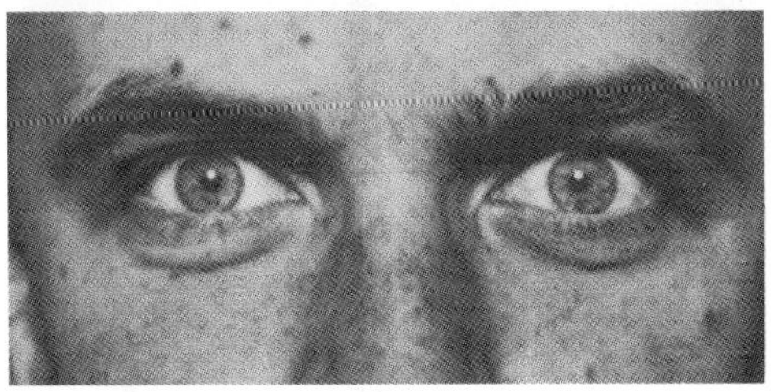

Fig. 17.10 *Contracted pupil allows less light to enter the eye*

CAMERA

The *shutter* controls the entry of light. A leaf shutter consists of a series of overlapping metal sections which move apart when the shutter release is pressed, and then spring back after allowing light to enter the camera. The time for which the shutter is open (exposure time for the film) may be adjusted to the required speed, which is indicated on a knob on the outside of the camera. Shutter speeds are given in fractions of a second, e.g. 1/1000 of a second

A series of overlapping metal plates act as the *diaphragm* (Fig. 17.11). The movement of the diaphragm, and thus the size of the *aperture* in the centre, is controlled by the diaphragm adjusting ring. The diameter of the aperture is expressed as a fraction of the focal length of the lens. Therefore the diaphragm adjusting ring is marked in f-numbers, e.g. f/2.8 (Fig. 17.12): the smaller the f-number the larger the aperture. On a dull day one would select a large aperture, f/2.8, with a slow shutter speed, 1/60 of a second. When we go up an f-number we halve the aperture, therefore only half as much light enters the camera

Fig. 17.11
(a) Aperture opened up to allow more light to enter the camera.
(b) Aperture closed down to allow less light into the camera

Fig. 17.12 *Diaphragm-adjusting ring and focusing ring*

TABLE 17.2 **continued**

FORMATION OF IMAGE

EYE

Behind the iris the choroid coat forms a thick ring of muscle called the *ciliary body*. *Suspensory ligaments* from this 'body' support the *lens*. The lens is like a magnifying glass made out of jelly. It is the suspensory ligaments and the ciliary muscles that act upon it to keep its shape. The lens shape is altered by the contracting and relaxing of the ciliary muscles. This is known as the eye's *power of accommodation*. It enables us to see clearly over a large range of distances. When the object is near (less than 6 m away) the ciliary muscles contract, the lens bulges and refracts the light more. Therefore it can bring the divergent rays of light to a focus on the retina (Fig. 17.13(a)). If the object is nearer than 25 cm the ciliary muscles cannot fatten the lens enough for it to focus the very divergent rays of light. The smallest distance at which an object can still be seen clearly is called the *least distance of distinct vision* or *near point* (on average 25 cm). When the object is more than 6 m away (the *far point*) the ciliary muscles relax, and the lens becomes thinner and weaker (Fig. 17.13(b)). Therefore the less divergent rays are brought to a focus on the retina. The lens adjusts only the focusing power of the eye: 75% of the refraction takes place at the cornea. The two humours also help to refract the light

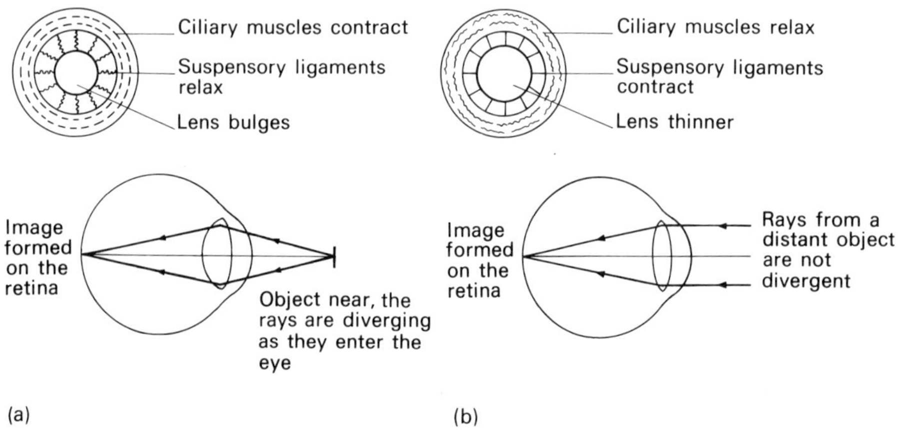

Fig. 17.13 *The eye's power of accommodation. (a) Focusing on near objects. (b) Focusing on distant objects*

CAMERA

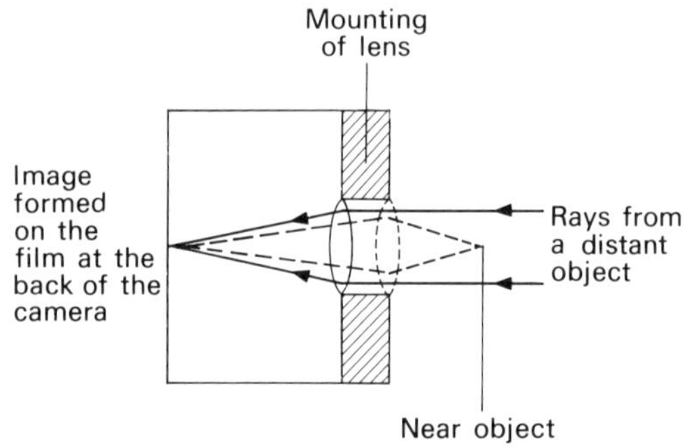

Fig. 17.14 *The focusing action of a camera*

When a picture is taken, the focusing ring is revolved so that it gives the required reading of the distance of the object from the camera. The rotation of the ring moves the lens within its screw mounting. Thus the distance of the lens from the film is adjusted. The movement of the lens enables near or distant objects to be clearly focused on the film (Fig. 17.14). When the camera is focused on a near object the lens is moved away from the film. The strength of the lens within the camera cannot be changed. However, additional lenses may be attached to the front of the camera to alter its total focusing power

TABLE 17.2 **continued**

THE SCREEN FOR THE IMAGE

EYE

The retina is the nerve tissue of the eye. It consists of ten layers of nerve fibres and nerve-cells. The nerve fibres are nearer the surface of the inner eye. The nerve-cells are buried deep in the lower layers of the retina (Fig. 17.15)

The nerve fibres are not sensitive to light and it passes through them. The nerve-cells sensitive to light are of two types: *rods* and *cones*. Their number and distribution within the retina differ. There are about 125 million rods and they are mostly found around the outer edges of the retina. They contain a purple pigment called *visual purple* or *rhodopsin*, which is bleached by light. This bleaching stimulates the rod cells. The brain interprets the signals from the rods as patterns of light and shade. When we enter a darkened room, over a period of half an hour our eyes become adjusted. This is because the sensitivity of the rods is increased by changes in the visual purple.

There are about 7 million cones and they are found in the central area of the retina. It has been suggested that there are three types of cones, each sensitive to a particular band of colour, such as red, blue or green. When different cones are stimulated to different degrees the brain interprets their signals as different tones and colours. Cones contain a pigment which is not sensitive to low levels of light. Thus in the dark rods and not cones are stimulated, so we do not see colours. Even in daylight 8% of man cannot distinguish between red and green. This is called red–green *colour-blindness* (page 14). It is thought to be caused by a lack of cones or by their inactivity

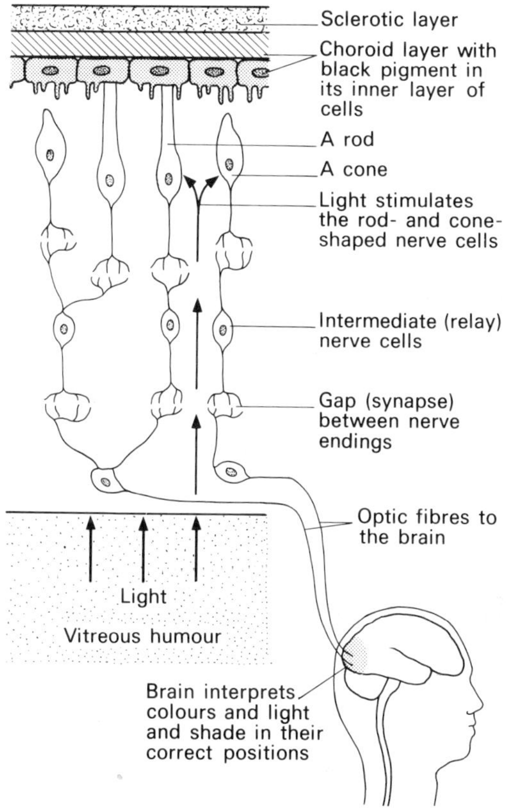

Fig. 17.15 *The structure of the retina*

CAMERA

There is a light sensitive film at the back of the camera. A black and white film consists of salts of silver nitrate (or silver bromide) on a gelatine base. These silver salts undergo a chemical change on exposure to light, and have to be chemically developed in order than we may obtain first a negative and then a positive picture. To obtain a colour picture we have to use a colour film

TABLE 17.2 **continued**

SENSITIVITY	
EYE	**CAMERA**
There is a high concentration of cones, but no rods, in the area known as the *fovea*. Because of this the fovea is an extremely sensitive area. Any image formed within this area (1.5 mm in diameter) is seen in fine detail. The least sensitive area is the *blind spot*, where the nerve fibres leave the eye as the optic nerve (see page 223)	A film is sensitive to the same degree all over its surface. However, we choose the type of film according to the light conditions. Some films are more sensitive than others, e.g. a 'fast' film is used in dark conditions
THE IMAGE	
EYE	**CAMERA**
This is formed upside-down on the retina. It is a *real image* for it is formed on a 'screen'. It is smaller than the object. The brain interprets the image as being erect. It does this using information from other organs of the body and our experiences of the visual world that have been built up from birth (page 22). The eye forms moving images on the retina. It also retains an image on the retina for about $^1/_{10}$ of a second – *persistence of vision*. Therefore if two still pictures are viewed one after the other within this space of time, the two images are superimposed. If the same figure is viewed in both pictures with only a slight change in position then that figure appears to move (see page 220)	A real upside-down image is formed. For general use the size of the image is smaller than the object. Additional lenses have to be added if we wish to magnify the object. The camera can take only one still picture at a time – it has no persistence of vision. A cine camera records a series of still pictures (frames) on a reel of film. These stills record slight changes in movement, and form a sequence which appears to move smoothly

17.5 **Common faults with vision**

The eye's power of accommodation gives us a good range of focus, normally from 25 cm (near point) to infinity (far point). However, this does not apply to everyone and it can vary with age. For instance, the retina needs to be only a fraction of a millimetre out of position to prevent the formation of a clear image. The wrong positioning of the retina can cause two common defects in vision to arise. These are *long sight* (*hypermetropia*) and *short sight* (*myopia*).

Long sight

A person is said to be long-sighted if they cannot see in detail things that are positioned at the normal near point (25 cm). As is shown in Fig. 17.16(a), it can be caused by the eyeball being too short. Very young children are generally long-sighted (at the age of seven this applies to about 80 per cent). Defects of vision can occur as growth and development occur within the eye. It is therefore important that eyesight should be checked regularly once a year during childhood. However, in most cases until the lens of the eyes harden in middle age long-sighted eyes can focus light rays on the retina.

Question 3 Why would it be unwise to give a five-year-old a book containing the size of print used here?

A weak lens which is unable to bend (refract) the rays of light to a focus on the retina is another cause of long sight.

To correct long sight a convex lens is positioned in front of the eye (Fig. 17.16(b)). This bends the rays inwards (converges) before they enter the eye. Thus these rays can easily be focused on the retina.

Short sight

A short-sighted person can see near objects clearly, but not those at a distance (Fig. 17.17(a)). This is because the rays of light coming from a distance are refracted to a focus in front of the retina. It occurs when the eyeball is too long, which may happen during

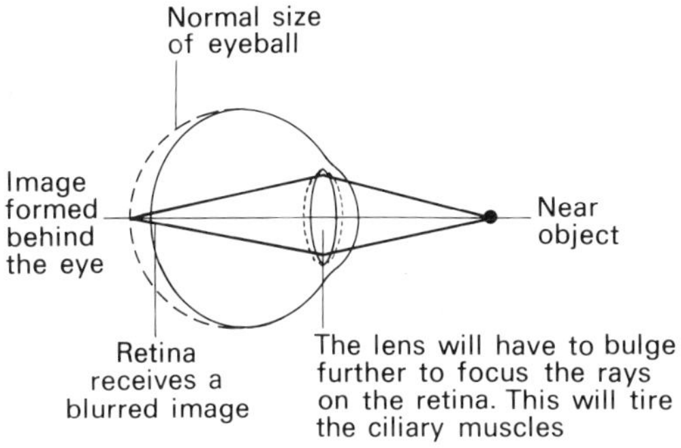

(a)

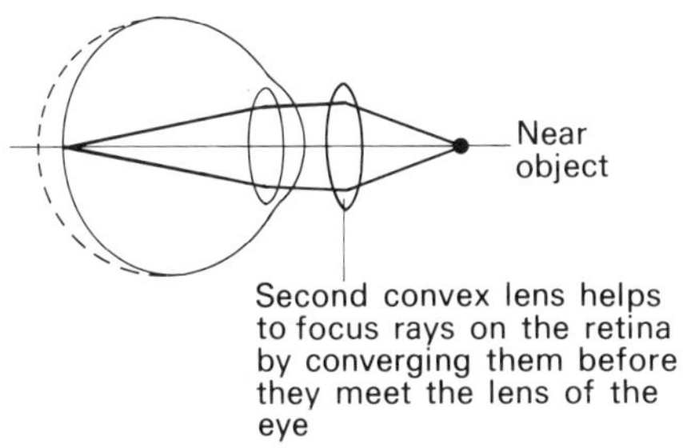

(b)

Fig. 17.16 *The defect of long sight (hypermetropia). (a) One cause of long sight is an eyeball that is too short. (b) A convex lens is used to correct long sight*

the process of growth after early childhood (Fig. 17.17(b)). Another possible cause is that the lens refracts the light too strongly.

No matter how hard a short-sighted eye tries it cannot focus the rays of light coming from a distance on to the retina. Thus to view things at a distance a *concave lens* must be placed in front of the eye (Fig. 17.17(c)). This spreads (diverges) the rays of light before they enter the eye. Thus these rays can easily be focused on the retina.

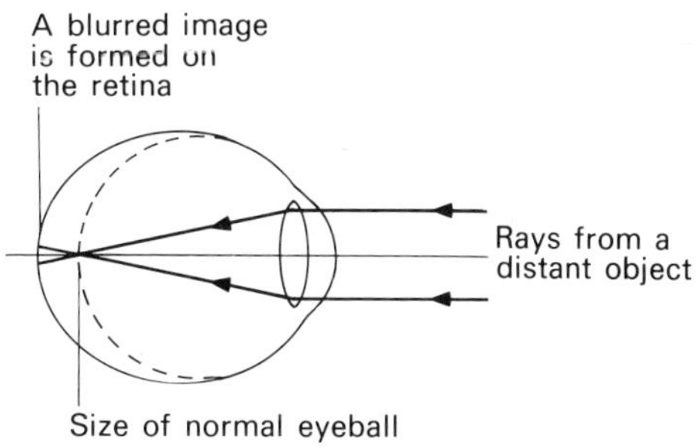

(b)

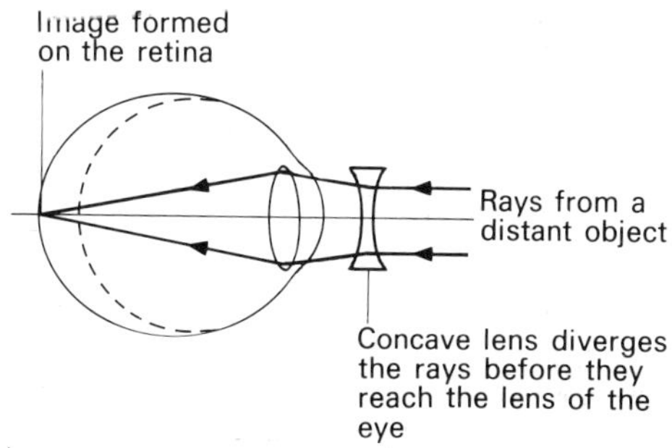

(c)

Fig. 17.17 *The defect of short sight (myopia). (a) A short-sighted person can see near objects clearly but cannot focus on distant ones. (b) One cause of short sight is an eyeball that is too long. (c) A concave lens is used to correct short sight.*

Fig. 17.18 *The presbyopic eye keeps good distance vision but is unable to focus at short range*

Presbyopia (old sight)
The lens within the eye, unlike the eyeball, continues to develop throughout life. By about forty-five it is no longer soft and elastic but has become hard and stiff. This happens to some people sooner than others, but eventually most elderly people have difficulty with near-range vision (Fig. 17.18) or suffer from headaches when they try to do close work. Anything that is within 40–50 cm from the face cannot be seen clearly. An optician will issue spectacles with convex lenses to produce the focusing power that the eye needs. If that patient already has a set of lenses to correct another condition, like short sight, two sets of lenses will be needed or a set of *bifocal* lenses. Bifocals are two types of lens in one: the bottom half is used for close work, the top half for seeing objects clearly at a distance.

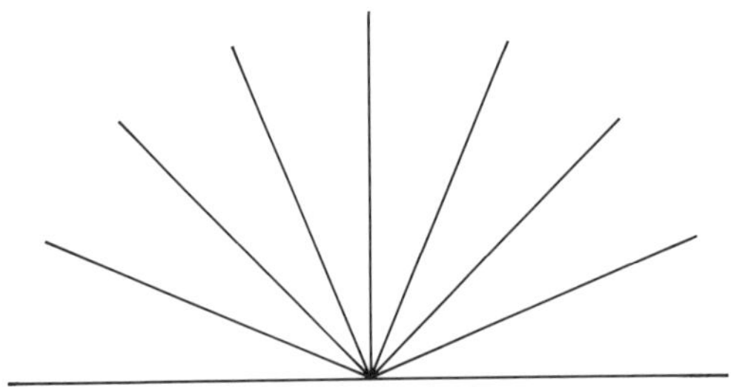

Fig. 17.19 *An astigmatic eye would be unable to see every line of this diagram clearly*

Astigmatism
The arrangement of regular lines in Fig. 17.19 can be used to diagnose astigmatism. A person with this optical defect will be unable to see clearly every single line. In extreme cases, vertical and horizontal lines will seem to slope. This is because the cornea or lens of the eye has an uneven curvature. To correct this condition cylindrical lenses are ground to provide the right curvature in the right places.

17.6 Eye tests and aids to good eyesight

A person may become aware that vision is poor in certain circumstances or it may become noticeable at a regular eye test. A familiar test is the 'distance acuity' test, where a person is asked to sit at a set distance in front of a chart. On this chart are lines of letters and numbers decreasing in size towards the bottom (Fig. 17.20). A person

Question 4 What is the distance acuity of a person who can read only as far as line 36 at 6 m?

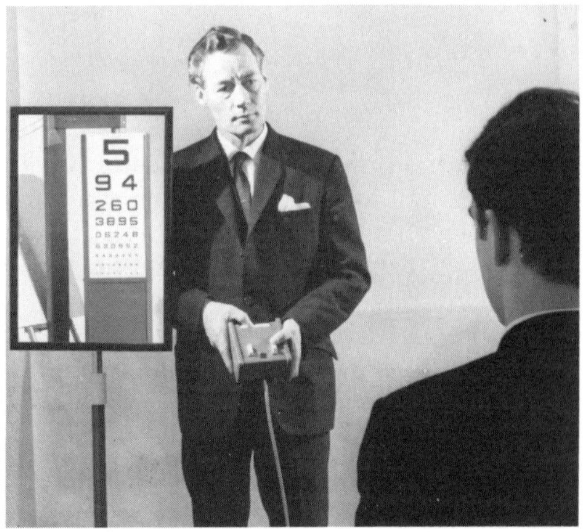

Fig. 17.20 *Person undergoing a distance-acuity test*

with what is judged normal sight should be able at 6 m to read line 6 of the chart. They have a distance acuity of 6/6. If a person can read only the first big letter (line 60), which can be read at 60 m with normal sight, then their distance acuity is 6/60 = 1/10. They have one-tenth normal distance vision. There is a large variety of lenses that can be prescribed to correct vision disorders. The ophthalmic optician's task is to make sure that each lens fits the person's eye requirements like a made-to-measure suit. Lenses mounted in an iron frame, i.e. *spectacles*, were known to be used during the last quarter of the thirteenth century. Today a wide range of frames is available because their importance to an individual's appearance is recognized.

The alternative to spectacles is *contact lenses* (Colour plate 10). These are thin, curved pieces of plastic which are placed in contact with the cornea of the eye. Some are fitted to cover the whole front surface of the eye. Nowadays most contact lenses cover just the area of pupil and iris. They are made from a material which absorbs water (hydrophylic) so they become softer when in contact with the eye and can be worn for very long periods of time.

17.7 Other eye disorders

Cataract and *glaucoma* are two disorders which can arise as we get older. A cataract occurs when the lens of the eye becomes dense and opaque, acting like the frosted glass in a bathroom window. An operation is carried out to remove the lens and spectacles can then be used to help the eye to focus. More than 90 per cent of people over the age of sixty-five show some signs of cataract. Often it is not widespread enough to affect seriously their clarity of vision.

Glaucoma is one of the most common causes of blindness. As with many conditions, its serious consequences can be prevented if it is spotted and treated early enough. The most common form gives no warning signs, but it can be spotted at an eye examination. Thus it is important to have regular examinations, at least twice a year after the age of forty-five. It is also clear that it is genetically linked and runs in families, so people with this disorder in their family history should ensure that they have regular eye inspections. Simple glaucoma occurs because an abnormally high pressure develops within the inner eyeball (Fig. 17.21). This affects the nerves of the retina and prevents the eye receiving the correct supply of blood. The high pressure develops because the normal drain of the aqueous humour gradually becomes silted up. Thus drainage becomes less and pressure within the eye gradually increases (Fig. 17.21). Treatment involves using drugs or surgery to lessen the inner eye pressure.

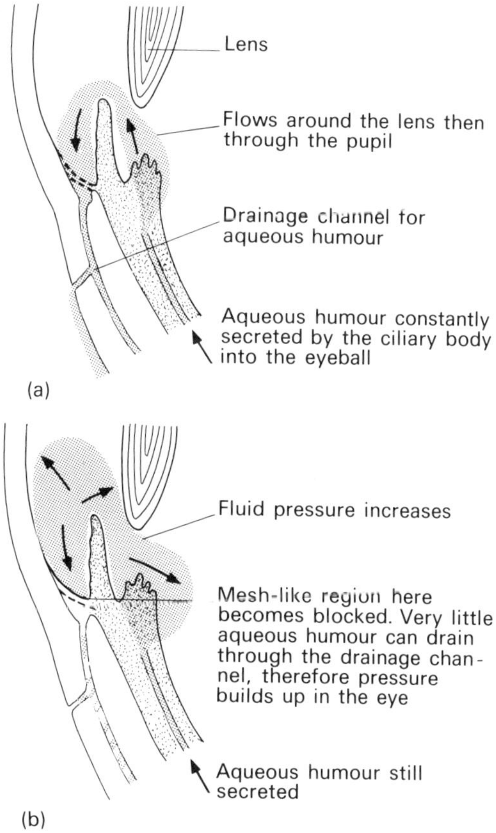

Fig. 17.21 *Simple glaucoma*

Children may be born blind through unknown causes or because of damage by an infection such as German measles. Blindness may develop in later life because of a physical disorder of the eye or through an accident. As with many physical disorders the nature and severity of the condition can vary, ranging from blindness to partial sight. Those who have no vision whatsoever need specialist teaching and aids to assist them in their studies. At centres all over Britain both blind and partially sighted people may learn to overcome their handicap and gain confidence in their own abilities. They may develop skills and receive training for a particular job. As we have stressed throughout this book, we are all individuals. A handicap does not make a person any less of an individual: their need to be treated as such is as great as ours.

17.8 Eye care

Children's eyes are growing organs and are therefore subject to tiredness and strain. Adults cannot strain the internal structure of their eyes. The dull ache that is associated with eye-strain is usually due to a slight defect in the muscular control of the eyes. It can also occur when the muscles become tired by trying to cope with vision in difficult circumstances, for example watching television in complete darkness and sitting nearer to the screen than 1.5 m. To prevent tiring the muscles of the eyes we should help by providing them with the best possible viewing conditions. Amongst other things, this means plenty of light to see by and from the right direction; for instance, for reading a light should always be directed over your shoulder so that it shines on to the book. Regular eyes checks are as important for adults as for children.

Answers to questions in the text

1 Motor cycles. It is also advisable not to hang out of car or train windows.

2 Look up and into the distance or close the eyes.

3 A very young child's ciliary muscles would have to work very hard to focus the lens on small print. This would quickly lead to eye fatigue.

4 Distance acuity = 6/36 = 1/6. This is one-sixth of normal vision.

THINGS TO DO

1 Stand on one foot with your eyes closed. Is it only the ears which assist your balance?

2 Obtain the eye of a bullock or sheep from your butcher. Your teacher will instruct you on its dissection.

3 Obtain a simple convex lens (a magnifying glass). Hold it up so that one curved surface faces a window. Behind the lens put a piece of paper. Move the paper backwards and forwards behind the lens. Observe the image on the paper.

4 Position yourself in front of a mirror. Cover one eye with your hand for about half a minute. Now uncover the eye and observe the size of the pupil.

5 Hold your hands at arm's length. Close one eye and try to bring your index fingers together so that their tips touch. Repeat the test with both eyes open. Which way is easier?

6 Find your right eye's blind spot. Draw an X on the left-hand side of a piece of paper. At a horizontal distance of 7.5 cm to the right of the X draw a black dot. Close or cover your left eye and look at the X, held directly in front of your right eye. Hold the paper at arm's length. Now bring the paper towards you, making sure that your right eye is focused only on the X. Observe that at a certain position the black dot seems to disappear. At this position the black dot is focused on the blind spot of your right eye.

7 Make a collection of diagrams that illustrate optical illusions.

8 Construct an illustrated chart to show some of the differences in eye structure and eye position that are to be found in other organisms.

9 Can we judge colours accurately at large distances? At the centre of a series of cards paint a dot about 1 cm in diameter. Use a different colour on each card. Use black on one card. Stand at a distance from another person and show them a card. Walk towards them and record the distance at which they identify the colour. Repeat the test for all the colours and compare the results. Try the test on a number of people.

10 Draw a letter of a certain size (e.g. 5 mm) on a piece of card. Stand at a distance from another person and show them the card. Record the distance at which they identify the letter. You can then calculate their visual acuity:

$$\text{Visual acuity} = \frac{\text{size of letter}}{\text{distance viewed}}$$

Both measurements must be in the same units.

11 Place a piece of graph paper at arm's length on an upright board. With one eye closed put a dot in the centre of fifty little squares with a sharp pointed pencil. Repeat the test with the other eye closed and then with both eyes open. When did you achieve the highest score?

12 On the corners of the pages of your notebook draw a minute-by-minute sequence of a simple action story (for example, a ship sinking). When you flick over the pages, you should see a moving picture.

13 Sit someone down and tell them to look forwards without any head or eye movement. Stand 1 m behind the chair. At a height that corresponds with the person's eye-level, hold a bright coloured ball. Move round from behind the person but at all times keep the same distance. Record on a plan the position at which the person (a) first catches sight of any movement, (b) identifies the colour of the ball.

14 Make a survey of the provisions for the blind or partially blind in your area. Are there any 'puppy-walking' families in your area? Find out about the training of guide-dogs.

THINGS TO WORK OUT

In Questions 1–3 write down the *one* answer which you think is correct.

1 The amount of light entering the eye is regulated by the
- **A** iris
- **B** lens
- **C** ciliary muscles
- **D** cornea
- **E** eyelashes

2 If you were to look up and out of the window the lenses in your eyes would
- **A** become thinner
- **B** become fatter
- **C** move towards the retina
- **D** move towards the cornea
- **E** not change their shape

3 A man who is colour-blind would not be able to become a
- **A** teacher
- **B** pilot
- **C** bricklayer
- **D** dentist
- **E** lorry-driver

4 Where are the following located in the eye?
(a) aqueous humour
(b) blind spot
(c) fovea (yellow spot)
(d) conjunctiva
(e) choroid coat

5 (a) What lens should be used to correct the following?
(i) long sight
(ii) short sight
(b) Draw two diagrams:
(i) to show the path of light through an eye with the defect of long sight
(ii) to show the correcting action of a lens

6 Describe *one* function of each of the following:
(a) tears
(b) tear-ducts
(c) eye muscles
(d) eyelids
(e) ciliary muscles

7 (a) Explain what is meant by stereoscopic vision.
(b) A phrase you may have often heard is 'I spotted you out of the corner of my eye.' Explain in biological terms what is meant by this phrase.

8 (a) Describe, using diagrams, how our eyes are able to see objects that are near and far away.
(b) What is this ability called?
(c) How may this ability of the eye alter as a person grows older?

9 (a) Name the light-sensitive screen of the eye.
(b) Describe how the eye distinguishes
(i) different intensities of light
(ii) different colours
(c) Explain why it is difficult for us to detect different colours when there is very little light.

18 The Nervous System

18.1 Communication and co-ordination

Fig. 18.1 illustrates the interior of a modern railway signal-box. The signalmen control the movement of trains on all the track served by this box. To be in control the signalmen need at least two paths of communication: they need to monitor what is happening on the tracks, and they need to be able to send out instructions to the train-drivers to set the trains in motion or to stop them. Communication between track and signal-box is achieved by means of electricity. A train on the track triggers an electric current, which flows along a wire and enters the signal-box. This incoming current causes a light to flash on a visual panel and indicates the position of the train. In response the signalman can operate any one of a number of electrical switches. These control the electricity flowing out to signals by the side of the track. Electricity flows through wires at a speed of approximately 300 000 km per second. In this way rapid communication and co-ordination can be developed within a complex system.

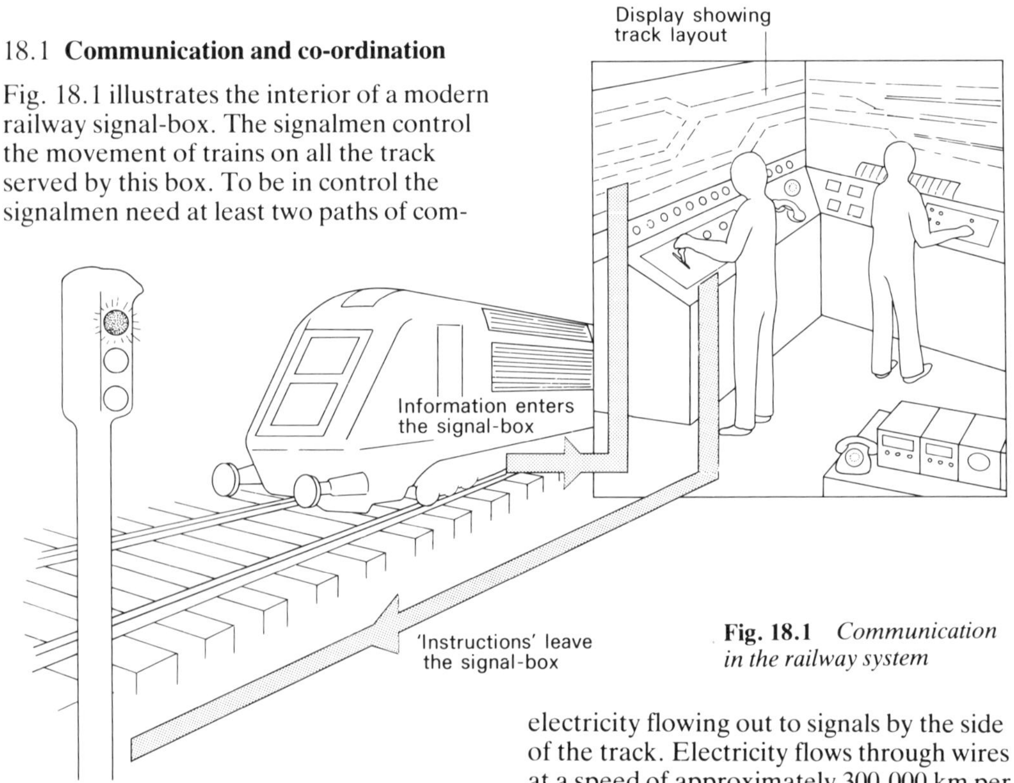

Fig. 18.1 *Communication in the railway system*

The human body needs a communications system that can work just as quickly and which can bring about *co-ordination*. Imagine you are climbing up a steep hill, when suddenly your feet start to slip. In a faction of a second your hands grab at the branches of a shrub, while your feet move rapidly to try and find their grip once more. Survival is dependent upon two things: the body receiving information telling it that it is in danger, and the ability of the body to respond to protect itself. The *nervous system* is responsible for monitoring changes and for organizing the response of the body to ensure its survival.

18.2 The monitoring of internal and external changes

In the previous chapters we discussed how the sense organs of the skin, eyes, ears, tongue and nose monitor changes in the body's external environment. *Sensory nerve endings* are the basic components in each of these sense organs. Throughout the internal structure of the body there are other sensory nerve endings. These are responsible for monitoring changes in tissues and organs. For example, in muscles there are small spindle-shaped structures (Fig. 18.2). These act like spring balances, recording the tension of the muscle fibres. When the

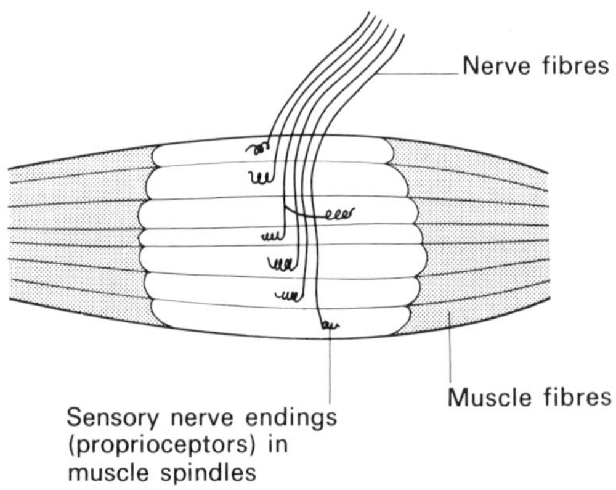

Fig. 18.2 *Sense organs in the muscle fibres (simplified diagram)*

muscle fibres are stretched these *muscle spindles* undergo a similar change. Sensory nerve endings (*proprioceptors*) embedded in the spindles keep the brain informed about the changes in tension. In this way the body is kept informed of the tension of muscles and the relative position of bones.

Question 1

(a) You are sitting in a chair, looking straight ahead. Do you need to look down to see whether your legs are crossed or laid out straight in front of you?

(b) With which sense organs must our muscular sense co-operate in order to help us to keep in an upright posture?

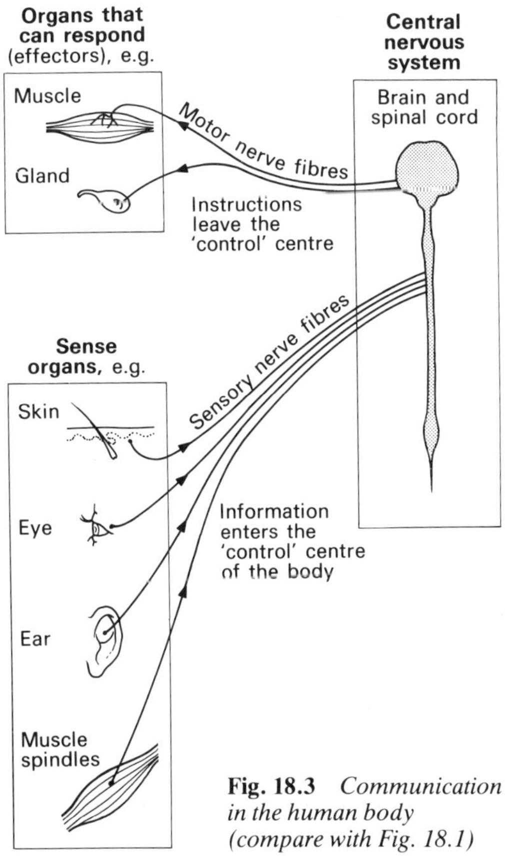

Fig. 18.3 *Communication in the human body (compare with Fig. 18.1)*

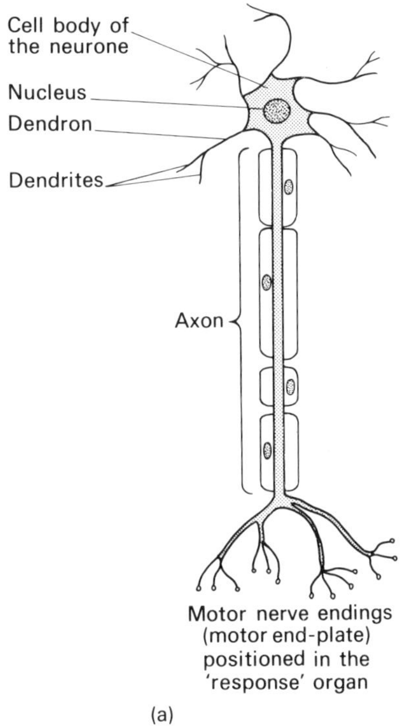

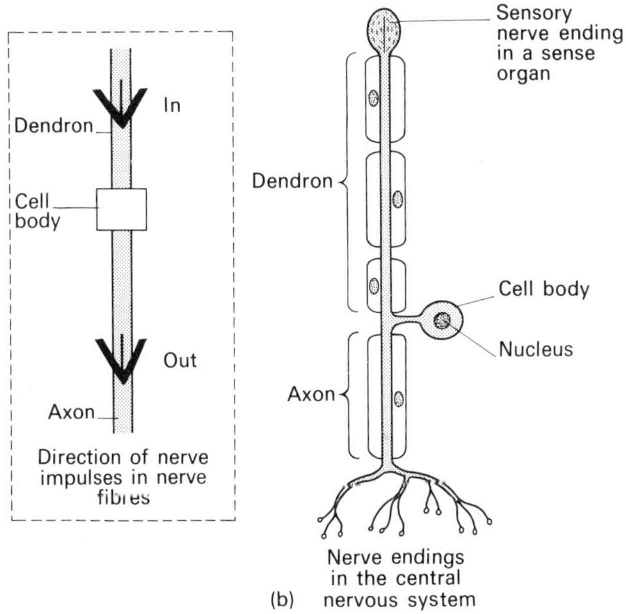

Fig. 18.4 *Types of nerve-cell (neurone) that are found in the body.*
(a) Motor neurone. (b) Sensory neurone

In Section 18.1 we discussed how information from the railway track is 'fed' into the signal-box. Similarly, in the human body the sensory nerve endings send information to a control centre. This is the *central nervous system*, made up of the *brain* and *spinal cord*. The 'wires' carrying the information INTO the central nervous system are *sensory nerve fibres*. Instructions from the central nervous system are sent OUT along *motor nerve fibres* to the organs of response, for example muscles or glands (Fig. 18.3). Both types of nerve fibre are fibres of *nerve-cells*.

18.3 Nervous tissue and nerve impulses

Nerve-cells or *neurones* make up the tissues of the nervous system. Fig. 18.4(a) and (b) shows their general structure. The region known as the *cell body* contains a nucleus, and is usually located in the brain or spinal cord. Several nerve fibres project from the surface of each cell body. Information is transmitted along these fibres as a *nerve impulse*. A *dendron* is a nerve fibre that carries nerve impulses towards the cell body. An *axon* is a nerve fibre that carries nerve impulses away from the cell body. Therefore nerve fibres can transmit nerve impulses in

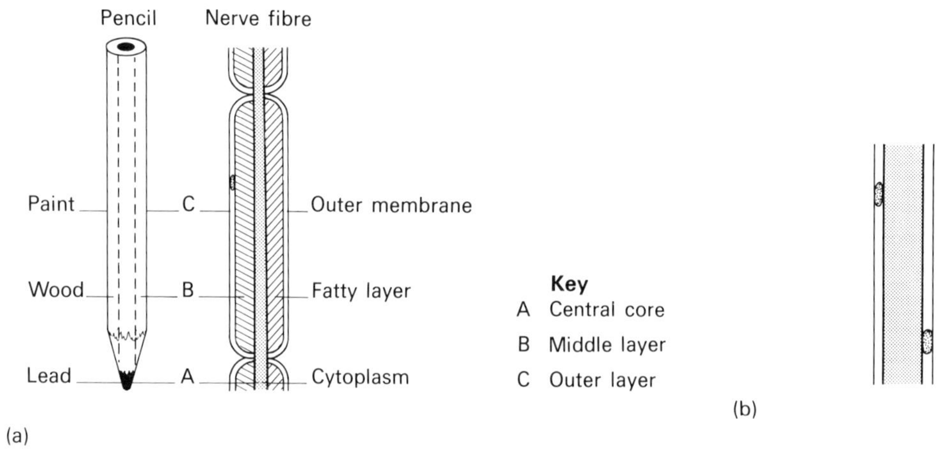

Fig. 18.5 *(a) Internal structure of a medullated nerve fibre. (b) Internal structure of a non-medullated nerve fibre*

only one direction. If you think of a pencil you can easily understand the internal structure of a nerve fibre (Fig. 18.5(a)). There is a central core of cytoplasm which is surrounded by a middle fatty layer that contains *myelin*. Around these two layers is a delicate outer membrane. Not all nerve fibres have a middle layer (Fig. 18.5(b)).

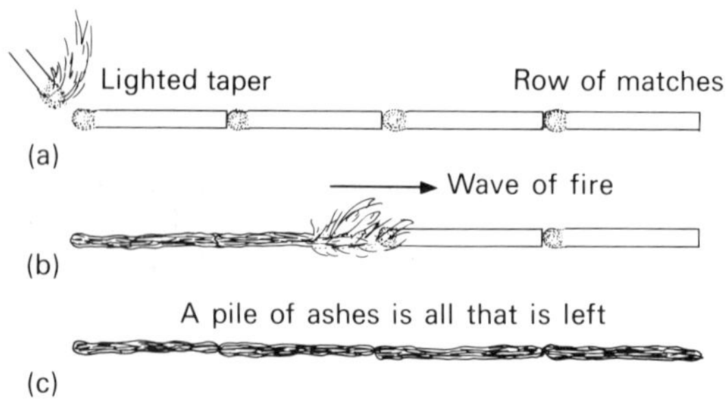

Fig. 18.6 *A wave of fire and a nerve impulse*

18.4 Transmission of an electrical impulse in a nerve fibre

Fig 18.6(a) shows a row of matches. Imagine that a light is applied to the match at one end of the row. Gradually a wave of flame progresses down the row of matches as shown in Fig. 18.6(b). Each match undergoes the same chemical change as the one in front, so the flame is a wave of chemical change. A *nerve impulse* is a wave of *electrical change* that progresses along the central core of a nerve fibre. Therefore we can detect nerve impulses or nervous activity in nerve-cells by detecting the electrical activity. As the flame progresses along the row of matches it leaves behind a pile of ashes (Fig. 18.6(c)). After burning a match cannot restore itself. However, a nerve fibre is not burnt out in this way. It quickly restores itself after every nerve impulse. A nerve impulse can progress along a fibre at speeds up to 160 m/s. There is no change in the strength or intensity of the impulse throughout the length of the fibre.

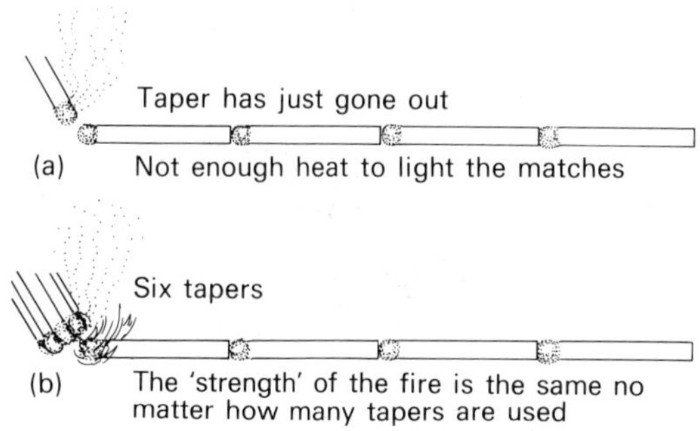

Fig. 18.7 *The all-or-none principle applied to matches*

18.5 The all-or-none principle

Think again about the row of matches in Fig. 18.6. If we fail to apply sufficient heat the first match will not ignite and there will be no wave of flame (Fig. 18.7(a)). Also, if we use a bunch of six tapers to light the first match we get a flame of the same strength as if we had lit it with one taper (Fig. 18.7(b)). The same principle can be applied to a nerve fibre: it is

Fig. 18.8 *All nerve impulses are of the same strength (intensity).*
(a) Weak (small) stimulus, e.g. small amount of light.
(b) Strong (large) stimulus, e.g. more light

called the *all-or-none principle*. A change has to be of the right intensity to create a nerve impulse in a nerve fibre, i.e it must be sufficiently strong to be above the *threshold of stimulation*. Such changes are termed *stimuli*.

Once it is above the threshold of stimulation a strong impulse produces exactly the same size of nerve impulse as a small stimulus. This can be shown by subjecting a single nerve fibre to stimuli of different strengths, and then measuring the level of electrical activity that is produced. The results may be presented visually on an oscilloscope screen or in the form of a graph as shown in Fig. 18.8(a) and (b). The graphs show clearly that there is no change in the size of the nervous impulse. However there is an important difference.

Question 2 How do the spaces between the peaks in Fig 18.8(a) differ from those in Fig. 18.8(b)?

In the previous chapter we discussed what was meant by the frequency of a sound – the number of vibrations per second. In Fig. 18.8(b) there are more nerve impulses per second than in Fig. 18.8(a). This shows that a strong stimulus produces nerve impulses of a greater frequency than a weak stimulus. However, there is an upper limit, the *optimum intensity of stimulation*. Once this level of intensity is reached no further increase in frequency is possible.

Fig. 18.9 *A multicore electric cable – a bundle of many conducting wires*

18.6 The connection of nerves to the central nervous system

The signal-box of the body is the central nervous system, made up of the brain and spinal cord. Nerve impulses are brought into the central nervous system through sensory nerve fibres, and are sent out through motor nerve fibres. In all of these nerve fibres the nature of the nerve impulse is the same, but the frequency varies. The organ that receives the nerve impulses must interpret the message from their frequency – a form of code.

If we look at a mains electrical cable we find that it contains many individual wires. Each wire is insulated from the rest by a plastic coating (Fig. 18.9). A *nerve* is like a multicore electrical cable. It is a bundle of nerve fibres held together by connective tissue and supplied by blood-vessels. Each fibre is insulated from the others by its fatty myelin sheath (Fig. 18.10). A nerve may contain fibres that are all motor fibres, fibres that are all sensory fibres, or a mixture of the

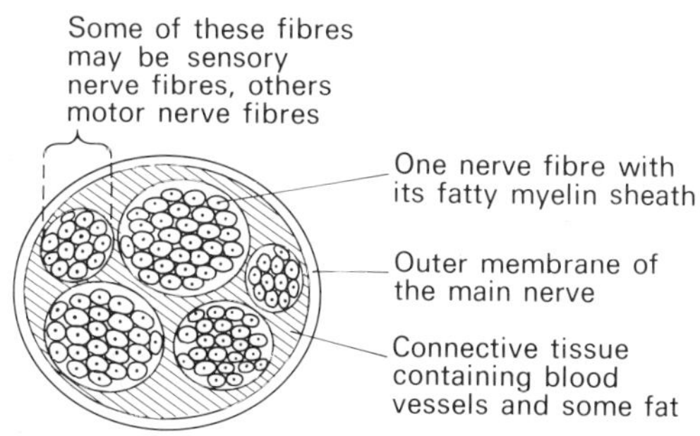

Fig. 18.10 *A nerve is a bundle of many conducting nerve fibres*

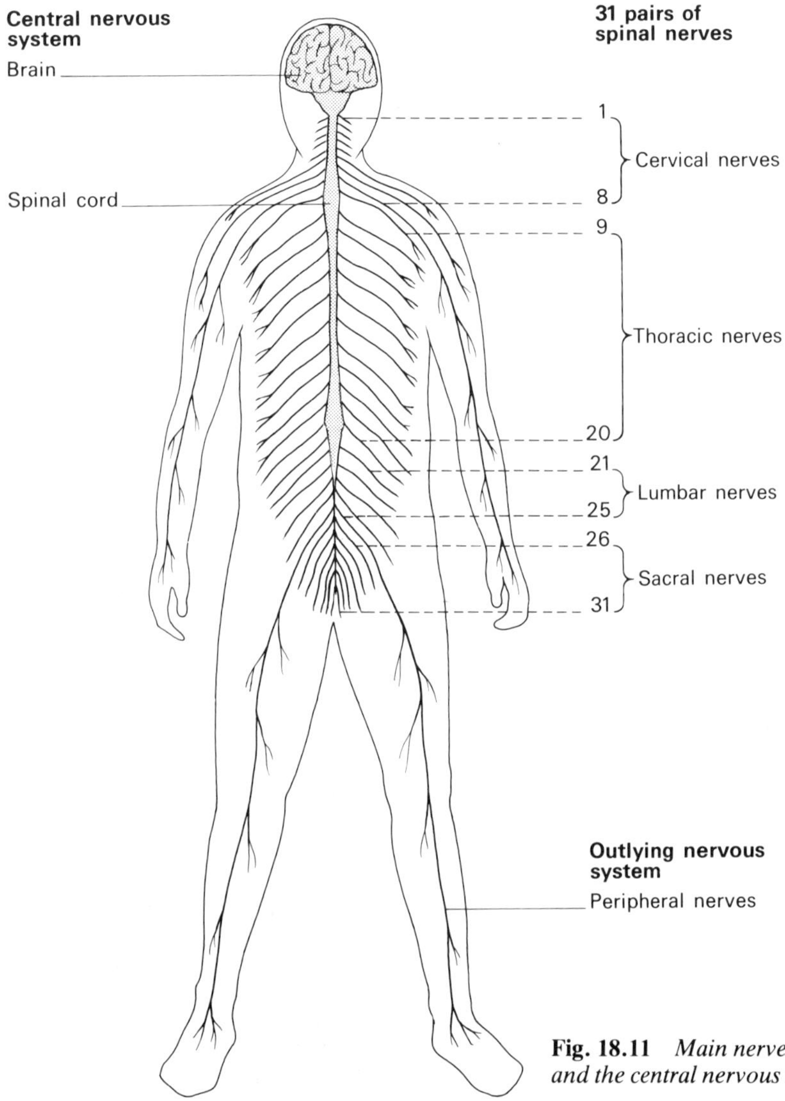

Fig. 18.11 *Main nerves of the body and the central nervous system*

two. Fig. 18.11 shows some of the main nerves of the body and the position of the central nervous system.

There are twelve pairs of nerves that are connected to the brain. These are *cranial nerves*.

Connected to the spinal cord are thirty-one pairs of *spinal nerves*. These are mixed nerves which form a double connection with the spinal cord. As Fig. 18.12 shows, the *ventral root* contains only *motor nerve fibres*, while the *dorsal root* contains only *sensory nerve fibres*. Thus the spinal cord has entrances for incoming impulses and exits for outgoing impulses. Nerve impulses travelling along sensory fibres in a dorsal root have to flow through the cell bodies of the nerve fibres before they reach the tissue of the spinal cord. The mass of cell bodies at this point forms a swelling on the dorsal root. On a nerve, this swelling is called a *ganglion*.

Spinal nerves from the cervical, lumbar and sacral regions of the spinal cord group together to form four knots of nerves. Each knot is called a plexus. Nerves are re-arranged within each *plexus* before they spread out to form a network of *peripheral nerves* that serve all regions of the body.

The distance covered by a nerve fibre varies considerably. For instance, the nerve which is found in an adult's big toe contains fibres which have their cell bodies in the lower part of the spinal cord, approximately 90 cm away! If you bang the point known as your 'funny-bone' on your elbow you stimulate the ulnar nerve. This nerve also serves the little finger of your hand, hence the tingling sensation!

18.7 **The protection of the central nervous system**

The brain is protected by the skull (Colour plate 11), and the spinal cord by the vertebral column. Nerves have to reach the brain through holes in the skull; they reach the spinal cord through the spaces between the vertebrae.

Question 3 How can a slipped disc interfere with a nerve root of the spinal cord?

The central nervous system is also surrounded by three protective membranes called *meninges* (Fig. 18.13). The innermost meninx is very delicate and is not easily distinguished from the surface of the brain or spinal cord. It contains many small blood-vessels which nourish the central nervous system.

You may have heard of a disease called *meningitis*. This is inflammation of the meninges caused by bacteria or virus infection. In mild cases it may produce symptoms of fever, headache and vomiting. When severe it can produce unconsciousness and

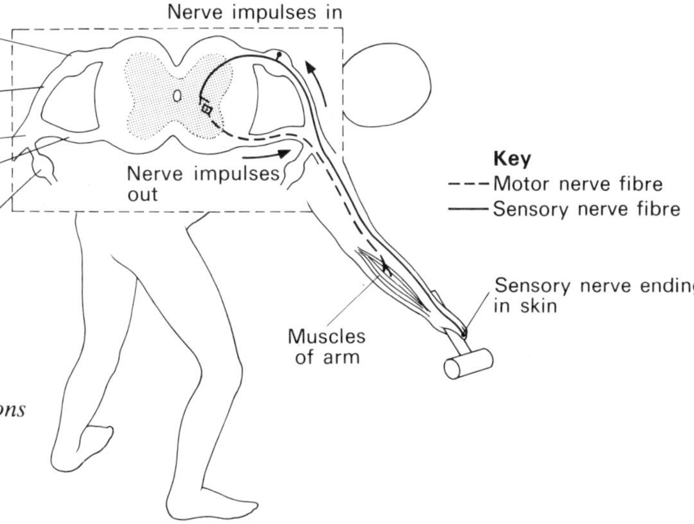

Fig. 18.12 *The connections of the spinal nerves with the spinal cord*

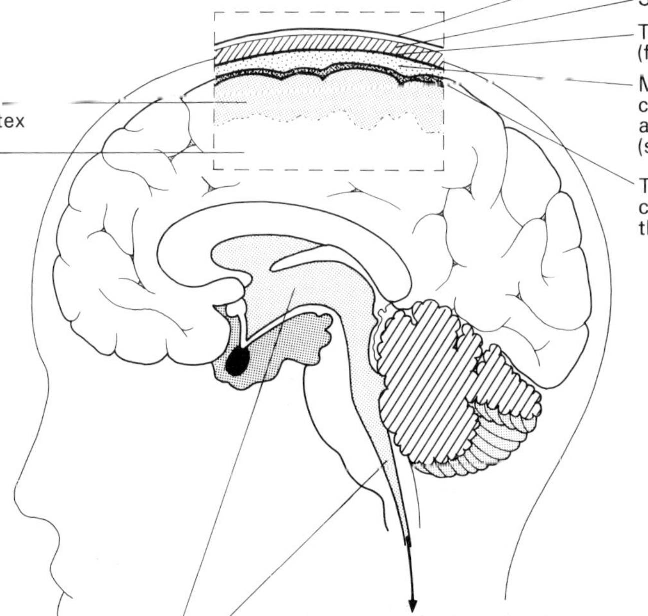

Fig. 18.13 *The three meninges and brain tissue*

convulsions. Meningitis can develop as a complication of other infections, such as poliomyelitis, mumps or even a serious middle-ear infection which, if left untreated, can spread through the bones of the skull. To diagnose meningitis a *lumbar puncture* may be carried out, which involves taking a sample of *cerebro-spinal fluid*. This fluid is present between the tough membrane nearest the skull and the middle membrane. It is a straw-coloured liquid very similar to lymph. It protects the nervous tissue against knocks as well as supplying nourishment and removing waste. To achieve this it circulates through cavities in and around the brain and spinal cord under pressure. To obtain a sample of the fluid a long needle is inserted through the skin of the lower back so that it enters the vertebral column. There is no danger of injuring the spinal cord itself because it ends at approximately the level of the diaphragm; in fact the spinal cord is only about 45 cm long. The spinal fluid is then examined for evidence of microbe infection.

18.8 The internal structure of the central nervous system

The tissue of the central nervous system is made up mainly of masses of cell bodies with many branches, seen as *grey matter*, and masses of nerve fibres, seen as *white matter*. The brain's outer layer, the *cortex*, is made up of grey matter (Fig. 18.13). Its depth varies from 1.5 mm to 3 mm and it lies in numerous folds, giving the crinkled appearance of a screwed-up piece of paper. Our brain is fully developed in size by the time we are two years old. At this stage it contains all the nerve-cells we need for future life. The cortex alone contains about 10 000 million nerve-cells. These cells cannot be rebuilt or regenerated by the body. Although some die off every day we have a sufficient supply for a lifetime. Brain cells die in a matter of minutes if their blood supply is cut off (page 76). So in emergencies it is essential to maintain a flow of blood to the brain if the

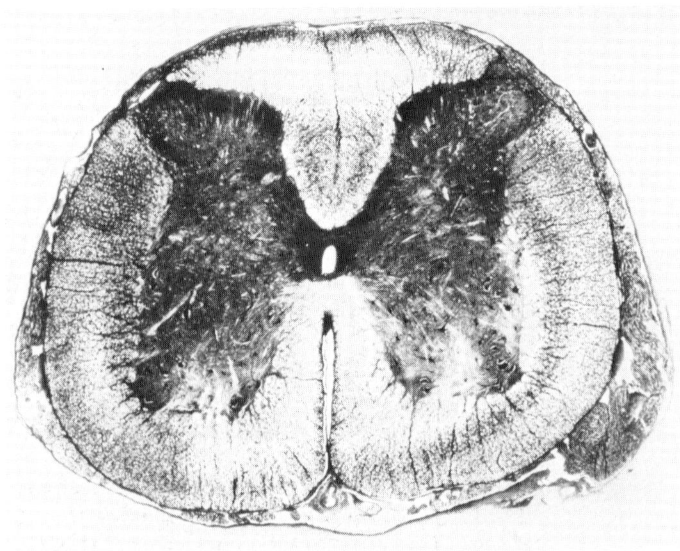

Fig. 18.14 *Transverse section through the spinal cord*

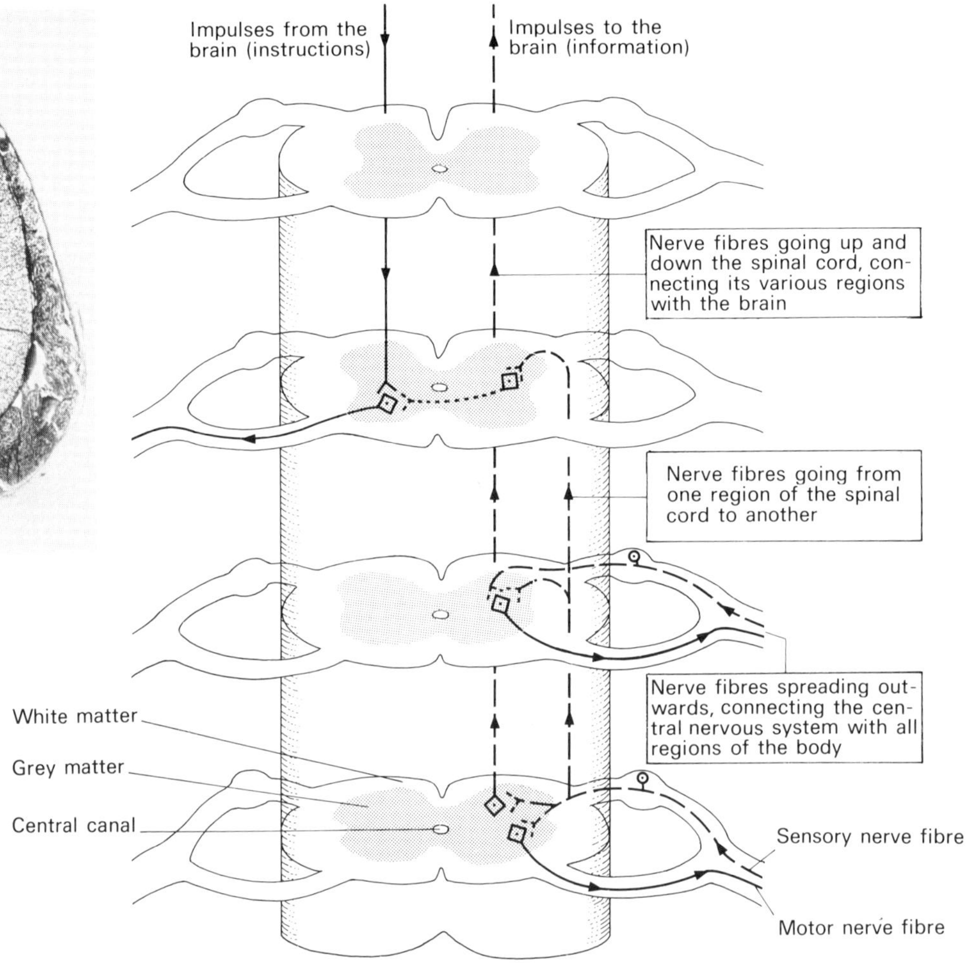

Fig. 18.15 (Right) *Nerve-cells and nerve fibres in the spinal cord. Arrows show direction of nerve impulses*

person is to live and not suffer brain damage.

Beneath the cortex there is a thicker layer of white matter. This surrounds many cavities within the brain, all of which contain cerebro-spinal fluid. The lowest cavity connects with the *central canal* of the spinal cord. You can see this central canal in Figs 18.14 and 18.15. You may also notice something different about the position of the layers of nervous tissue within the spinal cord compared with those of the brain. The grey matter, an H-shaped region, is contained within the white matter. The grey matter contains the spinal cord's own nerve-cells. The white matter contains nerve fibres which (a) travel outwards to form the spinal nerve fibres, or (b) travel along the length of the spinal cord (Fig. 18.15). The 'up-and-down' fibres transmit impulses from one region of the spinal cord to another. They also transmit impulses to and from the brain. The arrangement of these fibres is not haphazard. They are all grouped within the spinal cord according to the region of the body that they serve.

18.9 **Infections of the central nervous system**

Although the central nervous system is reasonably well protected against damage from external blows, it can still be damaged through infection. Inflammation of the brain tissue is known as *encephalitis*. It can be caused by a virus and sometimes occurs in epidemics. Another virus which invades nervous tissue causes *rabies* (page 99). The rabies virus is introduced through a break in the skin from the source of the infection, such as the saliva of a rabid dog. It remains near the point of its introduction for about two weeks and gradually multiplies within muscular tissue. The virus then invades the peripheral nerves and works its way along them to the central nervous system. After three to eight weeks the painful symptoms develop. There is a watery discharge from the nose, restlessness and painful spasms or convulsions. The person expresses fear at the sight or sound of water (*hydrophobia*) and dreads the onset of the next convulsion. Death usually occurs through heart failure or asphyxia (lack of oxygen in the blood).

The virus of *poliomyelitis* (page 96) reaches the central nervous system via the bloodstream. Within the central nervous system it multiplies and usually destroys the cell bodies of motor nerve fibres. The degree of paralysis caused by the loss of these motor nerve fibres depend upon the intensity of the infection (Fig. 18.16). In severe cases even respiration is affected.

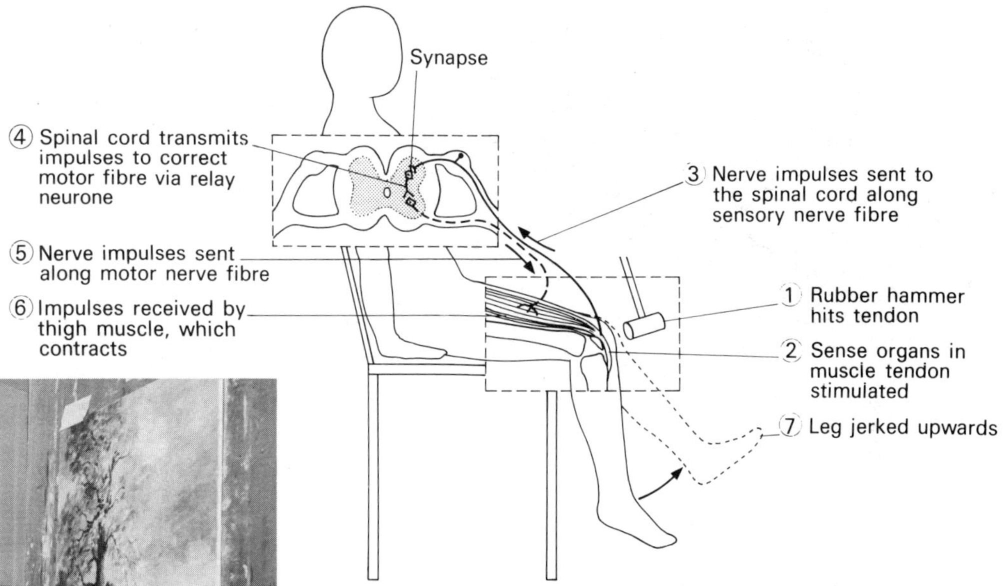

Fig. 18.17 *The simplest reflex action – the knee-jerk*

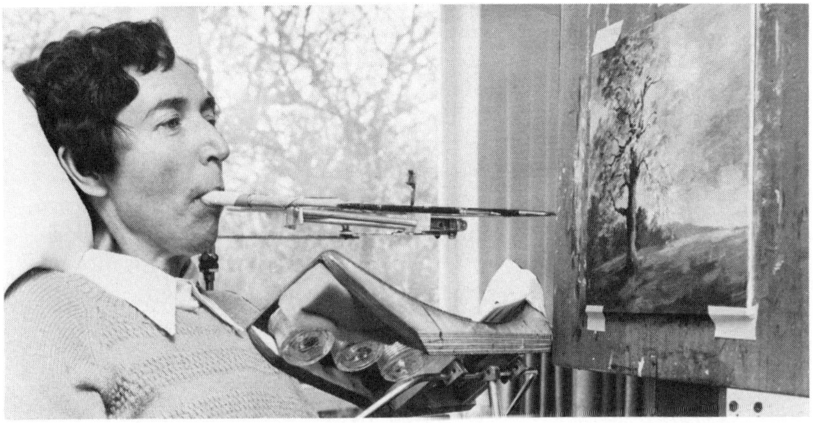

Fig. 18.16 *Elizabeth Twistington-Higgins was a ballet dancer, trained at Sadler's Wells. She became completely paralysed (apart from a slight movement of the head) by polio in 1953. She has been awarded the M.B.E. for her paintings, which she does by mouth.*

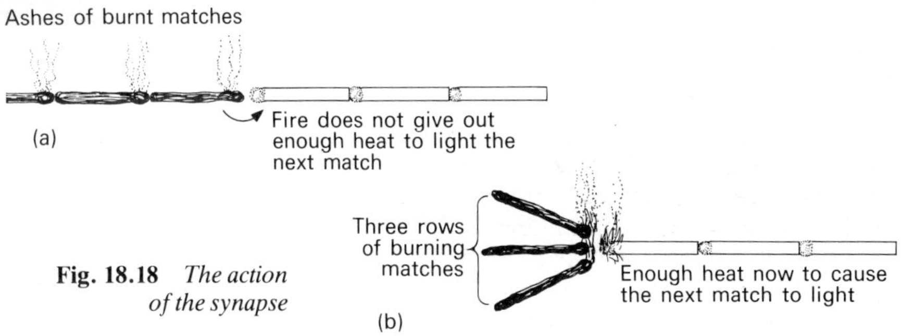

Fig. 18.18 *The action of the synapse*

18.10 The nervous system in action and reflex actions

Our survival is often dependent upon how quickly we respond to stimuli. The nervous system can easily cope with this need for speed in communication and co-ordination. Our bodies' most rapid response is a *reflex action*.

True reflex actions happen automatically. We are not conscious of thinking about what our response should be, nor can we control it. The simplest reflex action of our body is the knee-jerk (Fig. 18.17). It is the simplest because the sensory impulses enter a particular section of the spinal cord and the motor impulses are sent out from the same section. You can see in Fig. 18.17 that the nerve fibres that make up the pathway do not make physical contact with one another. There is a small gap between them called a *synapse*.

Action of the synapse

To see how the synapse works, imagine our row of matches once more, but this time with a gap half-way along (Fig. 18.18). If the flame is not hot enough it cannot cross the gap (Fig. 18.18(a)). However, if we increase the heat of the flame by adding several extensions (Fig. 18.18(b)), it is able to cross over. The synapse acts like this fire-break: the nerve impulse will not be transmitted across the synapse unless it is of a high enough frequency, or several impulses reach the synapse at the same time. Thus the all-or-none principle applies again. Actually the electrical charge does not cross the gap itself like a spark, but causes the end of the fibre to secrete a hormone. This hormone stimulates and thereby creates a nerve impulse in the nearby dendrite or cell body.

How the response is made

The nerve impulses leave the spinal cord in a motor fibre. This ends in the organ of response as a *motor end-plate* consisting of numerous branches. For example, in a muscle each branch of the motor end-plate serves an individual muscle fibre (Figs 18.4(a) and 18.19(a)). The all-or-none principle can be applied again to the way that the nerve ending stimulates the muscle fibre into action. The motor end-plate can be thought of as a detonator placed on a bundle of gelignite (Fig. 18.19(b)). The gelignite contains stored chemical energy like the muscle. The gelignite will explode, releasing its energy, when it receives a large enough impulse from the detonator. Similarly, a muscle fibre will contract when it receives a nerve impulse of the correct frequency (Fig. 18.19(c)). Once again the impulse is conveyed into the muscle fibre by the secretion of a hormone from the nerve ending. One bundle of gelignite will produce a certain number of joules of energy. It will produce the same quantity of energy whether a small detonator or a large detonator is used to fire it.

Question 4 What could be done to make the explosion larger, i.e. increase the energy released?

A muscle produces a stronger response when a greater number of its muscle fibres are 'fired' by nerve impulses.

Reaction time

The time it takes for the reflex action to occur after the stimulus is known as the body's *reaction time*. On page 234 there are some suggestions on how to measure a person's reaction time. On average you will find that this is somewhere between 0.15 and 0.18 seconds. The speed at which we react can make a lot of difference to our chances of survival.

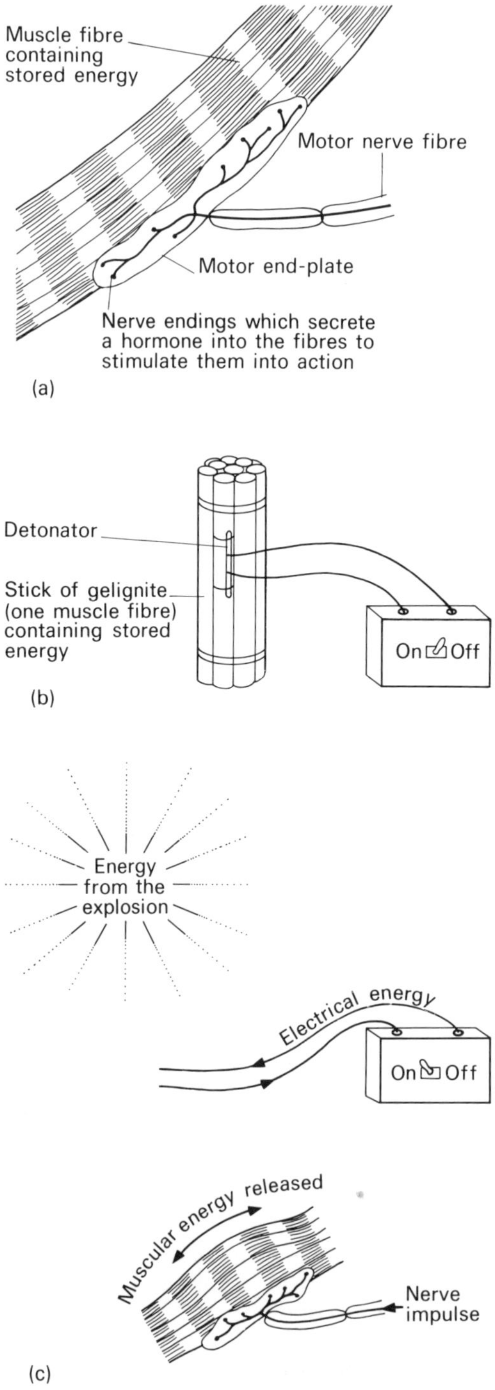

Fig. 18.19 *Action of the motor end-plate*

18.11 Why reflex actions are not as simple as they seem

The passage of the nerve impulse around the nerve fibres that make up the reflex pathway is very quick. However, even the simplest reflex arc is not just a single pathway made up of sense organ, sensory neurone, relay neurone, motor neurone and response organ (effector). There are other interconnecting pathways. For example, in the knee-jerk reflex, the lower part of the leg is moved upwards by the contraction of the front thigh muscle. But as we have seen in Chapter 3 muscles work in pairs. For the front thigh muscle to move the lower bone, the back thigh muscle must relax.

We have already seen in Fig. 18.15 that nerve impulses can travel along several pathways in the spinal cord. In the case of a knee-jerk the sensory impulse influences the motor neurone of the back thigh muscle as well as the motor neurone of the front thigh muscle (Fig. 18.20). In fact the sensory impulse suppresses the other impulses received by this motor neurone. Thus the back thigh muscle loses its tone a little, and its relaxation allows the front muscle to contract, jerking the leg upwards. A nerve impulse which suppresses the effect of another nerve impulse is said to *inhibit* the response. For example, some reflex actions can be suppressed or *inhibited* by the conscious control of the brain. If you pick up a very hot object your reflex action makes you drop it. However, if the hot object is a precious piece of the best dinner-service you consciously make yourself hold on to it against every impulse to let go!

Question 5 What reflex actions take place in the following instances?

(a) A piece of food goes down the 'wrong way'.
(b) You enter a darkened room.
(c) A finger tickles the side of your foot.

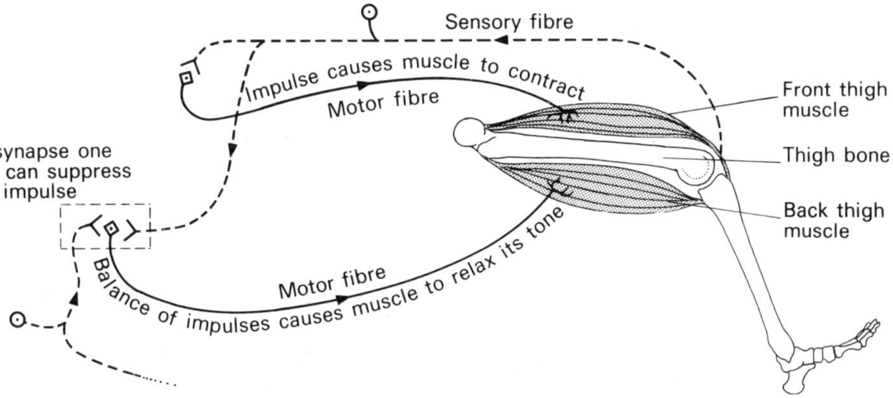

Fig. 18.20 *How nerve impulses affect one another at a synapse*

Mid-brain
Beneath the cerebrum is the *mid-brain*, which consists mostly of fibres connecting the various regions of the brain. In front and tucked under the 'peak' of the cerebrum is the *pituitary gland* (page 178). Above the pituitary gland is the *hypothalamus* (page 183).

Pons and cerebellum
The *pons* lies beneath the mid-brain and is made up of both grey and white matter. It forms a 'bridge' for nerve impulses between the two lobes of the cerebrum. The *cerebellum* lies beneath the rear portion of the cerebrum and has the same external appearance.

The cerebellum is mostly responsible for co-ordinating movement and muscular tone. It also helps us to keep our balance, because it receives impulses from the semicircular canals of the ear and receptors in the muscles. All of these activities are involuntary and not under our conscious control.

18.12 Regions of the brain

Cerebrum
The largest region of the human brain is the *fore-brain* or *cerebrum* (Fig. 18.21). This consists of two large *cerebral hemispheres* which are separated by a cleft on the top surface, but joined by a bridge below this. The cerebrum may be regarded as the high centre of the brain, responsible for conscious thought, interpretation, memory, behaviour and emotions.

The structural development of the cerebrum can be seen if we observe a range of life forms from the most simple to the most complex. Running parallel with this is the development of the organism's *conscious* or *voluntary* action as opposed to *automatic* or *involuntary* action.

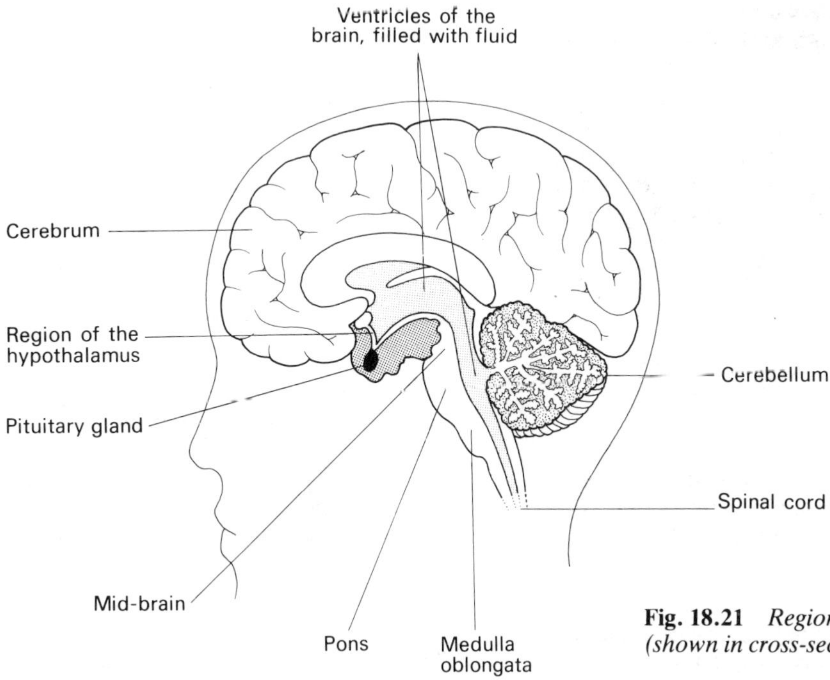

Fig. 18.21 *Regions of the brain (shown in cross-section)*

Medulla

The region of the brain that connects with the spinal cord is the *medulla oblongata*. This region is about 2.5 cm long and where it leaves the base of the skull it forms the spinal cord. It contains many nerve fibres going to and from the brain. In fact, sensory and motor nerve fibres cross over from left to right in the medulla. This explains why damage to the left cerebral hemisphere may cause paralysis in the limbs of the right-hand side of the body. An example of this is seen in a person who suffers a *stroke* caused by a brain haemorrhage. In such cases the brain can no longer communicate with the muscles that it used to control, so it cannot instruct on movement. Fig. 18.22 shows a person who is handicapped in this way and who has sought to maintain his independence. The article highlights the need for people with any handicap to live as fully as they can and shows how others can help them.

Within the medulla there are *vital centres* which control automatic and reflex actions. These deal mainly with the following:

(a) the rate and depth of breathing
(b) the dilation or constriction of blood-vessels
(c) heartbeat
(d) stomach movements, including vomiting
(e) secretion of saliva and gastric juice

The medulla can be part of a reflex action; it may also act on instructions from other regions of the brain, such as the cerebrum or hypothalamus. This happens particularly in times of stress, fear or emotion. Have you ever noticed how the skin of your face pales when you are frightened or angry? The emotion you feel in turn stimulates the vital centres of the medulla. These centres send out impulses causing the heart to beat more quickly and the walls of the capillaries to constrict so that less blood reaches the face.

The central nervous system works closely with the endocrine system. For example, in times of fear the nerve impulses sent through the automatic system stimulate the production of adrenaline (page 179).

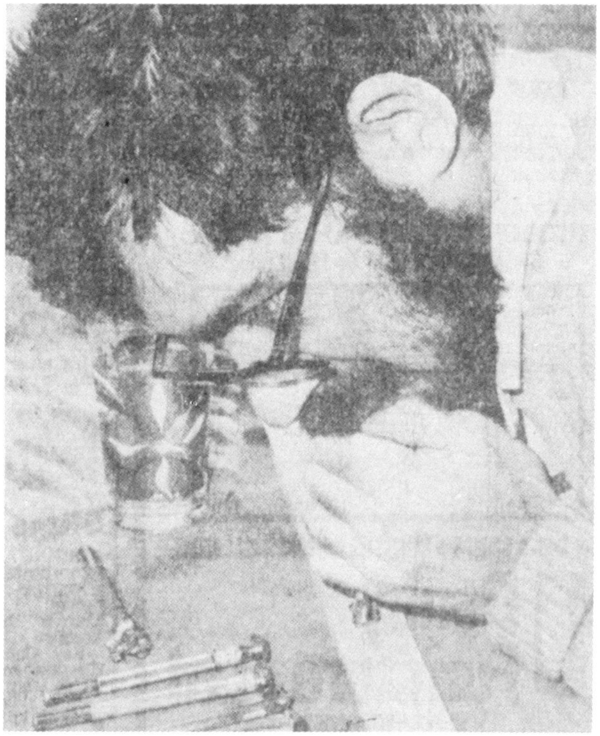

INDEPENDENCE is what most teenagers take for granted. At 18, Martin had as much as most. There were expectations of running the family business. He was full of the perpetual motion of being young and fit and his passion was horse riding. One day, in traffic, he was thrown and badly injured. His life almost stopped. When it started again, it was all in slow motion. After the accident came two strokes and paralysis. He went into residential care.

But his independent spirit survived. It has taken him several years to make it back to a sort of freedom. Now at 30, he has his own flat. One arm still hangs helpless down the paralysed side and every twisted step is an effort of will. But he manages, living alone except for his puppy.

His speech muscles still won't obey his brain. He can muster just a few well-rehearsed phrases and for the rest he uses his best hand to write a slow-motion word or two.

Marvellously, he manages to tool leather, producing attractive belts, purses and key rings with his crippled hands.

A friend

Valerie, who has become a friend, has been one of the mainstays of encouragement and practical help in giving Martin independence again. He is one of more than a score of young handicapped people Valerie is helping as organiser of Outreach, a Jewish organization set up to reach out to young handicapped people of any persuasion who need aid.

'We didn't even know if there was a need to reach out in the first place, but we wanted to offer practical help to young people who were handicapped physically, mentally or emotionally, and also to give support to their families,' said Valerie.

Once Valerie has discovered a need and made contact she has a small band of volunteer helpers she can call on. 'They are just ordinary people willing to give time and friendship, maybe taking a youngster out for an evening, delivering and collecting them from a disco or possibly getting involved in a hobby together.

'It often brings a streak of independence into the lives of the young handicapped, giving them a chance to make their own friends.

'However good-hearted a family may be, some independence is a vital ingredient in a handicapped life that often replaces frustration and depression.

There are others around Greater Manchester who could also be helped with social, educational, work and family matters . . . youngsters aged between 10 and 30 with a variety of handicaps. Some are spastic, epileptic, blind, deaf, emotionally disturbed or educationally sub-normal.

Confidence

'Because they are cut off they don't always get the chance to develop social and conversational skills. When we have taken them to youth clubs, discos and various organisations, they soon develop confidence,' said Valerie.

Fig. 18.22 *Life in slow motion*

TABLE 18.1 THE AUTONOMIC NERVOUS SYSTEM

Parasympathetic	Sympathetic
This consists of: (a) vital centres in the medulla region (b) end of the spinal cord	This consists of two chains of ganglia, one either side of the spinal cord
Function of nerve impulses from this system: Lower the body's state of activity 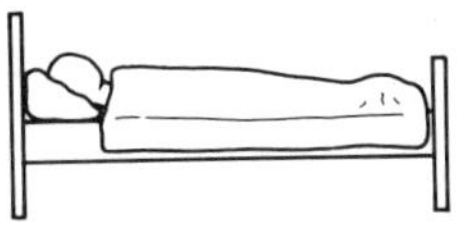 Examples of action: Heart rate slowed down Breathing rate slowed down Digestive juices secreted and digestive processes stimulated Blood flow increased to the digestive system rather than the muscles Stimulates the liver to store glycogen	Function of nerve impulses from this system: Prepare the body for action  Examples of action: Heart rate accelerated Breathing rate increased Peristalsis and digestive processes slowed down Blood flow increased to the muscles and withdrawn from the digestive system Increased blood flow to the brain Stimulates liver to release more blood-sugar Pupils of eyes dilated Adrenaline released from adrenal glands

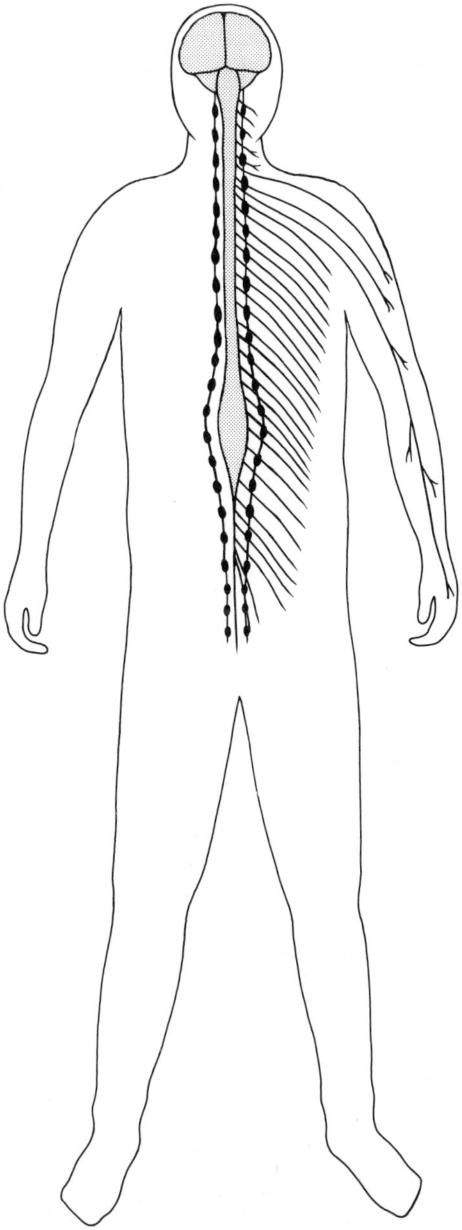

In the diagram (spinal nerves are shown on only one side of the spinal cord)

18.13 Autonomic nervous system

As we saw in the previous section there are vital centres in the medulla and hypothalamus that control the automatic processes of the body, such as heartbeat or peristalsis. If this were not so we would have to sit around all day consciously working out every action of our bodies' systems! This automatic control is exercised through a separate part of the nervous system known as the *autonomic* nervous system (Table 18.1). It consists of two chains of ganglia situated either side of the spinal cord, the automatic centres in the brain and the end of the spinal cord. Each autonomic ganglion has a smaller nerve connecting it with the spinal nerve (Fig. 18.11). Other nerves from the autonomic ganglia spread out to various organs of the body, generally to areas of involuntary or unstriped muscle (page 35).

The nerves of the autonomic system can be put into two groups which usually have opposite effects upon an organ. The group of fibres that forms the *sympathetic section* of the autonomic system prepare the body physically for action. The opposite group of fibres, which forms the *parasympathetic section*, prepare the body for inactivity and sleep. The activity of an organ is determined by the signals that it receives from these two systems and the balance between them. Naturally the autonomic system works closely with the central nervous system and endocrine system (page 177). For example the region of the brain known as the hypothalamus (page 183) exercises some control over heartbeat through the autonomic nervous system. Also the conscious regions of the brain can in some cases exercise control over this autonomic system, for example when we consciously control our breathing in singing or playing an instrument.

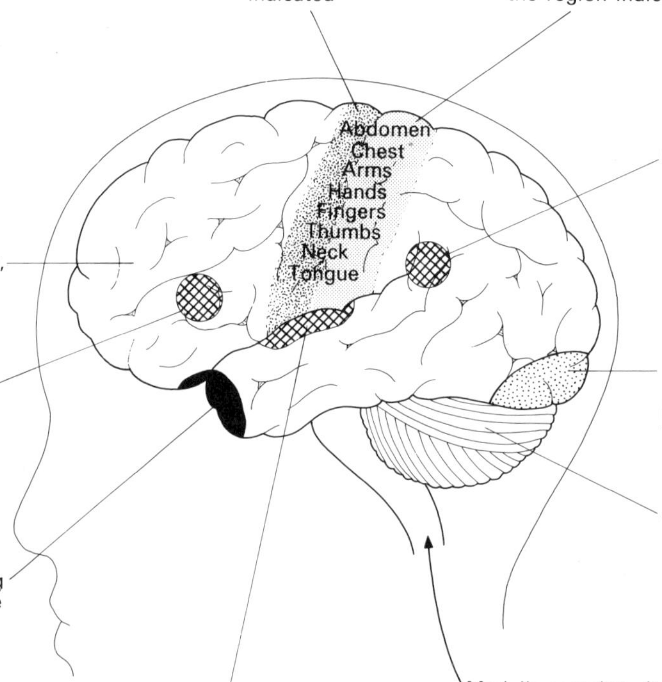

Fig. 18.23 *Mapping of the cerebrum*

18.14 The function of the cerebrum

Palmists claim that from studying the palm of a hand they can tell a person's fortune. In the nineteenth century phrenologists believed that the brain consisted of several 'compartments'. Each compartment was supposed to be responsible for a particular characteristic of the individual, for example firmness or cautiousness. These compartments gave rise to 'bumps' in the skull. The skilled hands of the phrenologist were supposed to be able to read these bumps to predict character, health and personality. We now know there are no such compartments, yet in a way the phrenologists were near the truth, for different regions of the brain do have different functions. However they deal with different sensory and motor functions, not different qualities of character (Fig. 18.23). The first ideas on this type of organization within the brain were developed during the last half of the nineteenth century. Observations were carried out on people suffering from epileptic fits. Their first sensations at the onset of the fit were noted. These were related to any known damage of the brain. Advances in medical techniques have allowed mapping of the brain to be carried out. Needles carrying small electric currents have been used to stimulate various areas of the brain in a conscious person. Notes were made of the part of the body where the person felt the sensation.

Nowadays it is possible to record the electrical activity of brain cells. The electrical impulses are picked up by special electrodes placed on the head (Fig. 18.24(a)). The electrodes feed the impulses into a machine which amplifies them and records them as a pen trace on paper (Fig. 18.24(b)). The recording is known as an E.E.G. or *electro-encephalograph*. The complex wave-forms on an E.E.G. have to be interpreted by an expert. An E.E.G. can be used, together with clinical evidence, in the diagnosis of an illness that involves abnormal electrical activity within the brain cells. For instance, the symptoms of *epilepsy* are caused by what might be termed an 'electrical storm' in the brain. If we understand this we can assist a person during an attack, as suggested in Fig. 18.25. At the same time we can recognize their need to live a normal independent life.

Further studies of the brain's sensory area reveal that legs, arms, trunk, etc., are 'mapped' in positions that resemble their relative positions within the body. However the amount of space occupied by each part of the body varies. The size of the space does not correspond to the size of the organ but to its sensitivity. For example, the lips occupy a much larger space than the whole of the trunk (see page 195). If we were to draw the body to the scale of the sensory brain map we should get something like Fig. 18.26.

Question 6 Why do you think the hands are so large in Fig. 18.26?

You can see the need for this organization within the brain if you think again about a railway network. If you had to meet an aunt who was coming up to London by train from Manchester you would go to Euston station. You would know where to meet her because Manchester trains always enter London at Euston. The arrivals board even gives the platform that the train will arrive at. In this way you can distinguish your aunt's train from identical ones by registering its point of arrival. Similarly the brain, because it receives an impulse at a particular point, can

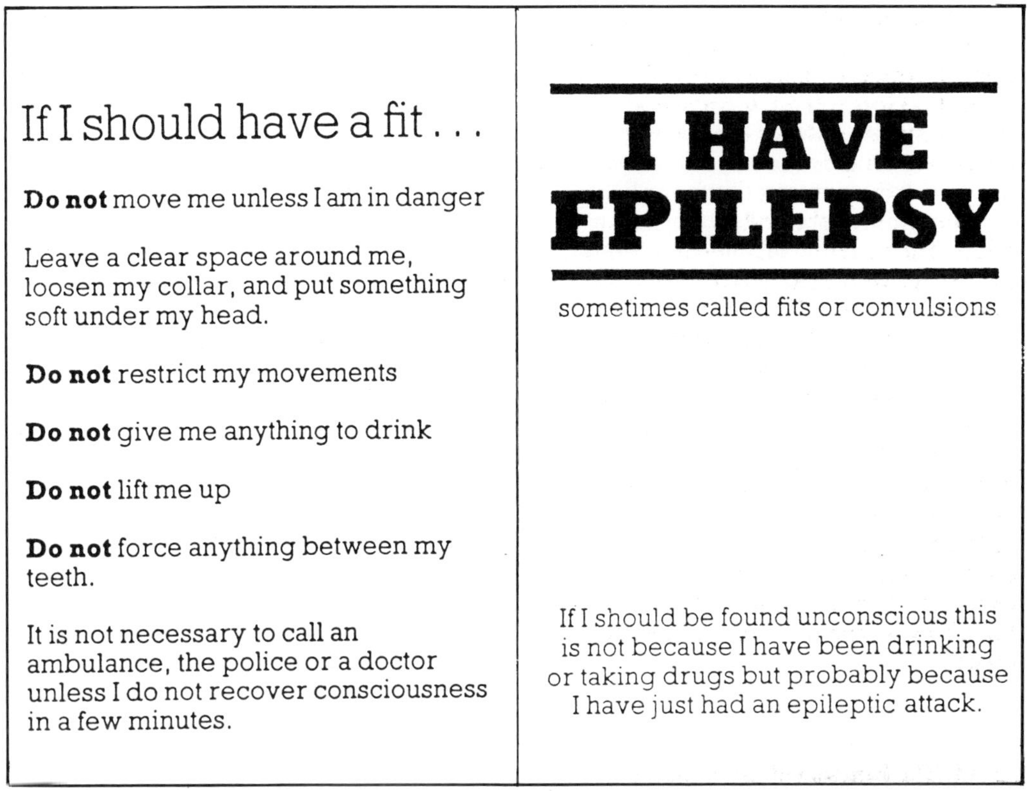
If I should have a fit . . .

Do not move me unless I am in danger

Leave a clear space around me, loosen my collar, and put something soft under my head.

Do not restrict my movements

Do not give me anything to drink

Do not lift me up

Do not force anything between my teeth.

It is not necessary to call an ambulance, the police or a doctor unless I do not recover consciousness in a few minutes.

I HAVE EPILEPSY

sometimes called fits or convulsions

If I should be found unconscious this is not because I have been drinking or taking drugs but probably because I have just had an epileptic attack.

Fig. 18.25 *Epilepsy card*

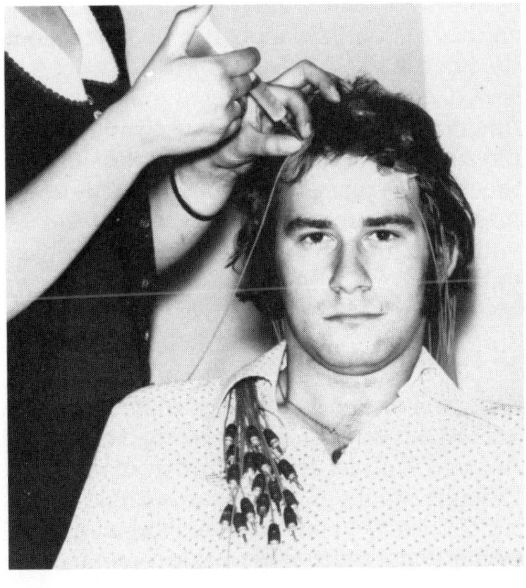

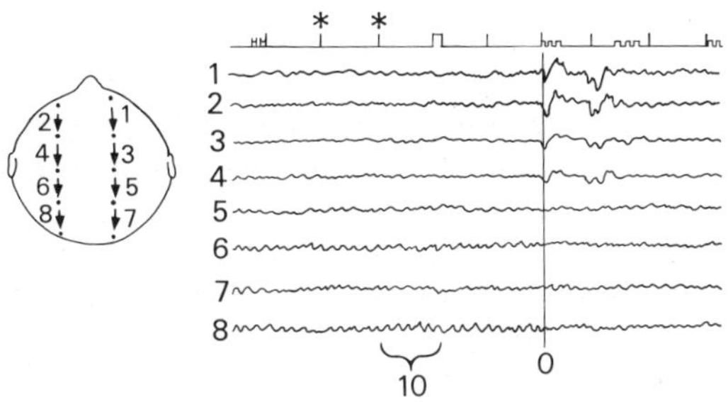

(b)

Fig. 18.24 *(a) Electrodes that are to record the brain's electrical activity being positioned on the head. (b) A normal E.E.G. In traces 1 and 3 you can see the change in the signal when the subject opens his eyes and (1 second later) blinks*

Fig. 18.26 *The body drawn to the scale of the sensory brain map*

identify the sensation. Think of the confusion if the railway lines of Euston and Paddington were crossed over! If the optic and auditory nerves were crossed so that auditory signals entered the brain at the point that normally received visual signals, the brain would interpret sound as light.

You may have noticed that there is no 'mapping' of internal organs on the brain. As they are not under our control we do not need to be aware of them. Often it is hard actually to pin-point the location of a sensation, such as pain from an internal organ. The pain may be general, spread over a wide area or referred, i.e. felt in a different region altogether. For example, in the first stages of a heart attack the pain may be felt across the top of the chest and along the arms.

Our senses give the brain a very quick feedback of information. They inform it of the results of every action, enabling it to change its response. For example, as we pick up a pen the skin receptors constantly record the pressure of the fingers. The brain knows whether to instruct the muscles to contract more to stop the pen slipping, or to relax if there is a danger of the pen being snapped.

The feedback of information concerning the results of our actions may be stored in the memory. It can then be used on other occasions to determine a course of action. Thus our previous experiences can influence the response of the body.

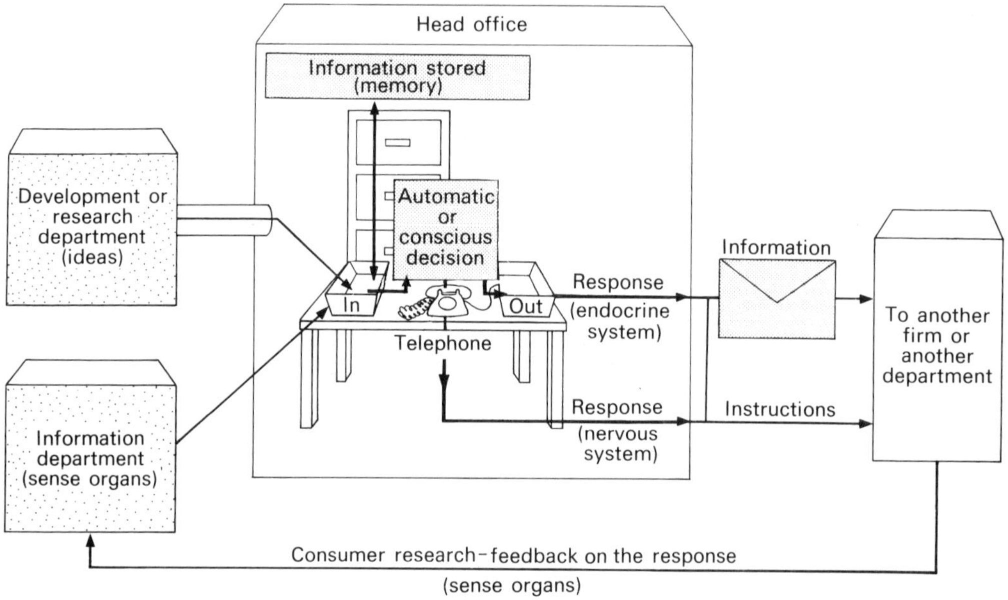

Fig. 18.27 *The brain as a 'head office'*

18.15 Behaviour

In the brain, like the head office of a business, there is constant communication between various departments. Our total response or *behaviour* is the visible result of co-operation between these departments (Fig. 18.27). Different head offices have different premises which receive and store particular information. Therefore their responses to a similar situation may not be the same. So it is that two people may react and behave differently in response to the same stimulus.

Reflex actions may be regarded as the simplest form of behaviour. We have discussed how some of these may be inhibited through conscious control, but there is another way in which they may be altered. This is called conditioning.

A *conditioned reflex* is a reflex which occurs in response to a stimulus that is quite different from the one that normally produces it. The classic example comes from the work on reflexes by the Russian, Ivan Pavlov. He found that, if he always rang a bell as he presented a dog with its food, he could eventually cause the dog to salivate just by ringing the bell when there was no food in sight. The dog now associates the ringing of a bell with the supply of food. Thus either the ringing bell or the food could on its own produce the same reflex action of salivation. The salivation caused when the bell rang is a conditioned reflex. It is suggested that such conditioning is possible because a different set of nervous connections is established within the brain, like cutting a new trail through a jungle. This could be regarded as a form of *learning*. However, conscious control may still be applied to such reflexes if the need arises.

A deeper understanding of the physical function of our bodies helps us to understand physical disorders. *Psychology*, the study of the mind, is just as important. It helps us to understand our mental development and its vital contribution to the total well-being of an individual.

18.16 Mental disorders

A mental disorder or illness is not the same as a mental handicap, which we shall discuss in the next section. It is a disturbance in thoughts, feelings and behaviour that affects the total well-being of the individual. It can occur at any stage in life. Its symptoms, severity and causes are mixed and variable. Generally there are two main groups of mental disturbances: *neurotic* and *psychotic*. We often use the word 'neurotic' in everyday life when we want to describe a person's nervous or exaggerated behaviour. Obviously all of us at some time or other feel tense, depressed, fearful or anxious. It is easy to accept that we are as subject to mental strain and stress as we are to physical strain and stress. However, when a person finds that such disturbances affect their general health and well-being they may seek medical advice and treatment.

A fear or phobia can seriously affect a person's life, as in the case of *agoraphobia*, a fear of open spaces. Understanding such disorders is as important as understanding the nature of physical disease and its effect on the well-being of the individual.

In psychotic illness the mental condition is such that the person loses touch with reality. Thoughts, feelings and actions become disconnected and the external world is rejected, as in *schizophrenia*. Treatment can involve the use of drugs or psychotherapy, where the person is encouraged to understand their illness and its possible causes.

Mental confusion, loss of memory, etc., may arise in conjuction with a physical disorder. Once the physical disorder is successfully treated the mental disorder usually disappears. Very elderly people may become mentally confused because of the gradual loss of brain cells (page 227). In such circumstances they may need assistance from other people. For example, if they fail to receive an adequate diet this could accelerate the mental deterioration.

18.17 **Mental handicap**

We can put on a blindfold if we want to get some idea of the problems and limitations that a person with the *physical handicap* of blindness faces every day. It is impossible for us temporarily to turn off an area of the brain to discover the problems and limitations that a person with a *mental handicap* faces every day. The main problem for many mentally handicapped people would appear to be that everything takes place too quickly. Like a physical handicap, a mental handicap can vary in severity. There are also many causes, ranging from genetic factors to physical brain damage. A genetic disorder in which the nuclei in a person's cells contain three number 21 chromosomes (page 14) is the cause of *Down's syndrome*. Physical brain damage can occur through such things as lack of oxygen (page 53). A mental handicap is usually established before or during birth or in the first few weeks of life. It is a disorder that will last a lifetime, yet with encouragement (Fig. 18.28) and practical support it is possible for many mentally handicapped people to achieve more than was in the past thought to be possible. They can acquire skills and a certain amount of independence (Fig. 18.29). The only limits are those related to the nature and severity of the handicap itself.

Fig. 18.28 *Young mentally handicapped children need a lot of help to learn*

Fig. 18.29 (Below) *Learning skills in the home*

Answers to questions in the text

1 (a) There is no need to look at our legs. Our 'muscular sense' informs us of their position.
(b) The ears and eyes.

2 The peaks are much closer together in (b) than in (a).

3 It causes pressure on the nerve, which irritates it.

4 More bundles of gelignite could be detonated at the same time.

5 (a) Coughing.
(b) Dilation of the pupils.
(c) The big toe curls inwards towards the side of the foot.

6 Our hands are extremely sensitive, with more nerve endings per square centimetre than many other organs. This sensitivity helps us to develop manual skills.

THINGS TO DO

1 Get a lot of friends to stand in a line holding hands. The first person looks at the second-hand on a watch and mentally records the time when he squeezes the next person's hand. He records how long it takes for the 'squeezing of hands' to pass down the line. When the last person's hand is squeezed, that person shouts out 'Stop'. Dividing the time by the number of people in the line should give the mean (average) reaction time. Try this experiment several times. Does the reaction time decrease with practice?

2 Study various unconditioned reflexes in your body.
(a) Stroke a pencil along the inner part of the sole of your foot. Note the curling action of the toes.
(b) Sit in a chair with one leg crossed over the other. Use the outer edge of your hand to strike the tendon just below the kneecap.
(c) Shine a torch in the eyes of another person and observe the contraction of the pupils. (This completely automatic reflex may be used as a diagnosis of concussion. A concussed person's pupils do not respond to changes in the intensity of light.)

3 Obtain a sheep's or calf's brain from your butcher. Observe its external and internal structure.

4 Throw a football rapidly at another person. They usually reach quickly and catch it. Observe what happens if, just as the ball is thrown, you order the person to either catch it or head it. Generally the response will be slower, as the person now has to think about his response. He has to associate the order (the stimulus) with the response. This requires mental activity and the reaction time becomes slower. Again with practice the reaction time may become faster. A lot of learning involves association, for example stopping the car when the traffic-lights are red, or recognizing the meaning of words.

5 Try to establish a conditional reflex in an animal. For example, just before you feed your pet make a particular noise. If you can do this every time, you will eventually find that when the noise is made the animal will respond as if it were feeding-time, although no food is actually present.

6 Collect together a number of illustrations, or do drawings, to show how in everyday life we are conditioned to respond to a certain stimulus. This is often a way of assisting our survival, for example the fire-bell at school.

7 Sit on a chair with one foot underneath you for some time. Eventually get up and try to walk around. In sitting in this way, you trap the sensory nerve which receives the signals from the proprioceptors embedded in the muscle spindles. The trapping of the nerve blocks the message to the brain, so we are not aware of the foot and it feels numb (it has 'gone to sleep'). As an awareness of the foot returns, the sensation in the sensory nerve fibres produces the familiar 'pins and needles'.

8 Select a skill and analyse how it is learned. Why do we say 'practice makes perfect'?

THINGS TO WORK OUT

In Questions 1–3 write down the *one* answer which you think is correct.

1 Nerve impulses pass from the central nervous system to effectors (e.g. muscles) along
- **A** sensory nerve fibres
- **B** motor nerve fibres
- **C** synapses
- **D** relay neurones
- **E** muscle fibres

2 The area of the human brain that controls speech is situated in the
- **A** medulla
- **B** cerebellum
- **C** cerebrum
- **D** pons
- **E** mid-brain

3 The vital centres which control the body's automatic actions (e.g. breathing) are contained in the
- **A** medulla
- **B** cerebellum
- **C** cerebrum
- **D** mid-brain
- **E** spinal cord

4 (a) Identify (*x*) and (*y*). In which direction does the nerve impulse travel along a neurone?

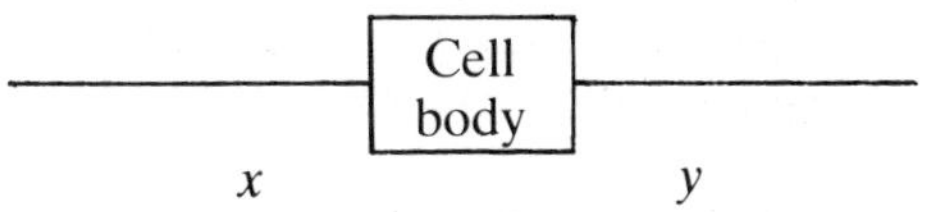

(b) What organs make up the central nervous system?

5 In which organ are the following nerve receptors located?
(a) a taste-bud
(b) a rod
(c) organ of Corti
(d) pain receptor
(e) a proprioceptor

6 Within the brain there are cavities filled with fluid.
(a) Name the fluid.
(b) Describe its function.
(c) Where else in the central nervous system is this fluid found?

7 Draw a simple outline of a human brain. Mark on this outline the region of the brain that is concerned with:
(a) bringing about movement
(b) receiving information from the eye
(c) reasoning
(d) balance and posture
(e) breathing

8 (a) Write out the following in the correct order and give the path of a reflex action: relay neurone, sensory neurone, muscle fibre, motor end-plate, motor neurone, pain receptor, spinal cord
(b) What is a synapse and how does it function?
(c) How is a nerve impulse created by a strong stimulus different from a nerve impulse created by a weak stimulus?

9 (a) Which of the following are automatic actions and which are learned?
(i) sucking
(ii) walking
(iii) curling the toes when a pencil is drawn across the sole of the foot
(iv) blinking
(b) Explain what happens to the pupil of the eye as you enter a darkened room. What hormone can cause a similar change?
(c) Explain what is meant by a conditioned reflex.

10 (a) If our bodies were not *not* sensitive to pain would this affect our chances of survival? Explain your answer.
(b) What is vital to the survival of brain cells? Can brain cells be regenerated?
(c) Distinguish between white and grey matter in the brain. What area of the brain is composed of grey matter?
(d) The surface of the brain has many folds. What are the advantages of this?

11 (a) Draw a simple diagram to show the spinal cord in cross-section.
(b) A child picks up a hot plate and immediately drops it. Use your diagram to show the path of the nerve impulses from the child's fingers to the muscle of the arm.
(c) If an adult had picked up the plate describe how the result might have been different.

19 Growing Up

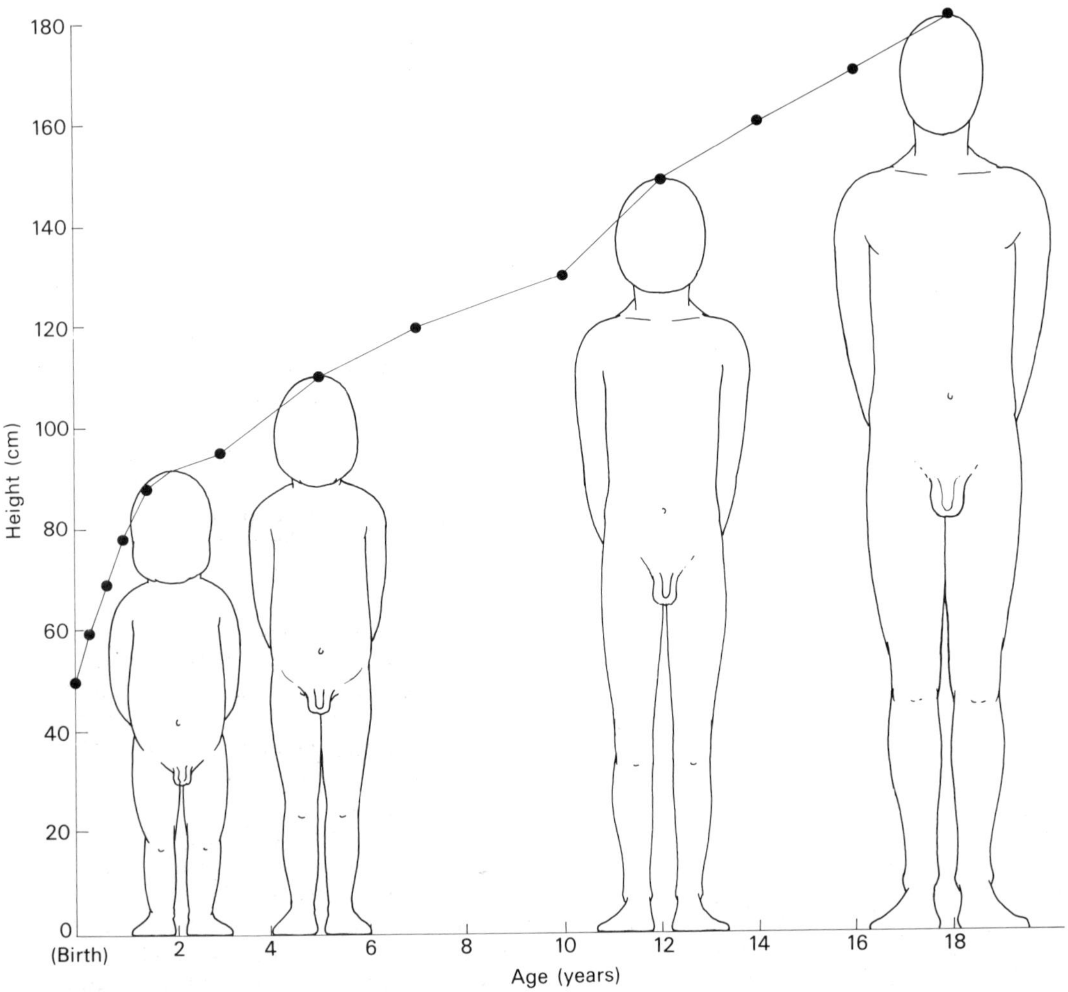

Fig. 19.1 *Physical growth in terms of height*

19.1 **Growth**

In the first chapter of this book we looked at life before birth and the characteristics of our bodies as we entered the world. We found out that we are all individuals, both physically and mentally. The blueprints of our bodies are laid down when an ovum is fertilized by a sperm. How that blueprint works out depends upon the raw materials that are available. So, to take care of her baby before birth, the expectant mother must consider her own health. Throughout life there is a basic pattern of physical and mental development. There will however be individual variations because of the influence of genetic and environmental factors.

The most obvious change that takes place in the first part of our lives is physical growth. Fig. 19.1 illustrates the general pattern of this physical growth in terms of height. Hormones, such as the growth hormone from the pituitary gland (page 178), influence the growth of bone and tissue. Growth in bones takes place by the conversion of cartilage to bone in the areas indicated in Fig. 19.2(a). This process is called *ossification*. New cartilage is laid down and is ossified at the ends of the bone shaft. Therefore the shaft gradually becomes longer (Fig. 19.2(b)). There is a similar growth in width, owing to action of the cells beneath the periosteum membrane that covers the surface of the bone. However this growth in width is less than the growth in length. The shape of the bone is maintained by the reabsorption of bone from areas where it is no longer needed (Fig. 19.2(b)).

Growth is very rapid during the first two years of life. For example, a baby's weight on his first birthday is about three times his

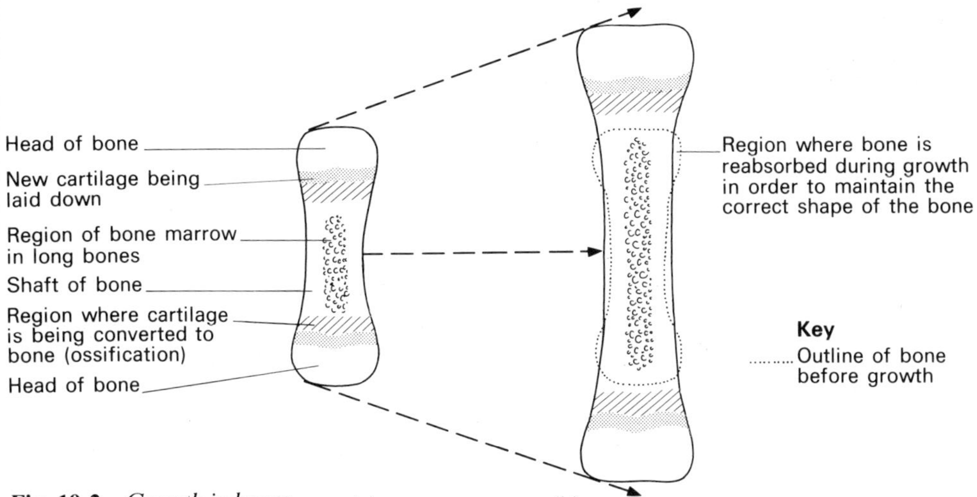

Fig. 19.2 *Growth in bones*

weight at birth. In fact up to the age of two 20 per cent of a baby's energy intake is used for growth. The growth rate then slows down; at eight to nine years only 4 per cent of a child's energy intake is used for growth (Fig. 19.1).

A further period of rapid growth starts at about the age of eleven. This is the beginning of the stage of life known as *adolescence*, the time between childhood and adulthood (Fig. 19.1).

If growth took place at the same rate in every region of the body we should have a larger, but identical form to that which we had at birth. For example, our head would make up a quarter of our total height (Fig. 19.3). Instead, growth takes place at a different rate in different regions of the body, which accounts for our change in form (Fig. 19.3).

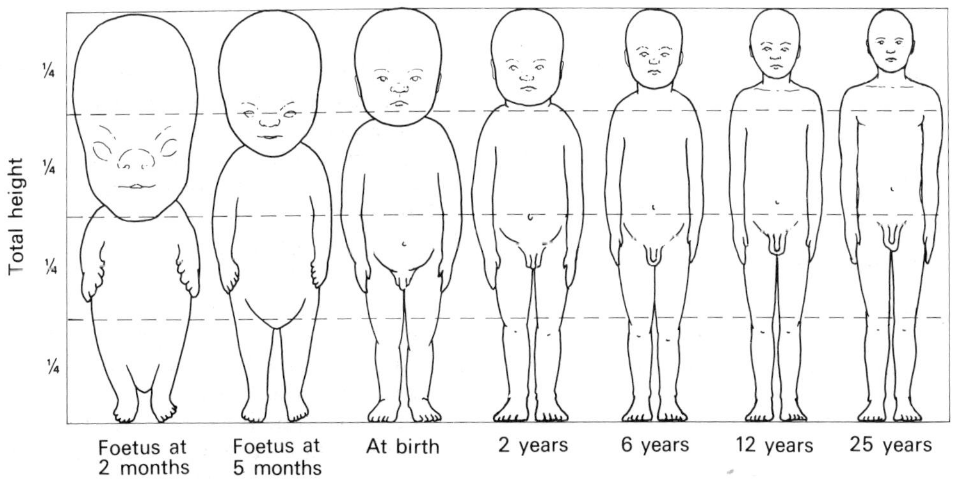

Fig. 19.3 *Growth takes place at different rates in different regions of the body*

Question 1 In an adult, what proportion of the body is accounted for by the legs?

Physiological growth of the body ceases in girls at about seventeen years and in boys at about nineteen years. For example, at this stage the cartilage regions between the shaft and heads of a bone have completely ossified. No further increase in the length of the bone takes place. Four per cent of the energy intake is all that is used for maintenance and repair of tissues, and none for growth. The body may be regarded at this point as having reached *physical adulthood* or *physical maturity*.

19.2 The male and female reproductive systems

At the birth of a baby the first question to be answered is usually 'Is it a boy or a girl?' The answer to this question was determined nine months before. It depends whether the sperm carrying a Y-chromosome fertilized the ovum or a sperm carrying an X-chromosome (page 12). Thus the part that an individual is going to take in the reproduction of the next generation is determined at the individual's own conception.

Human reproduction, like that of all vertebrates, is *sexual*. This means that a man's reproductive cell, a sperm, must fuse with a woman's reproductive cell, an ovum, to produce the next generation (page 251). The woman's reproductive organs known as the *ovaries* produce ova. The man's reproductive organs, *testes*, produce sperm.

The structure of a man's reproductive system enables him to place his sperm inside the woman during sexual intercourse. This internal fertilization makes sure that at least one sperm can reach an ovum. The structure of a woman's reproductive system enables her to receive sperm and to protect the developing fertilized cell. Tables 19.1 and 19.2 and Figs 19.4 and 19.5 show the internal structure of the male and female reproductive systems.

TABLE 19.1 MALE REPRODUCTIVE SYSTEM

Scrotum

A pouch of skin

Holds the testes outside the body. The development of sperm takes place at a lower temperature than body temperature. The testes descend into the scrotum just before birth. This is checked at birth and at medicals during childhood. If one or both testes fail to descend a small operation may be needed to release them. It is considered best to do this in early childhood

Testes

Two hard, oval bodies about 4 cm long, 2.5 cm wide and 2 cm thick. They can be seen in the scrotum when the body is in a hot bath

Produce the male reproductive cells – sperm. Produce hormones called androgens. The most important of these is testosterone, which regulates the development of secondary sexual characteristics

Epididymis

A finely coiled tube about 6 m long which is attached to one side of the outer surface of each testis

Stores sperms until they are mature

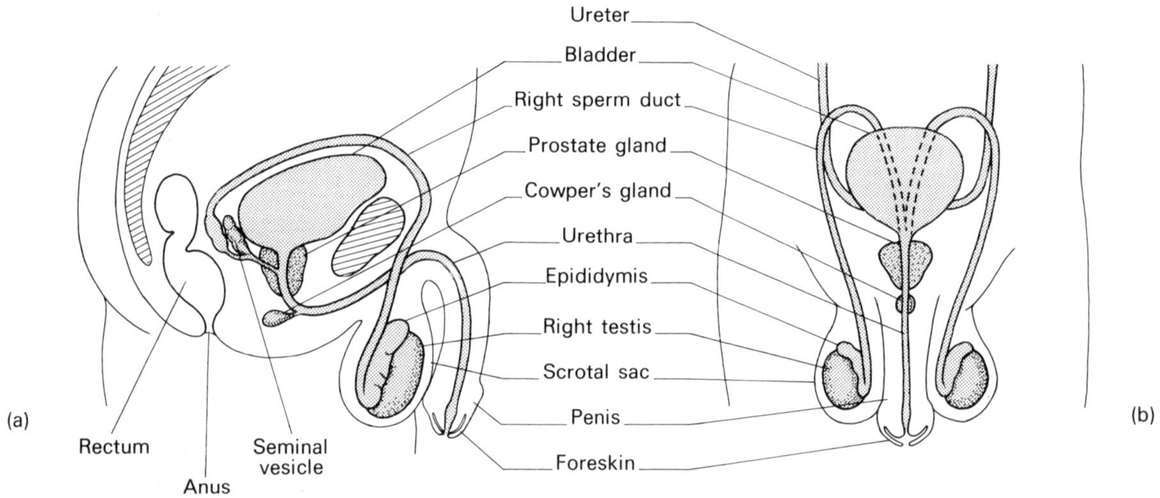

Fig. 19.4 *Male reproductive system. (a) Side view. (b) Front view*

Penis

This is made up of spongy tissue which surrounds the urethra. It has a very sensitive end which is surrounded by a ring of skin called the *foreskin*. This can normally be pulled back. In cases where it is too tight a portion of it may be removed. This is called *circumcision*. Circumcising may also be done as part of a religious rite

Under stimulation the spongy tissue of the penis becomes dilated with blood. It becomes hard and erect (page 251)

Urethra

The right and left ejaculatory ducts open into it

Urine is expelled from the body through the urethra. Sperms are released during intercourse or during 'wet dreams'

Sperm-duct

The epididymis from each testis connects with its own sperm-duct. This is a muscular duct. It enters the abdominal cavity and curves round behind the bladder to form the *ejaculatory duct*. This duct connects with the urethra

The muscular walls force the sperms along to the *seminal vesicles* at the top of the ejaculatory duct. Here there is a secretion of seminal fluid. Other glands, the prostate and Cowper's gland, also secrete fluids which nourish the sperm, neutralize fluids such as urine and 'activate' the sperms

TABLE 19.2 **FEMALE REPRODUCTIVE SYSTEM**

Ovary

Two ovaries situated to the right and left just below the 'rim' of the pelvic girdle. About 4 cm long, 2 cm wide and 0.8 cm thick. Creamy-white in colour

Produce female reproductive cells – ova. Produce hormones such as oestrogen and progesterone. Oestrogen regulates a woman's monthly cycle and the development of secondary sexual characteristics (page 180). Progesterone stimulates the uterus to prepare for the nourishment of a fertilized ovum

Oviduct or Fallopian tube

Each is about 10 cm long. One end is joined to the upper region of the uterus; the other end opens out into a funnel by the side of an ovary. Around the rim of the funnel are numerous finger-like processes. The inner linings of the Fallopian tubes consist of ciliated cells (page 4)

The funnel catches the ovum as it is released from the ovary. The ciliated cells then transport the ovum along the tube towards the uterus

Uterus or womb

In a woman who has had no children it is about 8 cm in length, 5 cm wide and 2.5 cm thick. It is protected by soft surrounding tissue and the bones of the pelvic girdle. The lower and narrower end of the womb is called the *cervix* or neck of the womb. The womb has thick muscular walls. These surround a small internal cavity. This cavity connects with the *vagina* by means of a narrow canal which passes through the cervix

The embryo is protected and nourished in the early stages inside the woman's uterus. The uterine walls are muscular so that at the right time in its development the baby can be pushed out into the outside world

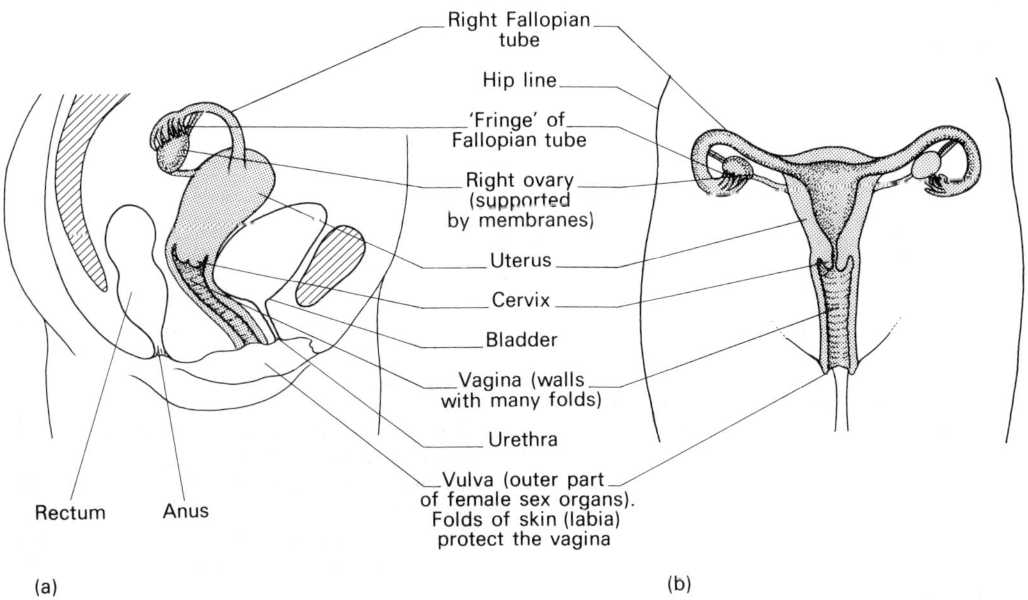

Fig. 19.5 *Female reproductive system. (a) Side view. (b) Front view*

Vagina or Birth canal

This is a very muscular passageway about 8 cm long. It has a soft mucous lining. When the vagina is examined internally, the cervix can be felt as a raised bump at the far end. Glands in the cervix secrete a clear mucus to keep the vagina clean and moist. This mucus is discharged from the vagina and dries, turning slightly yellow. The external opening of the vagina is between the anus and the urethra (page 171). This region is surrounded by folds of skin. Certain areas are extremely sensitive

This is the region of the woman's body which receives sperms from the man during intercourse (page 251)

19.3 **Puberty and secondary sexual characteristics**

At birth we possess all our reproductive organs, but at this stage they are not fully developed. For instance, a girl's ovaries at birth contain a total of approximately 10 000 000 ova waiting to mature. Only about 500 of these will eventually mature and be released. The remainder will disintegrate and disappear.

The stage at which the sexual organs mature is known as *puberty*. As with all developments within the body, the time when puberty takes place and the rate of the changes vary considerably from individual to individual. Girls on average experience puberty eighteen months to two years earlier than boys. We should expect these differences between sexes and understand that they are quite normal. Boys and girls may cover the course of maturity at different speeds, but in the end everyone covers the same course!

The organ that sets the development into motion and controls it is the pituitary gland (page 178). In the case of a girl it stimulates the ovaries into producing the hormone *oestrogen* (page 180). Oestrogen causes the laying down of fat under the skin in regions such as the thighs, buttocks and upper arms. Thus the girl's body fills out and assumes a softer, less angular appearance. Fat gradually accumulates around the mammary glands, which develop in the region of the second to sixth ribs. This leads to the development of *breasts*, which will vary in size from person to person according to the body's blueprint. As is shown in Fig. 19.6 the glands in the breasts are connected by *ducts* to the *nipple*. This is surrounded by other glands and an area of darker skin known as the *areola*. After a woman has given birth to a baby another hormone of the pituitary gland stimulates the mammary glands to produce milk (page 256).

During puberty both boys and girls experience another growth spurt. In girls this will reach its peak at about the age of twelve. A boy's growth is much greater than that of a girl. A boy's muscles develop considerably in size and strength. This extra growth is accounted for by the fact that it is stimulated by *testosterone*, the hormone from the testes. Although he goes on growing until nineteen or so, a boy reaches the peak of his growth spurt at about fourteen. Testosterone also stimulates the development of other secondary sexual characteristics, such as the growth of hair under the arms and between the legs (pubic hair). Facial hair develops later on (Fig. 19.7). The voice becomes deeper as the larynx changes shape. Puberty in boys is generally complete by sixteen.

Girls also grow hair under the arms and between the legs. At the same time development of the uterus and vagina reach their final stages. The pituitary gland begins to trigger the secretion of a hormone which instructs the ovaries to release an ovum. A girl may not be aware of her first ovulation but this will be followed by her first *period* or *menstruation*. This occurs on average between the ages of eleven and fifteen, when puberty can be regarded as complete.

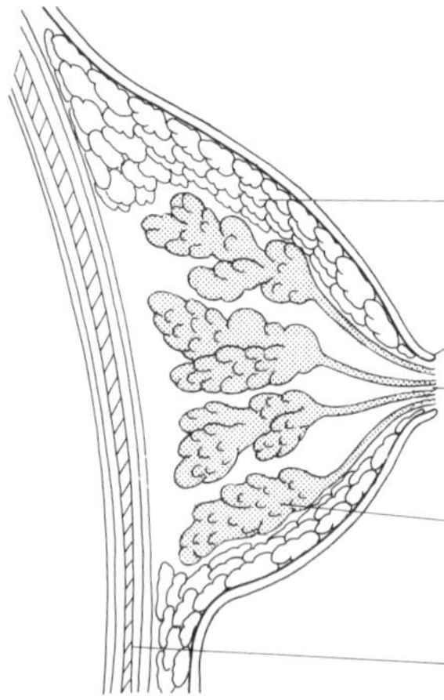

Fig. 19.6 *Internal structure of the breast*

19.4 **Ovulation**

Fig. 19.8 shows a simple diagram of a single ovary. It consists of a fibrous centre with an outer surface of special cubical cells called *germinal epithelium*. These are capable of constant division, and they build up cells inwards towards the centre of the ovary.

Question 2 In what other region of the body have we studied a similar germinative layer of cells?

When the ovaries are stimulated by a hormone from the pituitary gland (page 178), the germinal epithelium builds up special cells round some of the immature ova. These cells are called *follicle cells* (Stages 1 and 2 in Fig. 19.8). At a later stage a division occurs between some of the follicle cells and the space becomes filled with fluid (Stage 3 in Fig. 19.8). This fluid-filled follicle is called a *Graafian follicle*. It contains the ovum surrounded by follicle cells and fluid, like a castle surrounded by walls and a moat.

The Graafian follicles continue to develop for about six days, by which time one follicle is usually more advanced than the others.

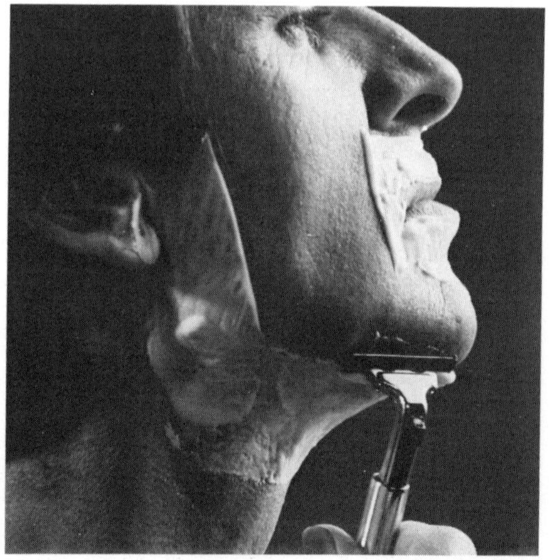

Fig. 19.7 *Facial hair*

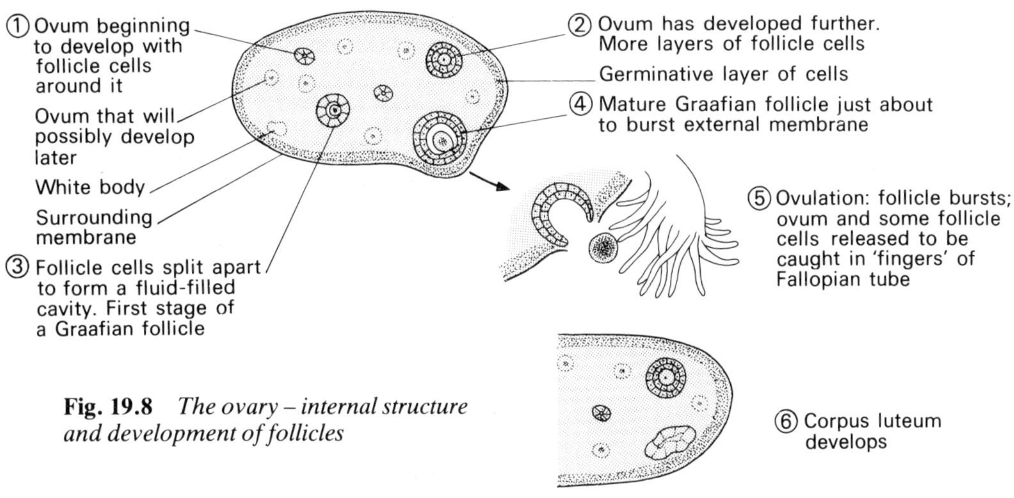

Fig. 19.8 *The ovary – internal structure and development of follicles*

This continues to grow while the others stop growing and return to their original condition. The single follicle continues to develop and become so swollen in size (about the size of a pea) that it stretches the external membrane of the ovary (Stage 4 in Fig. 19.8). The pituitary gland then secretes another hormone, which stimulates the follicle so that it bursts. The ovum, measuring about 0.1 mm and surrounded by a ball of follicle cells, is released and caught in the 'fringes' of the Fallopian tube. The release of the ovum is known as *ovulation*. Usually only one ovum is released at a time and each ovary takes it in turns. The 'fringes' of the Fallopian tube funnel the ovum into the inner depths of the tube. The ovum can live for between twelve and forty-eight hours before it dies and disintetrates.

19.5 **Menstruation**

The ovum is carried along by the cilia on the inner lining of the Fallopian tube. In this way it is transported towards the uterus. Meanwhile in the ovary the empty follicle that is left gradually fills with blood and then with a yellow fatty substance. This yellow body is called the *corpus luteum* (Fig. 19.8(6)). It acts like an endocrine gland, and secretes the hormone progesterone into the bloodstream. Progesterone stimulates the uterus so that it provides the correct environment for the ovum if it is fertilized. Thus the uterus lining becomes thicker and its blood supply is increased. Fig. 19.9 shows these changes in the uterus lining in relationship to the secretion of progesterone. If the ovum is not fertilized within forty-eight hours, it disintegrates and passes out of reproductive passages along with other normal secretions.

Without a job to do, the yellow body gradually disintegrates. This region then fills with a white tissue, which can be seen as a white spot on the ovary (Fig. 19.8). Without the corpus luteum the progesterone level in the blood falls. This happens at the same time as the ovary secretes less oestrogen. This lowering of levels of the two hormones triggers *menstruation* or a *period* (Fig. 19.9).

Fig. 19.9 *Hormone changes and developments within the ovary and uterus during a woman's monthly cycle*

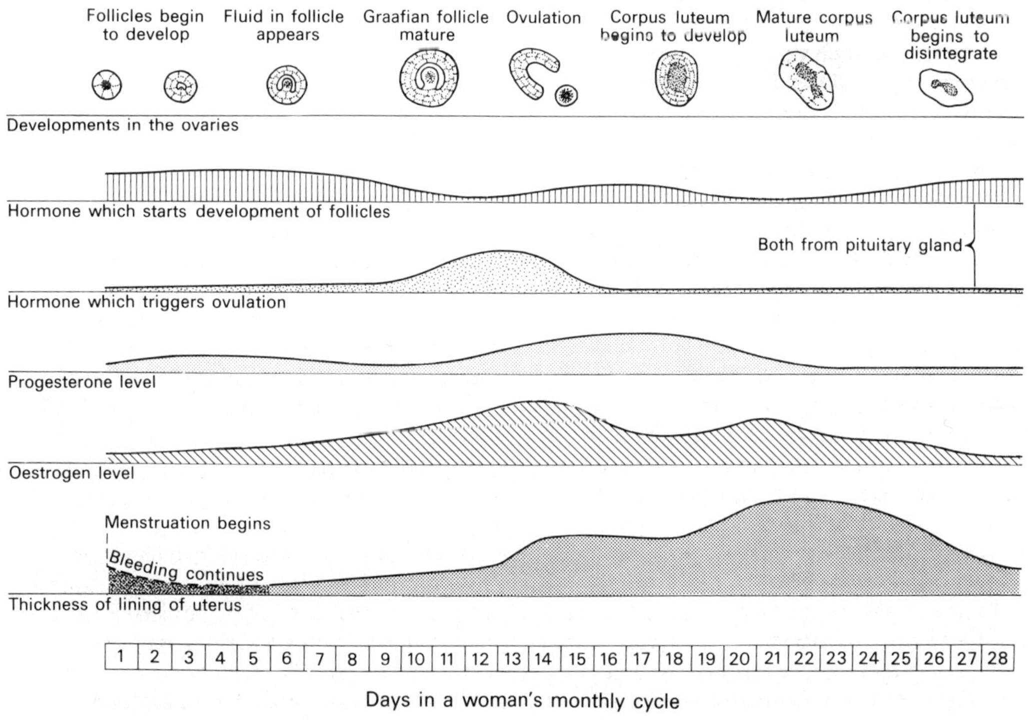

Have you ever seen old plaster crumbling, cracking and flaking off a brick wall? This is similar to what happens to the lining of the uterus. It disintegrates and is shed, together with about 200 cm^3 of blood, through the vagina. The blood is lost because the walls of capillaries are broken as the lining comes away. A period occurs on average on the twenty-eighth day from the date on which the follicles first began to develop. The menstrual flow lasts for about five days, and is a perfectly normal monthly occurrence. It need not hinder any normal activity. However it is important that at this time even more care should be taken over personal hygiene. Some form of protection has to be worn to cope with the flow and to prevent any staining of clothes. (Fig. 19.10). This protection can take the form of an absorbent pad worn either externally (a sanitary towel) or internally (a tampon). These both need to be changed regularly and should be disposed of in the way recommended by the manufacturer. Some many have to be incinerated or placed in a special unit, while others may be flushed away down a lavatory.

The first day of menstruation is taken as the first day of a woman's monthly reproductive cycle. It is then that the follicles begin to develop again. The whole process therefore repeats itself, unless interrupted by pregnancy. However the twenty-eight day cycle can vary: the pituitary is under the control of the hypothalamus, which is linked to the cerebrum of the brain; so an emotional strain can upset the timing of the cycle.

A woman will ovulate and menstruate until she reaches an age somewhere between forty and fifty years old. She is then said to have reached the *menopause* or *change of life*. This is a natural change that again varies considerably in timing, rate and the effect it can have on an individual. The menopause involves a change in the secretion of hormones, so a woman may be affected physically and emotionally. Such changes will require understanding from those around her, and medical advice can help considerably.

Fig. 19.10 (Below) *Products available for a woman's personal hygiene.*

19.6 The testes and sperm production

When the testes mature at puberty sperm are developed within them for the first time. As you can see in Fig. 19.11 the production of sperm takes place within the minute coiled tubes that make up the inner structure of the testis. The germinal epithelium constantly divides, and numerous layers of cells are produced within the inner lining. As the cells move towards the centre of the tube, they undergo a change (Fig. 19.12). The innermost layer of each tube consists of hundreds of mature sperm. Each sperm has a head section containing a nucleus, a neck region and a tail. At this stage the sperm are not active and are squeezed along by the muscular contractions of the tubes. It is only when secretions from glands, such as the prostate, meet the sperm that they become active and move by lashing their tails from side to side.

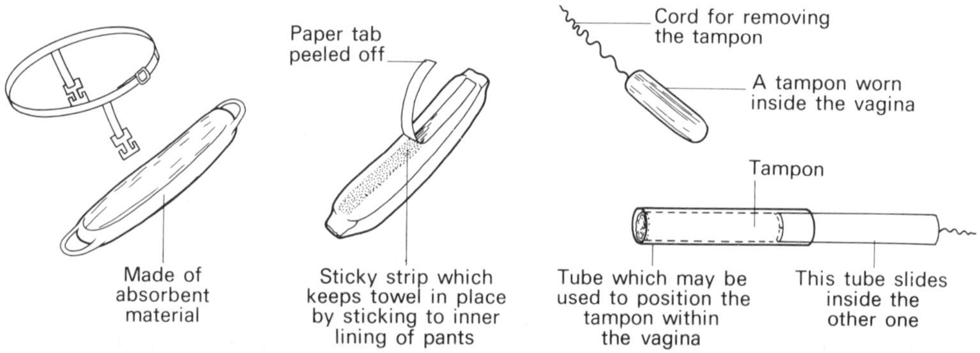

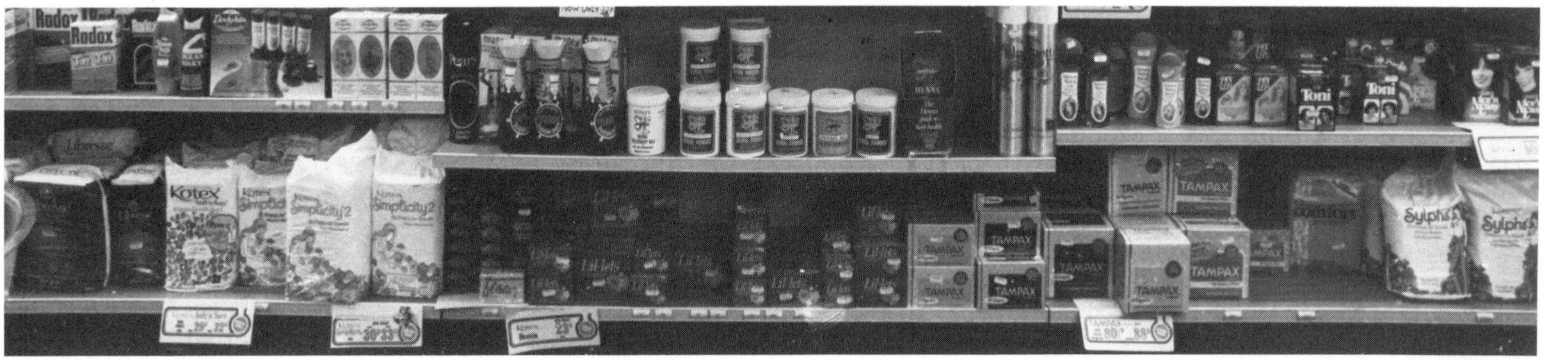

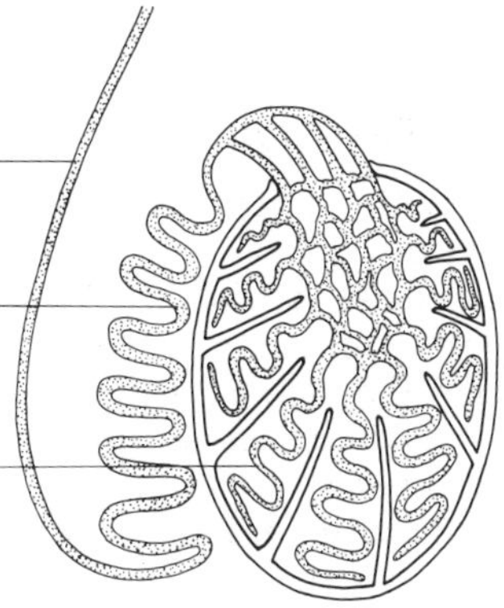

Fig. 19.11 *A testis sliced along its length to show the internal tubes (diagrammatic)*

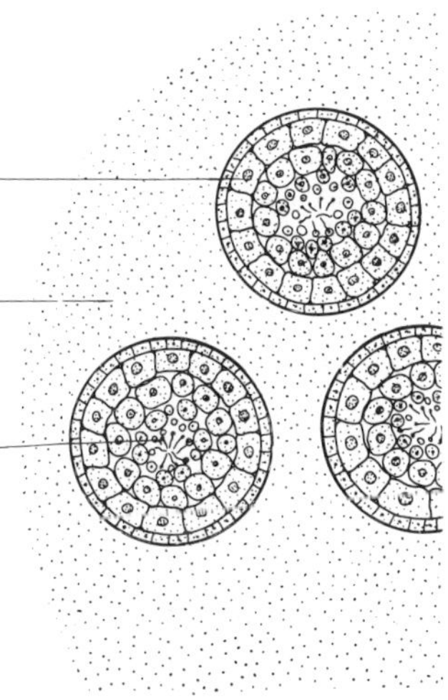

Fig. 19.12 *A testis sliced across its width to show the internal tubes and sperm production*

From puberty millions of sperms are made every day. Some are stored within the epididymis and in the sperm-duct near the seminal vesicles. There are times when no more can be contained within the body. At such times, and usually during a period of sleep, some of these sperms are released (*wet dream* or *nocturnal emission*). This is quite normal and needs to be understood as an unconscious reaction of the reproductive system.

19.7 Disorders of the reproductive passages

As we have seen throughout this book, most organs of the body are subject to disease. This can arise through an internal disorder or through infection from another person. The reproductive systems are no exception. It is essential to understand the nature and transmission of these infections.

In a woman the vagina is kept clean, moist and healthy by a clear or slightly milky discharge which can vary in amount at various stages in the monthly cycle. Changes in colour and smell of this discharge, perhaps with an itching or burning sensation, can indicate a possible infection. As the anus, urethra and vagina openings are so close together it is easy for a microbe to be transmitted from one system to another. It is also noticeable that infections such as 'thrush' (a form of fungus infection) are more likely to develop when a woman has been taking antibiotics or the contraceptive pill. It can also develop when a woman's general health is low (page 94). All of these mild infections usually develop without transmission from another person and are easily treated. In fact the microbes can be present without any obvious symptoms. Their presence may be detected when a woman undergoes a routine *cytotest*. A cytotest is a simple test carried out to see if there is any possibility of the development of cancer at the neck of the womb (*cervical cancer*). It involves scraping a few cells from the neck of the womb and placing them on a microscopic slide. Nothing is felt, as this is done quite gently. The *cervical smear* is then examined at a health laboratory and the woman is informed of the result. She may be recalled for further tests. This is often because a mild vaginal infection has been identified and will need treatment before a further test can be done. A cytotest is done as a routine check at post-natal examinations and at family-planning clinics. It is a test that should be done regularly, because if cervical cancer is discovered at a very early stage it can easily be treated and cured. Ideally a woman should have a cytotest every three to five years, especially if she is over thirty-five years old and has children.

19.8 Sexually transmitted diseases

Over the last ten years sexually transmitted diseases or *venereal diseases* (V.D.) have been on the increase. They are transmitted when there is sexual contact or intercourse with an infected man or woman. Venereal disease can be treated by drugs such as penicillin but, as with all diseases, it is essential that medical advice is sought in the very early stages of the infection. Such confidential advice can be sought from a family doctor or by attending a special V.D. clinic. Most large hospitals provide this service. Even in the later stages it is possible to treat the infection. However the problem here is that often organs have been damaged and cannot be repaired. If this happens it can result in much unhappiness later on because the person may have become infertile.

The two most common forms of V.D. are *gonorrhoea* and *syphilis*. Gonorrhoea is a bacterial infection which in a man causes inflammation of the testicles, penis and urethra. If left untreated the infection can spread to other parts of the body. The sperm-ducts can become permanently blocked, leading to infertility. In a woman the bacteria inflame the mucous membrane inside the reproductive passages. Any early symptoms can therefore go unnoticed and this is where the danger lies, for the infection soon spreads to the oviducts. The scarring of these fine tubes can block them permanently. If a baby is born to a mother with untreated gonorrhoea its eyes can become infected, and the child can go blind. Tests are carried out as a matter of course during antenatal care to detect this type of infection.

Syphilis is caused by a spirochaete (page 80). In the early stages, about three weeks after sexual contact with the infected person, a sore develops on or near the sex organs. In the case of a woman this may be inside the vagina. The sore contains thousands of spirochaetes and is extremely infectious. Eventually the ulcer clears up and if the infection is not treated the micro-organisms enter the bloodstream. In the next six to twelve weeks a skin rash may erupt, together with a high temperature. Again these symptoms disappear, but the microbes carry on and seriously affect vital organs such as the brain, heart and liver. It is this action of the microbes that in later years can cause a person to become crippled by paralysis, to go blind or even die prematurely. A woman who has syphilis and becomes pregnant can transmit the microbes to the embryo, across the placenta. In the later stages of infection syphilis can be detected in a blood test (page 18).

As we have seen in Chapter 7, infectious diseases are spread through contact with a person who is already infected. If we go out shopping we meet many people, and a few days later we may develop a heavy cold caught from an unknown contact. In this case we could not take any avoiding action, other than never going out of doors and mixing with people! In the case of venereal disease we can take avoiding action. If a sexual relationship is part of a permanent relationship with one person, then the partners know each other intimately and they need not worry about V.D. However, V.D. thrives when people have several sexual relationships and where there is little known about the other partner. In this sort of situation it takes only one person to be infected for V.D. to be transmitted to many people. Like all infectious diseases which can have serious consequences for the community, efforts are usually made to trace the source of infection. This presents each of us with a responsibility for our own protection and that of others from these infections.

Answers to questions in the text

1 Approximately half the total height of the body is accounted for by the legs.

2 The germinative layer of the skin.

THINGS TO DO

1 If possible, assemble a montage of photographs which show how you have grown from babyhood.

2 Measure your present height, weight, breadth of shoulders, hips and chest. Compare these measurements with those of others from your age group. Observe how much measurements can vary. Conduct a survey of clothing which is available for your age group. What variety of sizes is available? Is such a system of sizing successful? What might be one of the problems in buying a bikini?

3 Fig. 19.3 shows how the different systems of the body grow at different rates. Imagine you are advising a group of mothers with young babies. Write down how you would advice the mothers to support the normal pattern of growth within the nervous and skeletal systems. What action might a mother take to assist her baby's lymphatic system in its defences against infection?

4 Construct a chart using illustrations to show how the responsibility for our own welfare develops from babyhood. Try to indicate what is involved, for example physical welfare or behaviour or attitudes. Find out what is meant by 'the age of responsibility' within the law.

THINGS TO WORK OUT

In Questions 1–4 write down the *one* answer which you think is correct.

1 Low levels of both progesterone and oestrogen in a woman's blood triggers
- **A** implantation
- **B** ovulation
- **C** menstruation
- **D** breast development
- **E** the secretion of milk

2 The narrow entrance to the womb is called the
- **A** cervix
- **B** vagina
- **C** urethra
- **D** sperm-duct
- **E** oviduct (Fallopian tube)

3 Syphilis and gonorrhoea are transmitted through
- **A** not carrying out good personal hygiene
- **B** contact with a lavatory seat used by an infected person
- **C** sexual contact or intercourse with an infected person
- **D** kissing a person that has the infection
- **E** using dirty hand-towels in a communal wash-room

4 Sperms are stored within the
- **A** scrotum
- **B** epididymis
- **C** sperm-duct
- **D** urethra
- **E** penis

5 (a) What is meant by puberty and adolescence? Give *three* developments that take place during puberty in (i) boys, (ii) girls.
(b) What is meant by the menopause, and on average at what age does it occur?

6 (a) During which *two* periods of our lives is our growth rate faster than at any other time?
(b) Describe the growth rate of
(i) the brain and head
(ii) the reproductive organs

7 (a) What are oestrogen and progesterone?
(b) What secretes progesterone in (i) a man, (ii) a woman?
(c) What is the role of progesterone in a woman?
(d) A woman's monthly cycle is a regular twenty-eight days.
(i) On what day would ovulation usually take place?
(ii) When would menstruation take place and for how long?
(iii) During which days is sexual intercourse most likely to result in fertilization?

8 (a) Name *one* sexually transmitted disease.
(b) Describe how sexually transmitted diseases are spread within a community and why it is possible for them to reach epidemic levels.
(c) Give *two* dangers to health of sexually transmitted diseases.
(d) State the ways in which individuals and the community as a whole can guard against the transmission of these diseases.

20 A Lifetime of Change

20.1 We are all individuals

Many of you will have experienced a reunion where groups of people meet for the first time after spending several years apart. Gradually, during the excited round of greetings they become aware of changes that have occurred in each other since they last met. The changes that they observe in those first few moments are usually physical ones, for example the growth of the children, or the fact that one of the adults has gained weight. Other changes that have taken place over the years will become apparent as they spend the day together. It is likely that these changes will be concerned with character, behaviour and attitudes. For example, a child may be less shy or an adult less tolerant of people's mistakes. It is these types of development and change that are so easily influenced by a person's experience.

Fig. 20.1 *'Viscount Horatio Nelson' by L. F. Abbott*

We all know the physical changes that accompany the growth and development of the body. We only have to look in a mirror or at old photographs to observe them in ourselves. However, knowing an individual involves much more than just observing physical appearance. For instance, if we want to get a balanced impression of a certain person from the past, we need to take more than a glance at a portrait or picture (Fig. 20.1). We need to read what that person said and wrote and what others said or wrote about him. In this way we learn about his ambitions, achievements and failures and his relationships with other people. This gives us some hint of what the 'whole person' was like, his likes and dislikes, his strengths and weaknesses, in short his character. So if we really want to know someone, to get 'beneath the skin', it takes time and requires more than a casual relationship. As the first person you have to learn to live with is yourself you can start by taking time to get to know yourself. You can consider how you see yourself and how others see you.

20.2 Relationships

We are all individuals, yet we choose not to live in isolation. Our lives involve us in relationships with many other people, of both sexes and all ages, all of them individuals in their own right.

A child gradually develops relationships with others, often by observing and becoming involved with those that are closest to him. At first this usually involves parents, grandparents, brothers and sisters. As the child grows, the 'field of relationships' extends beyond the family to include other adults, and boys and girls of the same age in the same school or street. Friendships with contemporaries become very important. During adolescence these strong ties amongst a group of friends may conflict with other relationships, such as those of the family. Personal relationships are concerned with what we feel for other people and what other people feel for us. They affect our emotions, our likes and dislikes, our happiness and sadness, even anger and jealousy. At times our physical state can raise or lower our feelings; this is particularly true in adolescence, when emotional feelings and ideals tend to be at a peak.

In Chapter 4 we discussed how life on earth is physically interdependent. Personal relationships involve us in another form of interdependence. This makes us responsible at times for the total well-being of others, not just their physical welfare. You might like to discuss how the nature and extent of this responsibility changes throughout our lives.

During adolescence sexual feelings begin to be experienced. The sexual feelings that are aroused in a man or woman are, like other responses of the body, a response to stimuli. Styles of dress and the way we act can provide the stimulus which arouses sexual feelings in others. However, as we have seen in Chapter 18, very few responses of our body are beyond the conscious control of the brain. Sexual feelings affect our consciousness and can be controlled. Each individual has to make up his own mind whether his actions are responsible and caring or selfish and indulgent.

When two people enter into a permanent relationship such as marriage, they realize that their sexual feelings have an important part to play in a satisfying and interdependent partnership. Within the framework of a stable relationship sexual feelings can give pleasure and fulfilment. Also the man and woman will usually have given thought to the possible consequences of their sexual relationship, i.e. the conception of a child and the responsibilities that this will bring to each of them.

20.3 **Fertilization**

When a man is sexually stimulated the nerve impulses activate muscles in the pelvic region. These muscles constrict the vein carrying blood away from the penis, and at the same time cause other muscles to relax, allowing the spongy tissues of the penis to fill with blood. By this action the penis becomes enlarged, hard and erect, and can enter the woman's vagina. This coming together is referred to as sexual intercourse. The movements of the penis within the vagina stimulates both partners. At the peak of this sexual stimulation male sperms leave the penis and enter the vagina. This *ejaculation* is caused by the muscular contractions of the sperm-duct and epididymis. The sperms are pumped along in an intermittent flow. There are similar muscle contractions in the walls of the woman's vagina and at the cervix of the womb. This seems to speed up the movement of the sperm in the woman's reproductive passages. After the man has reached a climax the muscular contractions cease. A normal blood flow is resumed in the penis, it relaxes and returns to normal size.

About 300 000 000 sperms are released in about 3 cm^3 of semen. The sperms move through the cervix and into the uterus (Fig. 20.3(a)). Their life span averages 48 hours. Some sperm in this time will normally reach the oviducts. If there is an ovum travelling down one of the oviducts then fertilization is possible (see Fig. 1.7). Fertilization usually occurs at the top of the Fallopian tube, near the ovary (Fig. 20.3(b)). Only one sperm manages to fertilize the ovum. Its head penetrates the outer membrane so that the two nuclei fuse. When this has happened the membrane becomes tougher and prevents the entry of any other sperms. If for some reason two ova have been released and are present in the Fallopian tubes, then each one may be fertilized by a sperm. This will result in the birth of *fraternal twins*. Fraternal twins (Fig. 20.2(a)) are not necessarily of the same sex, nor are they more alike than any other brother or sister. *Identical twins* (Fig. 20.2(b)) are of the same sex and identical in looks. They develop when a fertilized ovum splits into two and both halves develop separately.

A woman does not detect that fertilization or conception has occurred. Only later will she realize she is pregnant.

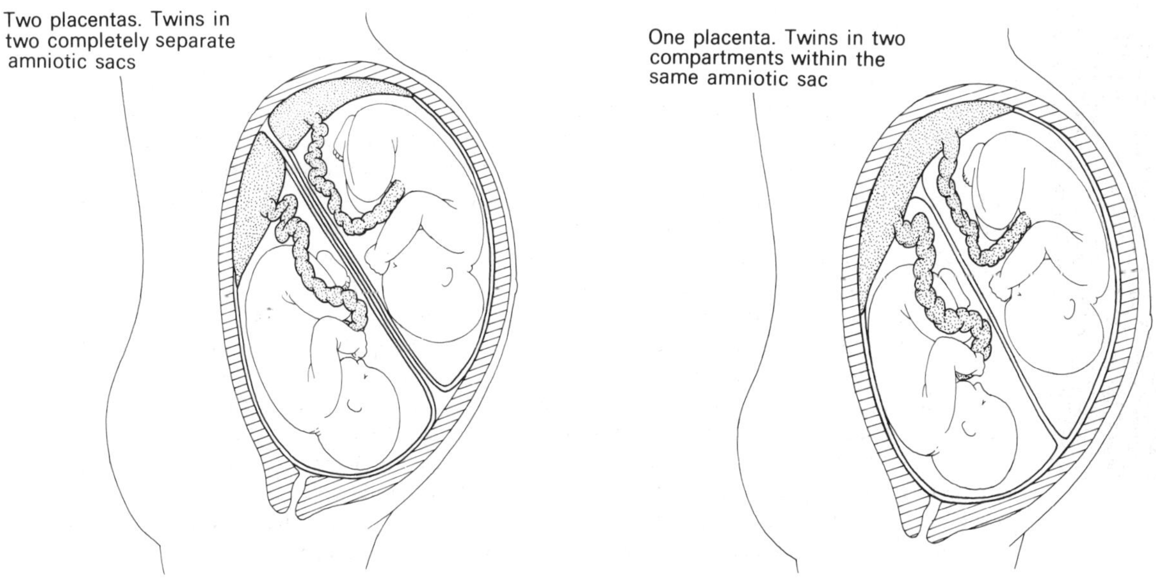

Fig. 20.2 *(a) Fraternal twins. (b) Identical twins*

20.4 Implantation and development

Within the next four to seven days the fertilized ovum (zygote) travels down the Fallopian tube towards the uterus. As it does so it continually divides (Fig. 20.3(c)), forming a ball of cells (see Chapter 1, pages 3 and 5). The *blastocyst* (Fig. 20.3(c)) enters the uterus. It can no longer nourish itself, so it fixes itself to the thick lining of the womb. This lining grows until it almost completely covers the top of the blastocyst. Finger-like projections develop as the placenta begins to form (pages 5 and 17). The attachment of the blastocyst to the lining of the womb is called *implantation* (Fig. 20.3(c) and (d)). During 1978 the world's first 'test-tube baby' was born. She was so-called because the sperm and ovum that were taken from her parents were fused together outside her mother's body. This technique was developed to help women who cannot conceive in the normal way. The fertilized and developing egg is then placed into the mother's uterus so that it can implant.

The woman becomes aware that implantation has occurred after the blastocyte secretes a hormone which enters the blood supply through the walls of the uterus. This stimulates the corpus luteum into action. It does not disintegrate but continues to secrete progesterone to keep the womb lining intact. (When the placenta is fully developed it too secretes progesterone to maintain the lining of the womb.) Thus the first indication of pregnancy is that a woman fails to menstruate.

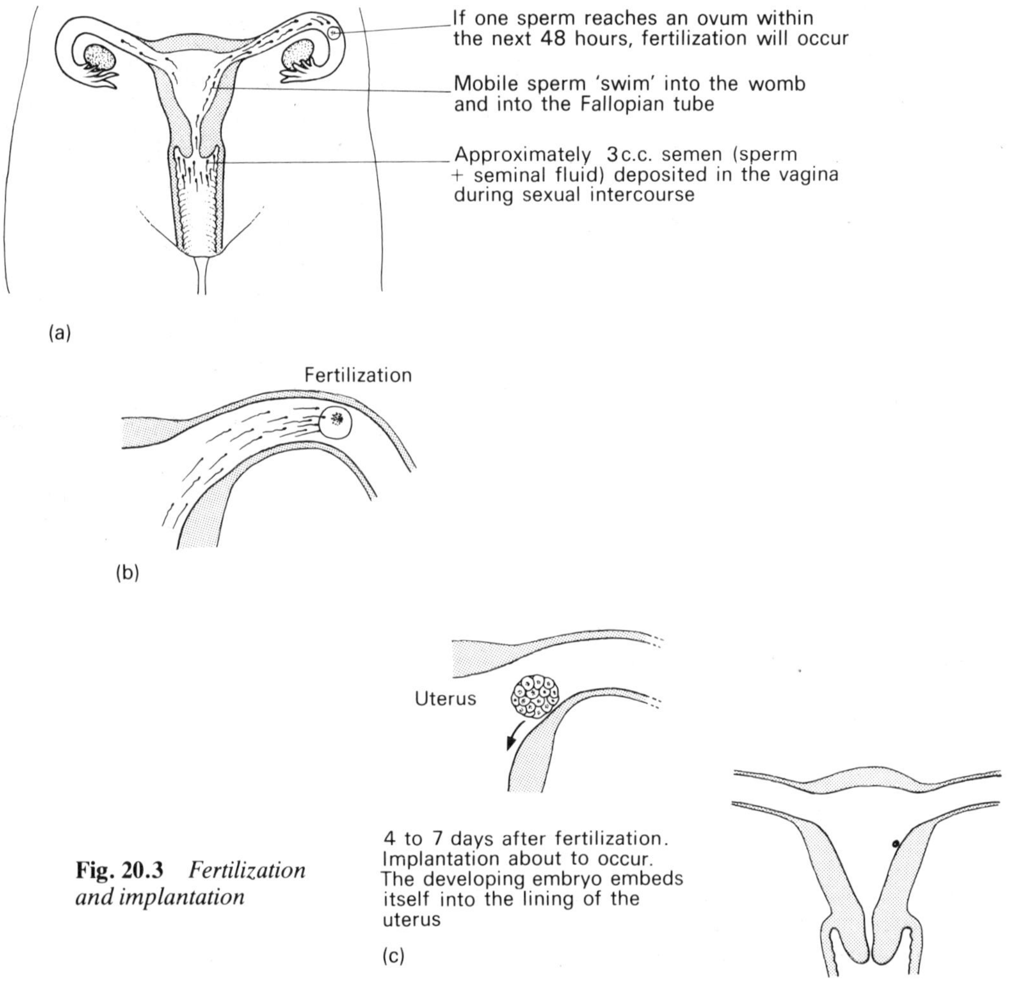

Fig. 20.3 *Fertilization and implantation*

20.5 Pregnancy

When a woman suspects that she is pregnant and has missed two periods she will see her doctor and may undergo a urine test to confirm her pregnancy. This involves giving a urine sample, which is then tested for a hormone known as HCG (human chorionic gonadotrophin). The outer cells of the blastocyte secrete this hormone into the mother's bloodstream. It leaves her body via her urine, so its presence is a reliable way of indicating that pregnancy has begun. The further development of the foetus during the following months has already been studied in detail in Chapter 1, page 5. The period between fertilization and birth is termed the *gestation period*. This is normally nine lunar months, although sometimes birth can occur before this (page 20). If it takes place before the baby can survive on its own, the mother is said to have had a *miscarriage*. This is usually before the twenty-eighth week of pregnancy. We have already discussed the ante-natal care of a woman during her pregnancy and why it is essential. However changes other than physical ones may need to be taken into consideration. When she is pregnant a woman may experience many emotional changes: on occasions she may become depressed. These changes are only natural and both the mother and those around her will need to expect and understand them and help her to cope with them.

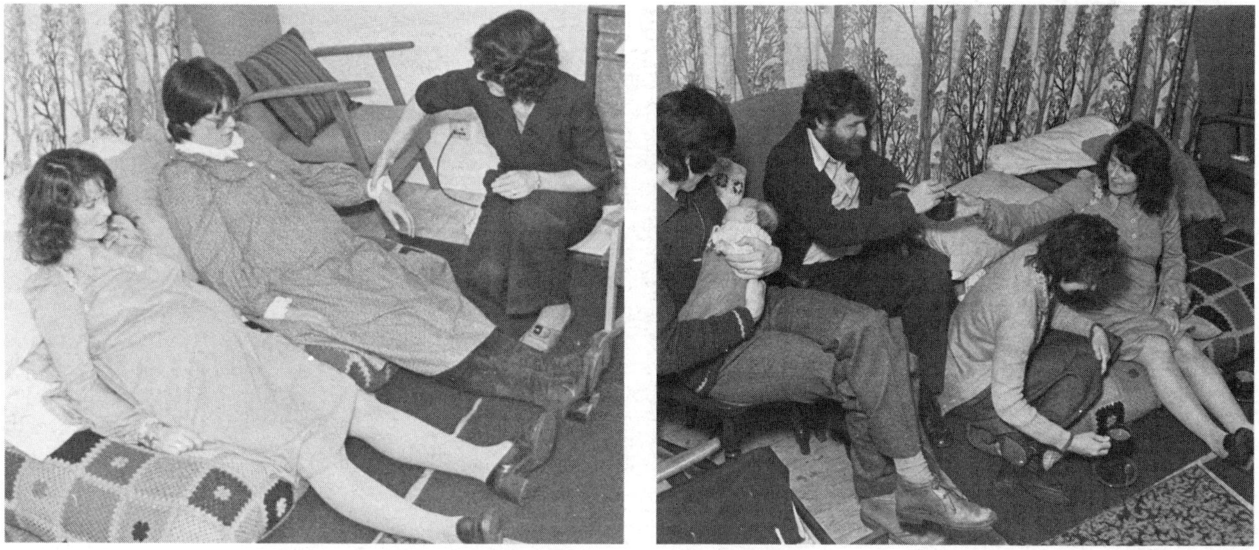

Fig. 20.4 *(a) The teacher of this antenatal class tests the mothers' relaxation as they breathe through a practice contraction. (b) Everyone is at ease in the happy, relaxed atmosphere of the class – even the baby, whose parents attended such classes before his birth. They are at this session to discuss their experiences of birth and parenthood with the 'expectant parents'*

Similarly, an understanding of the process of childbirth itself is highly desirable, if not essential, for the mother, particularly in the case of her first child. She can attend classes where she will be taught breathing and relaxation exercises (Fig. 20.4(a)). These will assist her and the baby during birth. The idea behind these exercises is to make the mother aware of the control she can have over her body during the baby's birth. This should enable the mother to be truly involved in the birth. She can make a positive contribution to help herself and her baby by relaxing certain muscles and so ease the passage of her baby into the world. She can learn how to work with her contracting muscles and not against them. Fathers may also be able to attend these classes. They need to understand clearly what is happening to their wives and how they can help, particularly if they wish to be present at the birth. Expectant mothers also find that these classes are happy social events. They get a lot of interest, enjoyment and encouragement from each other's experiences (Fig. 20.4(b)). They can ask questions and get informative answers which will dispel some of their anxieties.

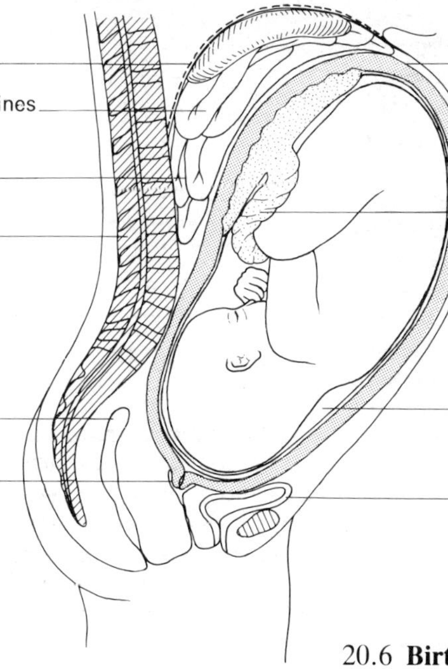

Fig. 20.5 *Towards the end of the ninth (lunar) month the baby is almost ready for birth. (If you compare this diagram with Fig. 19.5 you will be able to see how the other organs in the abdomen have been pushed together.)*

20.6 **Birth**

During the last part of the ninth month the baby 'settles down' ready for birth. Its head is usually pointing downwards and facing the mother's right side (Fig. 20.5). It is often said that the baby's head has 'engaged', which means it is right down in the pelvic region.

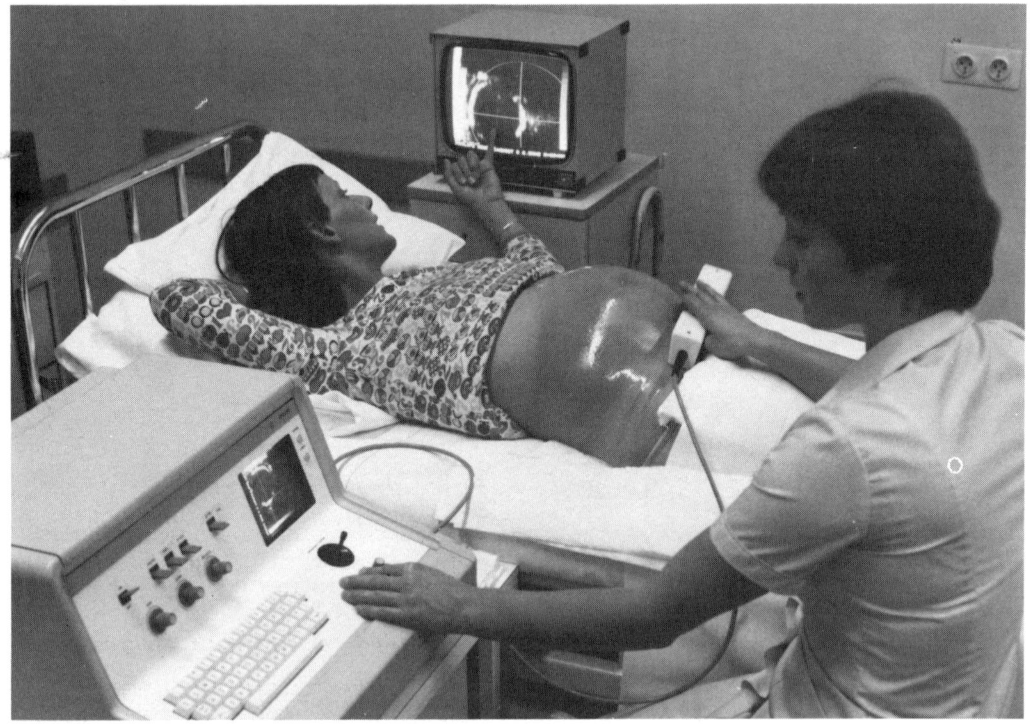

Fig. 20.6 *An ultrasonic scan will show the position of the foetus, its size and whether there will be a multiple birth*

The doctor will be able to feel its position. An ultrasonic scan can indicate the position more clearly, and show if one or more babies are to be expected (Fig. 20.6). The process of birth is called *labour*, because the mother's body has to work hard to push the baby out. This is achieved by the muscular contractions of the womb, which are triggered by the release of oxytocin from the pituitary gland.

To see how the uterus works place a tennis-ball inside the foot part of a sock (Fig. 20.7(a)). This resembles the baby inside the womb with its narrow passage, the cervix (the neck of the sock). If we squeeze the sock to push the ball out, it is like the muscular contractions of the uterus during the *first stage of labour*. The first contractions are usually short, with long intervals in between. However, every birth is different and the size and rate of contractions are the most variable features. If the mother is to have her baby in hospital she will have been told at what stage she should report for admittance.

The baby is pushed down against the cervix like the ball against the neck of the sock (Fig. 20.7(b)). As the muscles get shorter after each contraction the cervix becomes thinner and moves back over the baby's head (Figs 20.7(c) and (d)). At this stage the contractions are usually longer and come more rapidly. The mother will be regularly examined internally by nursing staff to see how far the cervix has dilated (Fig. 20.7(e)). When the cervix is fully dilated (about 10 cm in diameter) the mother is usually taken to the delivery room, unless she is having the baby at home. The amnion (membranes) will probably break at this stage, releasing the 'waters', i.e. the amniotic fluid. (This could even happen before contractions begin in the first stage of labour.) The baby has started to turn so that it can cram itself into the pelvic region.

Now the *second stage of labour* begins. The mother works with her abdominal and uterine muscles to push the baby through the vagina or birth canal. The urge to push comes with every contraction. When the baby's head can be seen stretching the outlet of the vagina (sometimes called 'crowning') the birth is near and labour almost over (Fig. 20.8). If it looks as if the area of muscle between the vagina and anus is going to tear the doctor may performan an *episiotomy*. A local anaesthetic will be given and a small cut will be made. After the birth this cut will be stitched, and when it heals there will be no after-effects. After the head emerges from

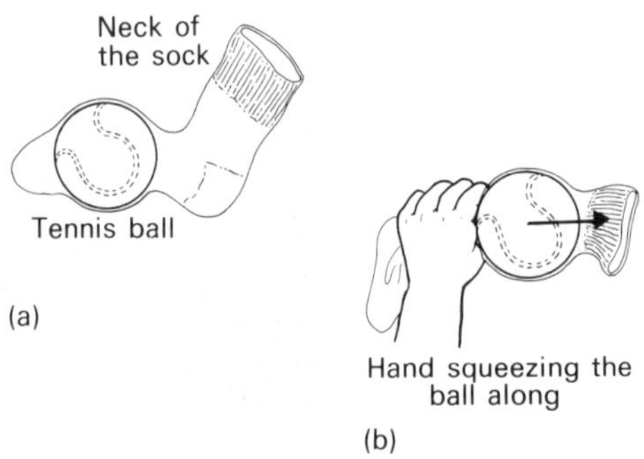

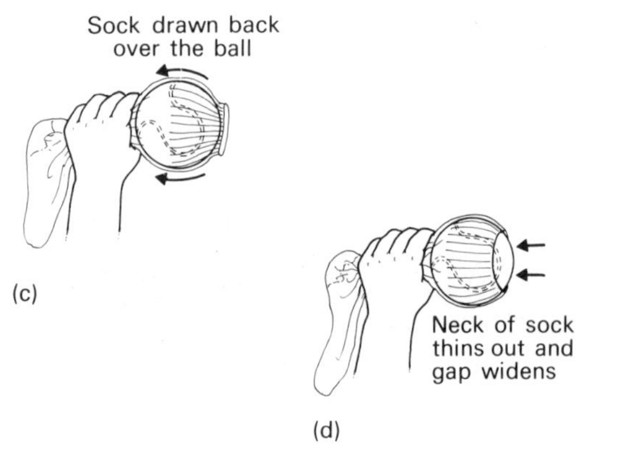

Fig. 20.7 *An illustration of how the uterus works to push the baby out. (e)* Far right: *End of the first stage of labour: the cervix is fully open*

the birth canal, the baby rotates. Its head is supported by the nursing staff, while the body is born one shoulder after another (Fig. 20.9). There is generally great excitement at this moment, even among nursing staff who have seen this situation many times before.

Now the *third stage of labour* must be completed. The baby is still attached by its umbilical cord to the placenta. The cord is clamped firmly on the side nearest the baby's abdomen and then cut. The nursing staff now encourage the mother to deliver the 'after-birth'; the placenta and remaining cord. The mother will feel the urge to push as the placenta loosens itself from the wall of the uterus. An injection may be given to stimulate this process and to control bleeding from the uterus wall. All the afterbirth is examined to check that nothing remains that might lead to infection of the uterus. The childbed fevers of years ago were due to lack of hygiene and care at this stage; cases are virtually unknown today. The whole birth takes about 14 to 15 hours for the first baby and 7 to 8 hours for following ones, but variations in these times are great indeed.

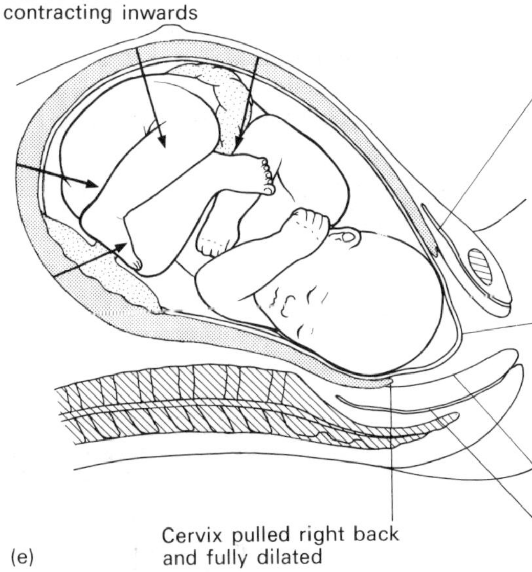

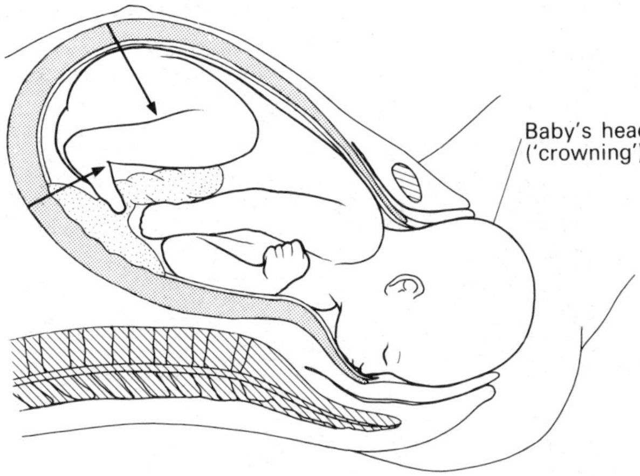

Fig. 20.8 *The baby is about to be born: labour is nearly over*

Fig. 20.9 (Below) *The baby is helped into the world by the nursing staff*

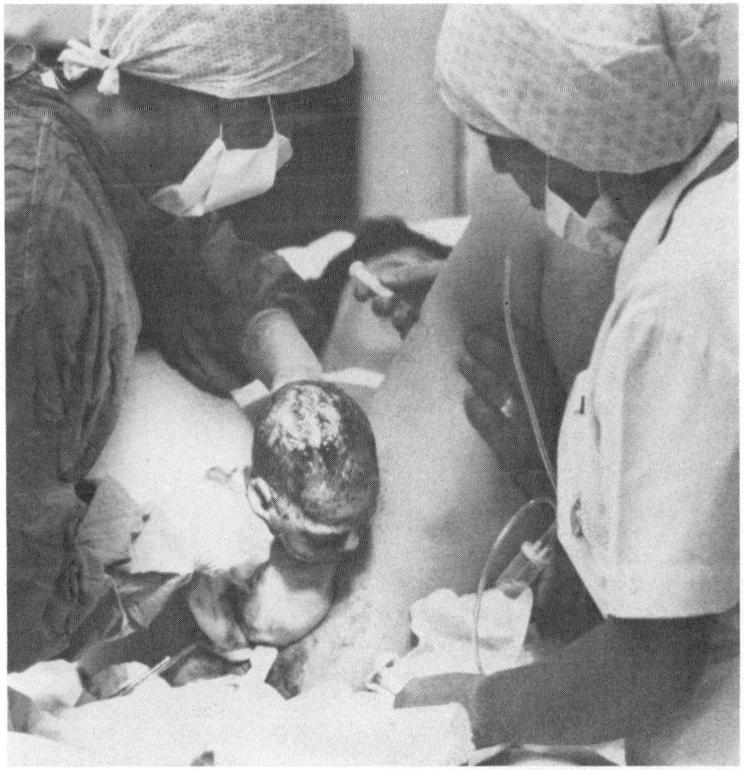

20.7 **Post-natal events**

The uterus continues to contract rapidly over the next few days until it returns to its normal size. Its condition will be checked at a post-natal examination six weeks later. The hormone prolactin is released from the pituitary gland and stimulates the breasts to produce milk. During the first three days a thick yellow fluid called *colostrum* is produced. This can give the baby a number of important nutritious substances and anti-bodies to prevent infection. Breast milk is considered best for babies as it contains the correct quantities of food materials and is delivered at the right temperature. The suckling of the baby at the breast stimulates further milk production (Fig. 20.10).

The birth of a baby may also bring about emotional changes in the mother. Physically babies are very demanding in the first few months of life. They require a lot of attention, need to be fed every few hours and can cry a good deal. The mother will need all the understanding and support that she can get. However, things gradually begin to settle down as experience of parenthood is gained.

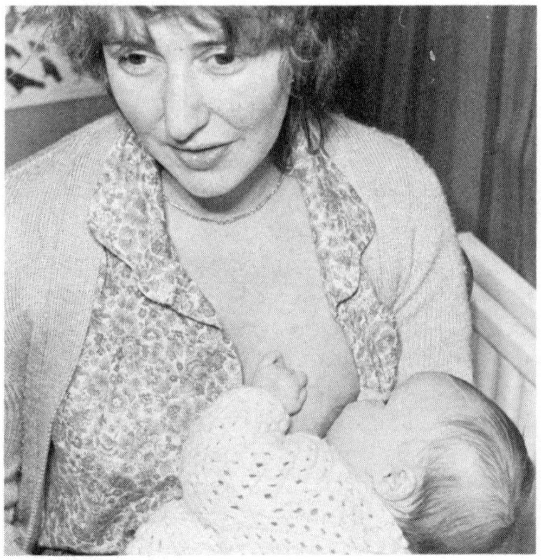

Fig. 20.10 *A mother breast-feeding*

The physical and emotional impact of caring for a young child can produce physical strains as well as times of great pleasure. It is most important therefore that secure personal relationships and affection should surround the baby when it is born.

20.8 **Contraception**

Contraception is the name given to any method of preventing conception. There are several contraceptive methods available nowadays to couples who wish to limit the number of children that they have. Expert advice is available to individuals on all of these methods from *family-planning clinics* and family doctors. Advice can be obtained about the suitability of each method and its effectiveness.

The least effective methods are those that rely on timing, such as the removal of the penis before ejaculation (*withdrawal* or *coitus interruptus*) or only having sexual intercourse when it is considered that there is no ovum in the oviduct (the *safe period*) (Fig. 20.11). The latter can be made a little more effective if a constant record is made of a woman's temperature every morning throughout her monthly cycle. Generally a woman's body temperature falls slightly at ovulation and then rises to a higher level immediately afterwards. Such a chart needs expert advice on its interpretation as the body temperature can fluctuate for other reasons, such as illness.

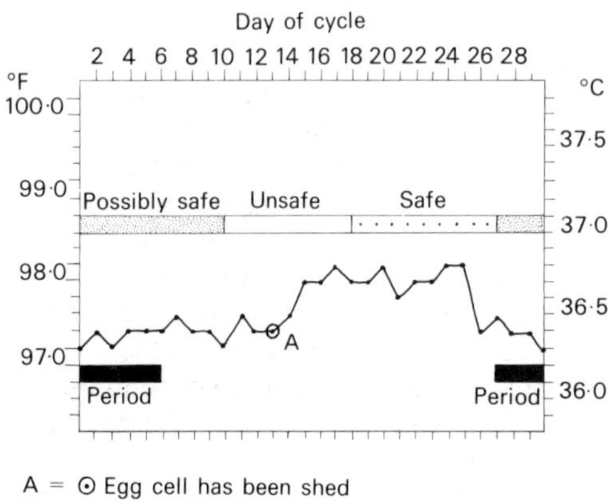

Fig. 20.11 *The safe period.*

Other methods present a physical barrier to the sperm. A soft rubber cap called a *diaphragm* can be placed in the vagina to cover the cervix. To be effective this needs a 'personal fitting' at a clinic and should be used with spermicides, which kill the sperm. A *sheath* (*condom* or 'French letter') is made of soft rubber and is fitted over the penis before intercourse to stop sperm entering the vagina.

The I.U.D. or *intra-uterine device* ('coil' or 'loop') is a small flexible plastic or copper object which is made in a variety of shapes (Fig. 20.12). It is inserted into the uterus by a doctor and is checked regularly. It works by interfering with the process of implantation.

The *pill* or *oral contraceptive* has to be obtained on a doctor's prescription. It is effective only if a course is taken regularly. There is a wide range of pills which contain different proportions of the hormones oestrogen and progesterone. If taken as recommended it is very effective, because the high levels of hormones that it introduces into the blood 'convinces' the pituitary gland that fertilization has taken place. As a result the pituitary gland stops stimulating ovulation.

Sterilization for both women and men can nowadays be a very simple procedure. In the case of men a *vasectomy* is performed, i.e. the sperm-duct is closed within the scrotum. Usually only a local anaesthetic is needed. Sterilization of a woman requires a general anaesthetic and a stay of a few days in hospital. The length of stay will depend on the technique that is used to seal off the oviducts. An abdominal operation may be necessary, but the simplest method used regularly nowadays involves the use of a laparoscope. A small cut is made below the navel and the laparoscope, containing optical fibres, is inserted through this small incision. The

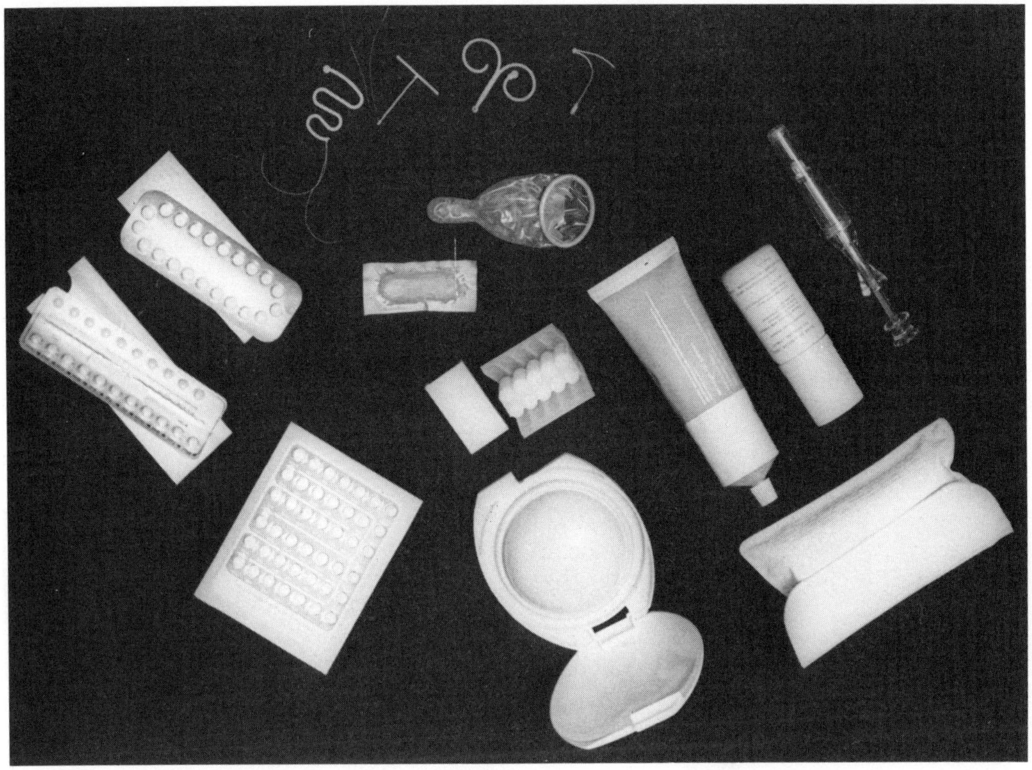

Fig. 20.12 *Methods of contraception*

Fig. 20.13 *The ceremonial dance that marks the coming-of-age of young men of the Masai*

laparoscope enables the surgeon to find the oviducts and to seal them. Sterilization should be considered as being irreversible. The woman's sterilization is effective immediately. A man has to take samples of semen for testing, until it is found that no more sperms are present in the fluid.

Abortion or termination of a pregnancy is the removal of an embryo from the woman's uterus during the very early stages, usually before the twentieth week. It is illegal except under certain circumstances, such as when medical opinion considers that the mother's well-being is at risk, or when there is a chance that the embryo is suffering from an abnormality. An abortion must be carried out on registered premises by doctors.

20.9 **A lifetime of change**

The young men in Fig. 20.13 are taking part in a ceremonial dance which signifies to them and to members of the tribe that they have reached adulthood. The elders consider that these young men now have the knowledge and skills that will enable them to assume the responsibilities of an adult member of the tribe. You may like to discuss what skills and qualities we need to possess in our society when we emerge from youth into the responsibilities of adult life. How far do you think our ability and readiness to face the responsibilities of adulthood depend on attitudes developed through the experiences of childhood?

Between the ages of twenty and forty most of us face many more responsibilities than we have previously experienced. These responsibilities can involve our work, personal relationships, marriage and possibly a family. Now the welfare of others depends upon how we manage our finances, our living conditions and how we respond and adjust to changing circumstances. If we are aware that changes and difficulties are inevitable we are more capable of coping with them.

Middle age is usually regarded as starting when we are forty. By now careers are more settled and children are less dependent upon us. Keeping our bodies in good health is important throughout our lives, but during middle age there is a tendency to begin to slow down physically. However, keeping fit and mobile can help us to remain healthy in later life. Also during middle age there may be more time for us to develop past hobbies and pursuits or develop new interests – physical, recreational or educational. Our physical well-being and our attitudes to life and health at this time will determine how we adapt to the last period of our lives, the over-sixty period or old age.

'I'm beginning to feel it is a disgrace to be disabled and a crime to be old. *Mr N.*'

'I was at one firm before I retired for fifty-six years. Why do I come back and work now? For the companionship. I couldn't sit lounging around. I enjoy it here, and I don't mind coming out and doing a bit of helping . . . If I can do my best, that's all I ask. I've always taken my share of the work. *Mr J. W., aged 81*'

'I don't know what I'd do without this place. I couldn't stand being on my own. I wish they had these things all over the country. *Miss B.*'

'I just want to move somewhere where there is company. I can't describe to you what loneliness is. If you've got your health it's something. But I haven't. Loneliness is an illness. It's the worst illness that you can ever have. *Mrs M.*'

'I come for the company. You're so lonely on your own. I've been on my own this last seven years. Coming here keeps you younger and on your feet. *Mrs S. C.*'

'I come here to get out of the house a bit. You get bored staying on your own. I'm a bit active and I'd rather be doing something. I don't think an elderly person should give up. You can make yourself an invalid by sitting about. *Mrs H. M. C., aged 75*'

The expectancy of life has increased during this century (see Chapter 7). Nowadays about 14 per cent of the population in the U.K. is over the age of sixty-five (Fig. 20.14). Physical changes take place in old age: arteries can harden, cells die or do not multiply or repair themselves so readily; bone tissue becomes thinner and more brittle, so that it is more susceptible to injury; flesh is easily bruised and becomes wrinkled in appearance. However, these changes do no necessarily mean that an elderly person has to become inactive, either mentally or physically. Elderly people may not have as much stamina as a young person but they can still tackle many of the physical activities of younger days. A good diet and regular exercise, like walking, is as important in old age as at any other stage in our lives. Some elderly people may develop a physical disability that could affect the way in which they have to manage on their own. Families, neighbours, doctors and many organizations can help to make sure that elderly people enjoy life to the full and do not suffer from loneliness or physical discomfort. If we understand the changes of old age we are more likely to derive every benefit from this period of our lives.

Fig. 20.14 *An old people's day centre*

20.10 Do we care?

We have only one body to see us through all these stages of our life. Unlike a motor car it has no trade-in value against a new model. Its value to us is our quality of life. Therefore it is in our interest to understand the mechanisms of our body and to take care of it. There is no doubt that our generation has the advantage of a health service and the technical advances in medicine and surgery. However, does a car owner immediately go out in his brand-new car and bump into a lamppost because he has a body repair shop around the corner? Every one of us has a personal responsibility for his own health and as members of a community we need to invest in the provision of a medical service that can respond to the needs of the people.

In this country, over two million old people live alone. A vast proportion of this number see virtually no one except the tradesmen from one week to the next. And the problem is growing. The number of people over the age of 60 has grown steadily from two and a half million at the turn of the century to ten and a half million today; the numbers are increasing by over 150 000 a year and are likely to go on increasing until the early 1980s.

Yet in our planning we still have not grasped the implications of this age explosion. It has been estimated that over seven million pensioners are not catered for in the provision of day centre places. Not every old person, of course, needs or wants the facilities a day centre offers. Some old people live a full and active life, completely integrated into the community around them. But for a significant number, day centres could provide a lifeline to friendship. They could provide interests, hobbies, and services like chiropody, cheap meals and bathing.

In many ways, we have got to exercise particular care, pay particular attention to the needs of the lonely, the frail and those at risk. But it is not just a case of doing things *for* the elderly. We have got to go much further, and restore a sense of worth and dignity. All of us want to feel we have something to give.

The truth is that many people *need* work. They are not happy unless they are busy making things and contributing to the world around them. It is significant that Norway, which has the highest retirement age in Western Europe, also has the highest proportion of men living to a ripe old age. It is significant that in war-time Britain elderly people lived longer than they do today, despite all our social services. They felt needed. More often than not they had work to do, whether their tasks were great or small.

One organization set up to help the elderly, Help the Aged is opening new Employment Fellowship Centres as fast as it possibly can – at a rate approaching one a week. But it cannot keep pace with the demand. The answer is obvious: if people *want* to work, and *want* to make a contribution to the world around them, we should ensure that this is possible. But it will not happen unless you, and people like you in your town, take the initiative.

THINGS TO DO

1 You are completing an application for a job which will involve meeting and working with a large number of people. You are to describe your physical appearance and then state how you think that your personality and attitudes will enable you to do the job well. You are also asked to state your other interests.

The application form also asks for a friend to give you a reference. Ask a friend to do this. Compare your friend's description of you with your own. Discuss the results.

2 Construct a chart, using illustrations, which shows how our relationships with others develop and change as we grow up from babyhood to middle age.

3 Collect together in a scrap-book a range of advertisements. Put them into groups according to the way in which you think they are influencing you to buy their product. For example, do they
(a) give factual information?
(b) appeal to certain attitudes?
(c) imply that the product can affect your relationship with others?

4 Find out about 'the family' as a unit and how it exists among different people all over the world. What marks the coming of age of the children? When do marriages take place? What form does the marriage ceremony take? Are the roles of individuals within the family different to the ones we know?

5 Conduct a survey of further-education and leisure facilities which are available in your area.

6 Conduct a survey of the social and welfare provisions for the elderly which exist in your area. Is there any help which you and your friends can give?

THINGS TO WORK OUT

In Questions 1 and 2 write down the *one* answer which you think is correct.

1 The method of contraception which stops sperm entering the vagina is generally known as the
A diaphragm (cap)
B coil (I.U.D.)
C pill
D sheath (condom)
E spermicidal foam

2 The most reliable indication of pregnancy is that the woman
A puts on weight
B feels sick every morning
C passes more urine
D feels giddy and faint
E misses a period

3 (a) Name *one* of the hormones that are present in the 'pill' (oral contraceptive).
(b) How does taking the 'pill' prevent conception?

4 Give the principles upon which the following methods of contraception are based:
(a) the diaphragm
(b) the safe period
(c) the I.U.D. (loop)
(d) spermicidal foam
(e) vasectomy
Comment on the effectiveness of each method.

5 (a) Why are the walls of the uterus formed mostly from muscular tissue? What stimulates the uterus into action?
(b) Describe the most significant occurrences during
(i) the first stage of labour
(ii) the second stage of labour
(c) What is the afterbirth? Why is it important that all traces of the afterbirth are removed from the mother's body?
(d) What are *two* advantages of being born head first? What does the term 'breech' refer to?
(e) What is a Caesarean birth?

6 (a) Draw a simple diagram to illustrate the production of sperm within the testis.
(b) Why do you think that part of the head region of a sperm contains chemicals which are capable of dissolving cell membranes?
(c) Why is the urethra in a man surrounded by spongy tissue?
(d) What is the name of the fluid that carries the sperm? Where is it produced?

7 (a) When fertilization takes place what happens to the lining of the uterus?
(b) Describe, with a diagram, how the foetus obtains food and oxygen from the mother.
(c) Why is it essential that the placenta provides a large surface area between the blood supplies of mother and foetus?
(d) How does the placenta protect the foetus? Give *one* instance of when this protection may not be effective.

Index